Advances in Nanoengineering

Electronics, Materials and Assembly

Royal Society Series on Advances in Science

Series Editor: J. M. T. Thompson *(University of Cambridge, UK)*

Published

Vol. 1: Advances in Astronomy: From the Big Bang to the Solar System
edited by J. M. T. Thompson

Vol. 2: Advances in Earth Science: From Earthquakes to Global Warming
edited by P. R. Sammonds and J. M. T. Thompson

PHILOSOPHICAL TRANSACTIONS OF THE ROYAL SOCIETY A
MATHEMATICAL, PHYSICAL & ENGINEERING SCIENCES

Royal Society Series on Advances in Science – Vol. 3

Advances in Nanoengineering

Electronics, Materials and Assembly

Editors

A G Davies
University of Leeds, UK

J M T Thompson
University of Cambridge, UK

ICP
Imperial College Press

Published by

Imperial College Press
57 Shelton Street
Covent Garden
London WC2H 9HE

Distributed by

World Scientific Publishing Co. Pte. Ltd.
5 Toh Tuck Link, Singapore 596224
USA office: 27 Warren Street, Suite 401-402, Hackensack, NJ 07601
UK office: 57 Shelton Street, Covent Garden, London WC2H 9HE

British Library Cataloguing-in-Publication Data
A catalogue record for this book is available from the British Library.

Royal Society Series on Advances in Science — Vol. 3
ADVANCES IN NANOENGINEERING
Electronics, Materials and Assembly

ISBN-13 978-1-86094-751-3
ISBN-10 1-86094-751-4

Typeset by Stallion Press
Email: enquiries@stallionpress.com

Printed in Singapore by Mainland Press Pte Ltd

PREFACE

Although researchers began to use the prefix "nano" more than thirty years ago, it is only in the last ten years that its use has spread to virtually every field of science, technology and medicine. Today it is used as much for fashion as it is for scientific classification, but the blossoming of interest nevertheless reflects a genuine explosion in the useful application of nanotechniques and nanomaterials to both science and technology. We have reached the point where it is possible to manipulate materials at the molecular and atomic level and create genuinely new materials and processes that are tuned for particular applications. Examples have emerged in fields as disparate as novel semiconductors for nanoelectronics and medicines for the treatment of hereditory illnesses. Capabilities are emerging in nanoscience and nanotechnology that could not have been imagined two decades ago and this book provides an invaluable underpinning for those genuinely interested in understanding their limits and capabilities so that they can apply them to the advancement of science and engineering.

When the prefix "nano" was first used in the 1970s, it genuinely referred to structures with dimensions that approached a nanometer or at least a few nanometers, and distinguished them from microstructures, but as its use spread, the definition was loosened to embrace structures up to 100 nanometer and that is where it has settled. It is important to preserve it at this level if the classification is to remain of value. This volume concentrates on the science and technology that underpins the genuine advances that have been made in manipulation and examination at dimensions below 100 nanometers. Starting with a chapter on carbon and its various molecular configurations it contains chapters written by experts on both man-made and naturally occurring structures, on nanodevices with potential application to information and communication technologies, and on the advanced analytical and microscopical techniques that have been developed to examine and assess these incredible small artifacts. There are chapters on molecular self-assembly and tunnel transport through proteins showing how science and technology can now operate at a level that probes the internal mechanisms of life itself. The nanoworld is so wide and diverse that no single volume is going to give comprehensive coverage of worldwide activity but this book covers as much as any and will long be useful as a reference to those entering the field or interested in its capabilities.

Lord Broers FREng FRS
Chairman, House of Lords Science and Technology Select Committee
Past President, Royal Academy of Engineering

CONTENTS

INTRODUCTION

Giles Davies

School of Electronic and Electrical Engineering
University of Leeds, Leeds, UK

You see things; and you say "Why?" But I dream things that never were; and I say "Why not?"
George Bernard Shaw

Of the volumes planned for this series of books from the Royal Society and Imperial College Press, this is the only one that is devoted to "engineering" rather than "science". The distinction between these broad disciplines is often blurred: scientists searching for the answer to their question "Why?" often need to develop technology to make progress, in effect becoming engineers. Similarly, engineers wanting to exploit science to answer their question "How?", or possibly "Why not?", often find they must understand better the underlying fundamental science and so, perhaps temporarily, become scientists.

The blurring between these disciplines occurs probably none more so than in the emerging field(s) of nanoscience and nanotechnology. Using the definitions established by the recent Royal Society/Royal Academy of Engineering wide-ranging report on nanotechnology,[a] nanoscience is the study of phenomena and manipulation of materials at atomic, molecular and macromolecular scales, where properties differ significantly from those

[a] *Nanoscience and Nanotechnologies: Opportunities and Uncertainties*, published on 29 July 2004 by the Royal Society and the Royal Academy of Engineering (see http://www.nanotec.org.uk/).

at a larger scale. Nanotechnologies are the design, characterization, production and application of structures, devices and systems by controlling shape and size at the nanometer scale. As such, the experimental and theoretical work of chemists, physicists, electronic and mechanical engineers, material scientists, biochemists, molecular biologists, *inter alia*, can all contribute to this cross-disciplinary field, making it, in my (perhaps biased) opinion, one of the most exciting and challenging research activities to pursue.

Broadly speaking, nanoscience and nanotechnology are concerned with materials that have at least one dimension less than 100 nm, or one-tenth of a micron. To put this into context, a carbon Buckminsterfullerene molecule ("Bucky Ball"), which comprises 60 carbon atoms arranged into a spherical soccer-ball-shaped structure, has a diameter of 1 nm — this is about 200 billion times smaller than the diameter of a real soccer ball, which itself is about 200 billion times smaller than the diameter of the earth. A nanostructure can be categorized as zero-, one-, or two-dimensional according to whether its features are confined to the nanometer scale (nanoscale) in three, two, or one dimensions, respectively. The fullerene molecule, for example, can be regarded to be zero-dimensional owing to its size being on the nanoscale in all three dimensions. Other zero-dimensional nanostructures include metal and semiconductor particles that are a few nanometers in diameter, and are sometimes called "quantum dots". A one-dimensional nanostructure (a "quantum wire") is confined in two dimensions, and extended in the third. Carbon nanotubes, for example, which can be visualized as rolled up sheets of graphene, can be regarded as quantum wires, as indeed can many molecules and biomolecules, particularly if they are polymeric. Finally, there are two-dimensional nanostructures, which are confined on the nanoscale in one dimension but are extended in the plane, and can manifest as coatings or thin films, or electron layers buried inside semiconductor devices, for example.

A further broad categorization is often made according to how the nanostructure is fabricated. The "top-down" approach, as the name implies, involves defining the nanostructure out of a larger macroscopic material perhaps by chemical etching, milling, or electrostatic confinement, *inter alia*, and crudely speaking, has predominantly lay in the remit of the physical and material scientist, or the electronic and mechanical engineer. The "bottom-up" approach, on the other hand, fashions the desired nanostructure from smaller, constituent parts, perhaps by chemical synthesis, and has its provenance in the laboratories of the chemist or biochemist, for example. These characterizations emphasize how nanotechnology is a convergence of a vast range of disparate science and technology, and is inherently a multidisciplinary field.

However, the focus of nanoscience and technology is not with materials that are simply small; the properties of the structure or material must be different from those exhibited in the bulk. There are two main reasons that this can be the case. Electrons, the fundamental particle central to most of the physical and chemical properties of materials, and in particular their electronic and optical characteristics, have a size. This size is related to their wavelength, a consequence of the wave–particle duality inherent in the quantum mechanics that governs electron behaviour, and this wavelength can be on the nanoscale. If the dimension of a material approaches the electron wavelength in one or more dimensions, quantum mechanical characteristics of the electrons that are not manifest in the bulk material can start to contribute to or even dominate the physical properties of the material. This allows fundamental quantum mechanical properties to be accessed for their study and potentially for their exploitation.

The second main reason that the properties of nanoscale materials can be different from those exhibited in the bulk is associated with their increased relative surface area. By reducing the diameter of a quantum dot from 30 to 3 nm, the number of atoms on its surface increases from 5% to 50%.[b] Therefore, for a given mass of material, nanoparticles will have a greater surface area compared to larger particles, and hence will be much more reactive, as chemical reactivity, catalytic activity, and growth reactions occur at a material's surface. Similarly, the high grain boundary area in materials comprising nanoscale crystalline grains can instill enhanced mechanical properties.

It is probably becoming clear that the field of nanotechnology is vast, and this book can only hope to give a taste of the immense activity currently taking place. A significant part of this book is devoted to the fundamental nanotechnology building blocks — the nanostructures themselves. In Chapter 1, Humberto and Mauricio Terrones describe carbon-based nanostructures and, in particular, carbon nanotubes and carbon fullerenes. The authors review the fabrication and properties of these fascinating structures, and discuss their emerging and potential applications. Moving from the organic to the inorganic world, Caterina Ducati discusses the growth of nanowires made of inorganic materials such as silicon, ruthenium oxide, and nickel sulphide by a number of physical and chemical processes in

[b] *Ibid.*

Chapter 2. And in Chapter 3, material scientists Jon Molina-Aldareguia and Stephen Lloyd describe multi-layered inorganic materials, which have a range of potential applications including hard coatings and data storage systems. In his fascinating chapter (Chapter 4), Simon Hall draws on nature for inspiration and techniques to fabricate inorganic nanowires, and illustrates how fruitful the adoption or exploitation of processes, techniques or systems from traditionally distinct disciplines can be. In particular, we see how chitosan, a derivative of chitin (one of the main components in the cell walls of fungi and insect exoskeletons), can be used to template the fabrication of high-temperature superconductor wires.

Equally important to the fabrication of nanostructures is the development of techniques to assemble them onto surfaces, or into appropriate geometries or circuits, or to interface them with the outside world. This is particularly necessary for the exploitation of nanotechnology to produce useful applications, since no matter how fascinating the physical, chemical, or biological properties are of any given nanostructure, it is likely that they will need to be organized into some kind of functional device to employ their properties. In particular, there is a need for directed assembly tools, in which the nanostructures can self-assemble or be programmed to self-assemble into their desired final configuration. In Chapter 5, Philip Gale discusses progress in the field of supermolecular chemistry, concentrating on how molecular subunits can be designed to assemble into larger chemical complexes, which allows one to engineer new molecular knots and chains, and even nanoscale molecular machines, that could not be made previously. Probably the best known self-assembling molecular system is DNA (deoxyribonucleic acid), which in its physiological state comprises two polymeric molecules of complementary chemical structure entwined around one another — the famous double-helix structure. If the individual strands are not chemically complementary, they remain separate and the double helix does not form. This has led a number of researchers to propose that DNA, and other (biological) systems with analogous lock-and-key recognition properties, could form the basis of a nanostructure assembly procedure. In Chapter 6, Christoph Wälti reviews this field and describes a number of experiments designed to exploit DNA to this end, including the selective attachment of molecules to surfaces at a nanometer-scale resolution, the manipulation of surface-tethered molecules by electric fields, and the fabrication of branched DNA constructs.

Chapter 6 approaches nanotechnology from the broad perspective of developing molecular-scale electronic devices — the natural evolution of the progressive miniaturization of semiconductor electronics over the past 50 years. This is a theme shared with the chapters that immediately follow. In Chapter 7, Jason Davis continues the discussion of the integration of biological molecules into electronic circuitry, and describes a range of experiments on metallo-proteins, proteins containing transition metals, including studies of their electrical conduction properties. In Chapter 8, John Cunningham returns the discussion to the top-down methodology and reviews a number of nanoscale electronic devices formed by electrostatic confinement, including devices that control individual electrons or operate by the action of individual electrons. Rolf Crook continues with this theme in Chapter 9 describing an innovative technique to pattern electronic nanostructures in an erasable fashion, providing a flexible approach for investigating and optimizing such devices. Moving sideways from nanoelectronic systems to nanomagnetic systems, Robert Hicken's chapter (Chapter 10) is concerned with the development of nanomagnetic materials and how they can be exploited for future data storage applications. Indeed, the ongoing parallel miniaturization of the electronic and magnetic components integral to consumer products such as personal computers is one example of how relevant this technology is to everyday life; nanotechnology is not just an esoteric research field that might find application in the future, it is in use, all around us, now.

The analysis and characterization of nanostructures is a crucial part of their fabrication, assembly, and understanding, and all of the preceding chapters describe the techniques employed to study and assess the specific systems under discussion. The last two chapters of this book, however, particularly concentrate on sophisticated analytical techniques. In Chapter 11, David Richards discusses new scanning-probe technology developed to address the nanoscale optically, while in Chapter 12, Mark Osborne describes fluorescent techniques to investigate single molecules and how they interact with their immediate environment.

The authors of these chapters are young researchers, many of whom hold or have recently held prestigious Royal Society University Research Fellowships or Advanced Research Fellowships from the UK's Engineering and Physical Sciences Research Council (EPSRC). They are working at the cutting edge of their fields, and these articles describing their research and setting it into a wider context provide an excellent overview of these topics and demonstrate the infectious enthusiasm of young people passionate about what they do best — asking the questions "Why?" and "Why not?"

Acknowledgments

I would like to thank all of the authors for their contributions. I am also very grateful to Ms. Katie Lydon at Imperial College Press, and to Prof. Michael Thompson of the University of Cambridge and Editor of the *Philosophical Transactions of the Royal Society A*, who is the series editor of these books. Finally, I would like to express my appreciation to Lord Broers, former President of the Royal Academy of Engineering, for kindly agreeing to write the preface to this volume.

Giles Davies studied at Bristol University where he graduated with first class honors in Chemical Physics in 1987, and obtained his PhD in 1991 from the Cavendish Laboratory, University of Cambridge, in Semiconductor Physics. He spent three years as an Australian Research Council Postdoctoral Fellow at the University of New South Wales, Sydney, before returning to the Cavendish Laboratory as a Royal Society University Research Fellow in 1995. He took up the Chair of Electronic and Photonic Engineering at the University of Leeds in 2002, becoming Director of the Institute of Microwaves and Photonics in 2005, and has built up large research teams studying high-frequency (terahertz) electronics and photonics, semiconductor device growth and processing, and bio-nanotechnology. He is especially interested in cross-disciplinary research and, in particular, the combination of biological processes with micro- and nanoelectronics. He is an associate editor of the *Philosophical Transactions of the Royal Society A*.

CHAPTER 1

THE SHAPE OF CARBON: NOVEL MATERIALS FOR THE 21ST CENTURY

Humberto Terrones* and Mauricio Terrones

Advanced Materials Department, IPICyT
Camino a la Presa San José 2055, Lomas 4ª Sección
78216 San Luis Potosí, SLP, México
**E-mail*: *hterrones@ipicyt.edu.mx*

Carbon is one of the elements most abundant in nature. It is essential for living organisms, and as an element occurs with several morphologies. Nowadays, carbon is encountered widely in our daily lives in its various forms and compounds, such as graphite, diamond, hydrocarbons, fibers, soot, oil, complex molecules, etc. However, in the last decade, carbon science and technology has enlarged its scope following the discovery of fullerenes (carbon nanocages) and the identification of carbon nanotubes (rolled graphene sheets). These novel nanostructures possess physico-chemical properties different to those of bulk graphite and diamond. It is expected that numerous technological applications will arise using such fascinating structures. This account summarizes the most relevant achievements regarding the production, properties and applications of nanoscale carbon structures. It is believed that nanocarbons will be crucial for the development of emerging technologies in the following years.

Keywords: Carbon, nanotubes, nanoelectronics, nanodevices, curvature.

1. Introduction

Various forms of carbon including graphite, diamond, and hydrocarbon molecules have been intensively studied since the beginning of the 20th century. In 1924, J. D. Bernal successfully identified the crystal structure of graphite and in the 1940s developments of carbon alloys such as spheroidal graphite (SG) in cast iron were carried out. Rosalind Franklin distinguished graphitizing and nongraphitizing carbons in the early 1950s. From the 1950s to 1970s carbon fibers were produced and developed for industrial applications. Diamonds have been successfully grown synthetically from 1955 and diamond thin films by chemical vapor deposition have also become a 21st century material. However, by the end of last century, the discovery of a third carbon allotrope Buckminsterfullerene (C_{60})[1] had opened up a novel and distinct field of carbon chemistry. As a result, in the early 1990s, elongated cage-like carbon structures (known as nanotubes) were produced and characterized. This gave a tremendous impetus to a new, multidisciplinary field of research pursued internationally.

Carbon possesses four electrons in its outer valence shell; the ground-state electron configuration is: $2s^2\ 2p^2$. Graphite and diamond are considered as the two natural crystalline forms of pure carbon. In graphite, carbon atoms exhibit what is known as sp^2 hybridization, in which each atom is connected evenly to three carbons (120° bond angles) in the xy plane. The C–C sp^2 bond length is 1.42 Å. The sp^2 set forms the hexagonal (honeycomb) lattice typical of a sheet of graphite. The p_z orbital is responsible for a weak bond, termed a *van der Waals "bond"*, between the sheets. The spacing between these carbon layers is 3.35 Å (Fig. 1(a)). The free electrons in the p_z orbital move freely within this cloud and are no longer local to a single carbon atom (*delocalized*). This phenomenon explains the reason why graphite can conduct electricity.

In diamond sp^3 hybridization takes place, in which four bonds are directed toward the corners of a regular tetrahedron (Fig. 1(b)). The resulting three-dimensional cubic network (diamond) is extremely rigid and is one reason for its hardness. The bond length between sp^3 carbons (e.g.,

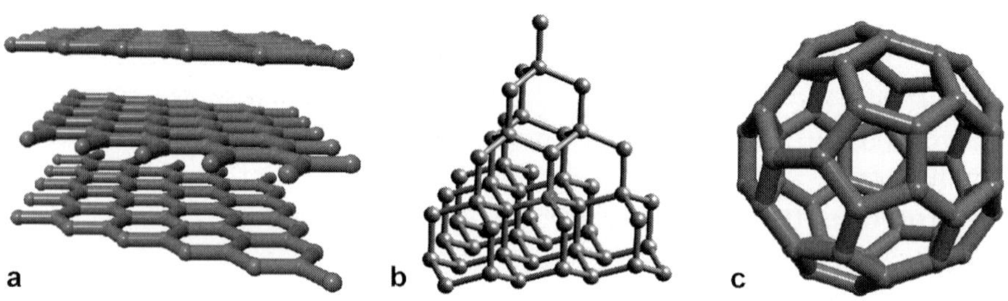

Fig. 1. (a) The crystal structure of diamond; (b) graphite; and (c) C_{60}: buckminsterfullerene.

diamond) is 1.56 Å. A hexagonal, wurtzite form of carbon has been found in meteorites and in shock-loaded graphite and has been named "lonsdaleite" in honor of Kathleen Lonsdale, who studied this system.

Diamond, on the contrary, behaves as insulator, because all electrons are localized in the bonds within the sp^3 framework.

2. New Carbon Nanostructures: Fullerenes, Carbon Onions, Nanotubes, Etc.

2.1. *Fullerene discovery and bulk synthesis*

Research that resulted in the Fullerene discovery originated in the 1970s, when Harry Kroto and David Walton (Sussex University) studied cyanopolyynes, molecules of the type $H(C{\equiv}C)_nC{\equiv}N$. The Sussex team succeeded in preparing HC_5N and HC_7N, and together with Takeshi Oka, an astronomer, and co-workers, detected radio waves emitted from cyanopolyynes HC_nN ($n = 5, 7, 9$) in the center of our galaxy.[1] In 1984, Robert Curl introduced Kroto to Richard Smalley (Rice University), who was then carrying out cluster experiments by vaporizing solid Si targets with a laser. Kroto wanted to vaporize graphite with the idea of proving that longer cyanopolyynes chains could be formed in the interstellar media.

In late August 1985, during the Rice experiments, Kroto and colleagues noted the dominant role played by the 60-atom cluster (the most intense in the spectra), and ascribed the structure of this 60-atom molecule to a truncated icosahedral cage, consisting of 20 hexagons and 12 pentagons, all carbons having sp^2-like hybridized carbon bonds, without dangling bonds and all atoms identically situated (Fig. 1(c)). The authors named the new cage molecule *Buckminsterfullerene,* in honor of the American architect Richard Buckminster Fuller, who had designed geodesic domes with similar topologies.[1]

Five years later, in 1990, Wolfgang Krätschmer, Donald Huffman and colleagues[2] found that C_{60} could be produced in macroscopic quantities, forming crystals, using an electric carbon arc discharge apparatus.[3] Almost simultaneously, the Sussex team became the first to isolate C_{60} and C_{70} molecules and confirmed that the structure of C_{60} was indeed that of a truncated icosahedron with I_h symmetry (nanosoccer ball).[3]

Nowadays, C_{60} molecules have been used as the basis of a new type of chemistry (fullerene chemistry), in which various types of organic, inorganic, and organometallic molecules have been reacted with these carbon cages.

When C_{60} crystals are doped with alkali metals, such as potassium, cesium or rubidium, it is possible to obtain superconductors at $< 33\,K$.[3] Among the strange properties of C_{60}, inhibition of the HIV has even been detected.[4]

2.2. *From giant fullerenes to graphitic onions*

In 1980, Sumio Iijima reported for the first time the existence of nested *carbon nanocages* (now known as *graphitic onions*) seen by using high-resolution transmission electron microscopy (HRTEM) (Ref. 5 and references therein). Eight years later, Harry Kroto and Ken McKay proposed also for the first time, the model of graphitic onions consisting of nested icosahedral fullerenes (C_{60}, C_{240}, C_{540}, $C_{960}, \ldots$) containing only pentagonal and hexagonal carbon rings.[1,5] In 1992, Daniel Ugarte observed the transformation of polyhedral graphitic particles into almost spherical carbon onions,[6] when he irradiated the specimens with fast electrons inside an electron microscope. Theoretical researchers proposed the idea of introducing additional pentagonal, heptagonal, or octagonal, carbon rings into icosahedral carbon cages, to form spherical onions (Refs. 5 and 7; Fig. 2). At present, the fabrication of electronic devices using spherical carbons waits in the future, but it is clear that some applications will arise in the nanotechnology field.

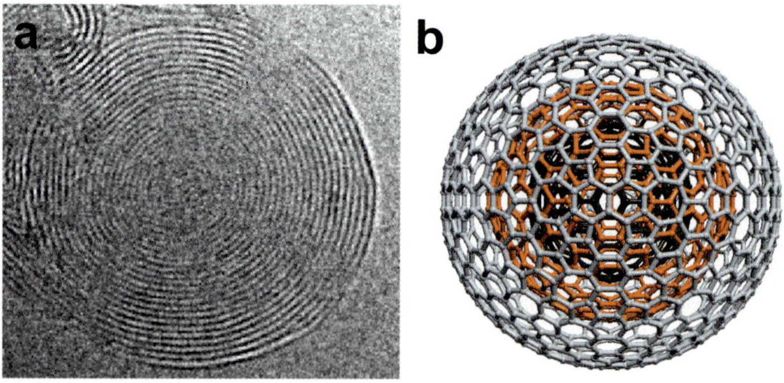

Fig. 2. (a) Spherical carbon onion produced in a TEM at 700°C and (b) model proposed by Terrones and Terrones for spherical carbon onions based on the introduction of additional heptagonal and pentagonal carbon rings (Terrones and Terrones, 1996).

2.3. *Carbon nanotubes*

2.3.1. *Identification and structure of carbon nanotubes*

Carbon nanotubes can be considered as elongated fullerenes (Fig. 3(a)). There are two types of tubes: single-walled (SWNTs) and multi-walled (MWNTs). In 1991, Sumio Iijima reported the existence of MWNTs, consisting of concentric graphene tubes (Ref. 8; Fig. 3(b)). These nested tubes (2–10 nm outer diameter; $< 5\,\mu$m in length) exhibited interlayer spacings of *ca.* 3.4 Å, a value that is slightly greater than that of graphite (3.35 Å). Iijima also noted that the tubes exhibited different helicities or chiralities. This refers to the way hexagonal rings are arranged with respect to the tube axis. Thomas Ebbesen and Pulickel Ajayan published the first account of the bulk synthesis of MWNTs using the arc discharge technique (see Ref. 9) only a few months after Iijima's publication. It is also important to note, that probably the first HRTEM images of carbon nanotubes (SWNTs

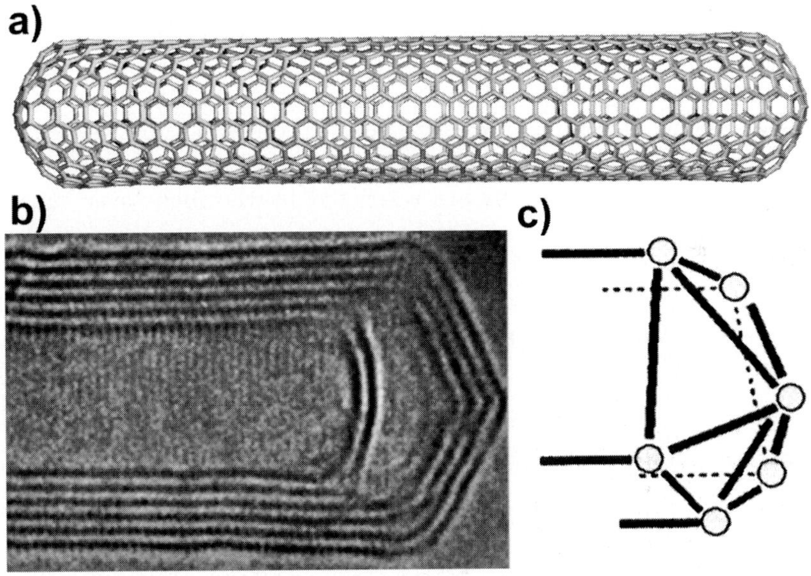

Fig. 3. (a) Molecular model of an SWNT (rolled hexagonal carbon lattice), which is capped due to the introduction of six pentagons in each nanotube end; (b) HRTEM image of one end of an MWNT (nested graphene cylinders; courtesy of P. M. Ajayan); and (c) model of a nanotube tip exhibiting the locations of the six pentagonal rings (open circles; courtesy of P. M. Ajayan).

and MWNTs) were obtained by Morinobu Endo in the mid-1970s.[10] He observed that tubular graphite of nanometer scale could be produced using this thermolytic process, and imaged the first ever-observed SWNTs and MWNTs.

Based on an hexagonal carbon honeycomb sheet, it is possible to create SWNTs of different chiralities described by two indices (m, n) (Fig. 4). These indices describe precisely how the carbon honeycomb sheet is rolled up into the final tube configuration, determining the direction the sheet is rolled in as well as the final diameter of the tube. Therefore, with the (m, n) indices, one can construct chiral and nonchiral nanotubes (Fig. 4). There are two types of nonchiral tubes: (1) armchair-type tubes, occurring when "$m = n$" (m, m) and (2) zigzag configurations, occurring when "$n = 0$" $(m, 0)$.

In chiral nanotubes "m" and "n" are different (m, n). In 1992, two groups predicted theoretically that the electronic properties of carbon nanotubes would depend on their diameter and chirality: in particular, all of the so-called armchair-type nanotubes could be metallic (see Fig. 4), and zigzag nanotubes could be semiconductors except for the cases in which "$m - n$" is multiple of 3 (see Fig. 4).[3] These results amazed the scientific community because bulk graphite behaves only as a semi-metal, and bulk diamond does not conduct electricity.

The unique electronic properties of carbon nanotubes are due to the quantum confinement of electrons normal to the nanotube axis. In the radial direction, electrons are restricted by the monolayer thickness of the graphene sheet. Consequently, electrons can only propagate along the nanotube axis, and so their wave vector distribution has points. These sharp intensity spikes shown in the density of states (DOS) of the tubes are known as "van Hove" singularities, and are due to this one-dimensional quantum conduction, which is not present in an infinite graphite crystal.[11]

2.3.2. *Carbon nanotube production methods*

Arc discharge method: The technique is similar to the one used for obtaining fullerenes developed by Krätschmer and Huffman,[2] with two main differences: (a) the pressure is higher, around 500 torr (for fullerenes the pressure is around 100 torr) and (b) MWNTs are grown on the cathode and not in the chamber soot. This method produces highly graphitic MWNTs with diameters ranging from 2 to 30 nm (separation between the concentric cylinders is *ca.* 3.4 Å). The length of these nanotubes can be up to 30 μm. Since the electric arc reaction is too violent, it is very difficult to control the

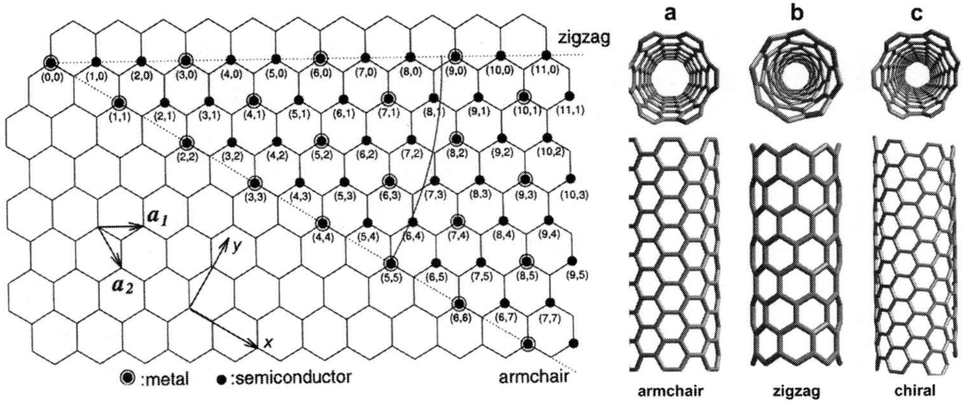

Fig. 4. Molecular models of SWNTs exhibiting different chiralities: (a) armchair configuration; (b) zigzag arrangement; and (c) chiral conformation. Left: Indexed graphene sheet (courtesy of M. S. Dresselhaus). Unitary vectors $\boldsymbol{a}_1$ and $\boldsymbol{a}_2$ are necessary to determine the rolling direction expressed by vector $\boldsymbol{C_n}$. Note that all armchair tubes are metallic, as are all tubes in which $m - n$ is an integer multiple of 3. All other tubes are semiconducting tubes.

formation conditions. In addition to MWNTs, polyhedral particles (giant nested fullerenes) are also formed as a subproduct.

The first reports on the production of SWNTs using the arc appeared in 1993.[3] These authors produced SWNT material by arcing Fe-graphite or Co–Ni-graphite electrodes in a methane–argon or helium atmospheres. Nowadays SWNTs can also be produced using the carbon arc method if metal catalysts (Gd, Co–Pt, Co–Ru, Co, Ni–Y, Rh–Pt, and Co–Ni–Fe–Ce) are introduced into the graphite anode (see Ref. 9). SWNTs produced by the arc discharge are deposited on a "collaret" as a rubbery soot formed around the cathode.

Pyrolysis of hydrocarbons: This technique consists of heating a hydrocarbon or any organic precursor containing carbon, in the presence of a transition metal catalyst such as nickel, cobalt, or iron. Two mechanisms have been proposed for the formation of carbon fibers which can be extended to SWNTs and MWNTs: the first, proposed by Baker *et al.*, consists of the diffusion of carbon through the catalytic particle and the deposition of the carbon material on the other side to form the fibers.[9] In the second, proposed by Baird *et al.* and Oberlin,[10] the filaments are formed by diffusion of carbon around the surface of the catalytic particle to generate the fiber.[9] The carbon diffusion parameters depend critically on the dimensions of the metal particles, the physico-chemical characteristics of the metal used as catalyst, the temperature, the hydrocarbons and the gases involved in the process. Using this method it is now possible to generate patterns of aligned carbon nanotubes[9,12] (Fig. 5).

It is also possible to produce SWNTs via pyrolytic methods. In this context, Dai *et al.*[13] reported the generation of SWNTs via thermolytic processes involving Mo particles in conjunction with CO (in other words, CO disproportionation) at 1200°C. In 1998, Cheng *et al.*[14] demonstrated that it is possible to obtain large amounts of SWNT ropes using experiments similar to those reported by Endo.[14] A novel production method involving the thermolysis of $Fe(CO)_5$ in the presence of CO at elevated pressures (< 10 atm) and temperatures (800–1200°C) was reported to be extremely efficient and nowadays bulk amounts (g/h) can be produced using this method. The latter was named HipCo process, and was developed by the Rice group.[15]

Laser vaporization: For producing MWNTs, the process involves firing a high-power laser (YAG type) toward a graphite target inside a furnace at

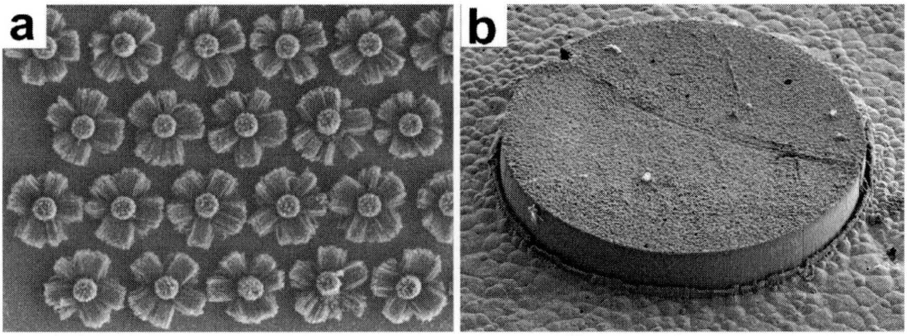

Fig. 5. (a) SEM images of nanoflowers created with patterns of aligned carbon nanotubes grown perpendicularly, by pyrolyzing xylene–ferrocene mixtures over SiO_x substrates (courtesy of P. M. Ajayan) and (b) SEM image of a "nanocake" produced by pyrolyzing benzylamine-toluene-ferrocene solutions over round SiO_x patterns.

1200°C in the presence of Ar.[5,9] Thess *et al.* demonstrated that if nickel or cobalt is added to the graphite target, SWNTs are obtained (see Ref. 9). The SWNTs exhibit diameters around 14 Å forming a two-dimensional crystal with a lattice constant of 17 Å. It is also possible to produce SWNTs using a CO_2 laser focused on a graphite–metal target[16] in the absence of any furnace. Eklund and co-workers[17] reported that ultrafast (sub-picosecond) laser pulses are able to produce large amounts of SWNTs (1.5 g/h).

Electrolysis: The method (developed by Wen Kuang Hsu) involves graphite electrodes immersed in molten LiCl (contained in a graphite crucible) under an argon or air atmosphere (see Refs. 9 and 18), applying a DC voltage between the electrodes. Under optimal conditions, it is possible to generate up to 20–40% of MWNTs using this liquid phase technique. The depth of the cathode and the current (3–10 amp) play an important role in the formation of nanotubes. Other studies carried out at Cambridge University indicated that the nanotube production strongly depends on the molten salt and the temperature of the electrolyte.[19]

Solar vaporization: The method, first developed by Laplaze and co-workers, is able to generate SWNTs and MWNTs, when solar energy is focused on a carbon–metal target in an inert atmosphere.[20] Laplaze and co-workers have been able to scale up this process using solar flux densities in the range of 600–920 W/cm^2, thus resulting in the generation of soot (2 g/h) rich in fullerenes and nanotubes.[20]

2.3.3. *Mechanical properties of carbon nanotubes*

The carbon–carbon bond observed in graphite is one of the strongest in nature, and therefore carbon nanotubes are excellent candidates to be the stiffest and the most robust structure ever synthesized. The first attempt to determine the stiffness of carbon nanotubes was made by Treacy and colleagues (see Ref. 9) using a TEM to measure the amplitudes of vibrating tubes at different temperatures. These authors found that MWNTs exhibit a Young's modulus of the order of 1.2–1.8 TPa, which is higher than that of conventional carbon fibers. Direct measurements using atomic force microscopy (AFM) revealed that the Young's modulus of MWNTs is around 1.28 TPa (see Ref. 9). Subsequently, Richard Superfine's group (see Ref. 9) observed that MWNTs could be bent repeatedly through large angles using an AFM tip, without undergoing catastrophic failure. Morinobu Endo and co-workers also observed by breaking vapor-grown carbon fibers (VGCFs) in liquid nitrogen,[21] that an inner tubule (diameter *ca.* 2 nm or more) could survive this usually catastrophic bending.

However, the values for Young's moduli could decrease considerably due to the presence of defects present within the structure (e.g., pentagon–heptagon pairs, vacancies and interstitials usually present in pyrolytically grown nanotubes).

2.3.4. *Electronic properties of carbon nanotubes*

Individual MWNT conductivity measurements have demonstrated that individual MWNT exhibits unique conductivity properties (resistivities at 300 K of *ca.* 1.2×10^{-4} to 5.1×10^{-6} Ω cm; activation energies < 300 meV for semiconducting tubes) (these measurements were perfomed mainly by Ebbesen's and Lieber's groups; see Ref. 9). Cees Dekker and co-workers (see Ref. 9) carried out the first transport measurements on individual SWNTs (1 nm in diameter), and demonstrated for the first time that the SWNTs behave as quantum wires, in which electrical conduction occurs via well-separated, discrete electron states that are quantum-mechanically coherent over long distances (see Ref. 9).

Using scanning tunneling spectroscopy (STS), it has been possible to determine the electronic local density of states (LDOS) on various individual SWNTs, showing that SWNTs can be either metals or semiconductors, depending on small variations in the (m, n) indices or diameter.[9] The possibility of superconduction in carbon nanotubes

($T_c = 0.55\,\mathrm{K}$) was demonstrated experimentally by Kasumov *et al.*[22] More recently, Tang and co-workers reported that SWNTs of 4 Å in diameter, embedded in zeolite matrices, superconduct below 20 K.[23]

2.3.5. *Thermal properties of carbon nanotubes*

McEuen's group determined that the thermal conductivity for individual MWNTs is higher than that of graphite (3000 W/K) at room temperature, and two orders of magnitude higher than those obtained for bulk aligned MWNT carpets.[24] Hone *et al.* have measured the thermal conductivity of: (a) mats of randomly oriented SWNTs (35 W/m K) and (b) aligned SWNTs (> 200 W/m K). The thermal conductivity of SWNTs at low temperatures exhibited linear acoustic bands contributing to the thermal transport at the lowest temperatures and optical sub-bands entering at higher temperatures.[25]

2.3.6. *Carbon nanocones*

It is also possible to create graphite-like cones, when five or less carbon pentagons are inserted in a graphene sheet (see Ref. 9; Fig. 6(a)). Various authors reported the existence of graphite cones and conical fibers using thermolytic approaches involving carbon precursors.[26] Charlier's group[27] has calculated the electronic properties of nanocones, thus finding that there is a charge accumulation toward the tip and that there are localized states near the Fermi level. Thus these structures may be suitable for electron field emitters. In addition, the synthesis of coalesced graphitic nanocones has been achieved by heating conical nanofibers up to 2800°C;[28] the electronic properties of coalesced graphitic cones have been calculated showing that the coalesced edges with heptagons and pentagons play a crucial role in the LDOS.[28]

2.3.7. *Negatively curved graphite: Helices, toroids, and schwarzites*

In 1992, Alan Mackay and Humberto Terrones[29] predicted that the presence of heptagonal or octagonal rings in sp^2 carbon would induce structures with negative curvature (Fig. 6(b)). A year later, Iijima *et al.*[30] reported this curvature phenomenon in carbon nanotubes due to the presence of an extra pentagon–heptagon pair within a tubule. By combining heptagons and pentagons in the predominantly hexagonal carbon framework, helicoidal graphite or helically shaped carbon tubes can be generated (Fig. 6(c)) (Ref. 18 and references therein).

It is also possible to produce toroidal nanocarbons when a different arrangement of heptagonal and pentagonal rings is embedded in the hexagonal carbon network (Fig. 6(d)). The first researchers to observe hemi-toroidal nanotube caps (axially elongated concentric doughnuts) were Iijima and Ajayan,[31] when studying the morphologies of nanotube caps of material produced in the arc discharge generator. Subsequently, Endo and co-workers found similar hemi-toroidal nanotubes produced in pyrolytic processes.[21]

The groups of Dekker and Smalley reported the existence of SWNT bundles forming tori (see Ref. 18 and references therein). In these structures, the presence of heptagons and pentagons is not necessary, because a tubule

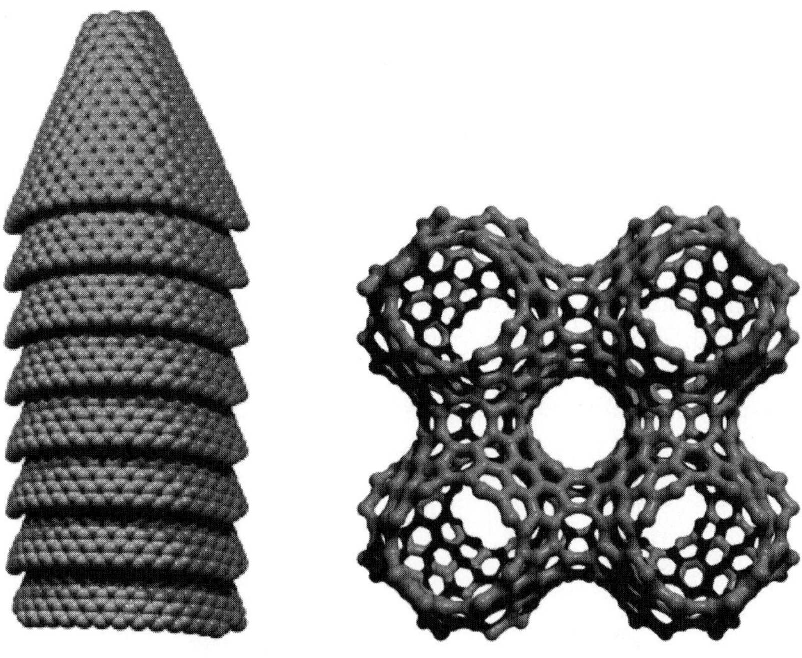

(a) (b)

Fig. 6. (a) Stacked cone carbon nanofiber. (b) Four cells of a periodic negatively curved porous carbon. (c) Helically coiled carbon made with heptagons, hexagons, and pentagons. (d) Toroidal carbon obtained by joining C_{60} molecules along the five-fold axis (the molecules coalesce and get distorted by the presence of rings with more than six carbon atoms which are necessary to join them).

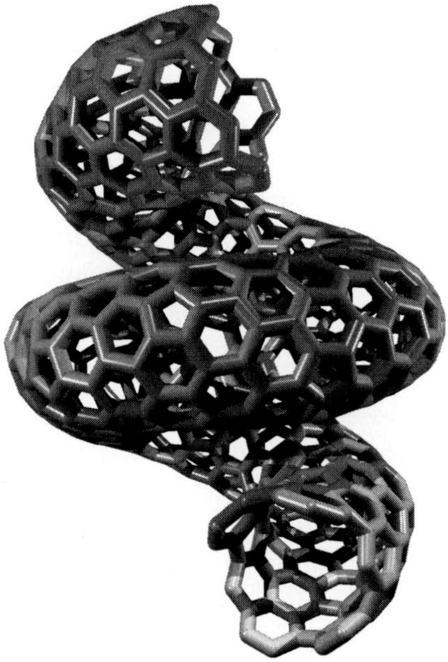

(c)

(d)

Fig. 6. (*Continued*)

can close in a doughnut configuration by curving so that its ends join. More recently, IBM researchers headed by Ph. Avouris were able to produce large quantities of bundles of SWNT tori.[32] More complicated negatively curved graphitic structures, based on the decoration of minimal surfaces (structures termed *Schwarzites*, analogous to zeolites) have also been proposed by Mackay and Terrones.[33] Recently, the magnetic response of carbon nanotori has been studied theoretically, and it is shown that depending on the arrangement of hexagons, pentagons, and heptagons it is possible to obtain different behaviors which can be paramagnetic or diamagnetic.[34]

2.3.8. *Haeckelites*

It has been proposed that there may exist a new hypothetical type of graphene sheets which admit pentagons, heptagons, and hexagons, in which the number of heptagons and pentagons should be the same, the negative curvature of the heptagons compensating for the positive curvature of the pentagons.[35] These arrangements have been called "Haeckelites" in honor of Ernst Haeckel, Professor of Zoology who produced beautiful drawings of radiolaria (a type of planktonic organism) in which such heptagonal, hexagonal, and pentagonal rings can be observed.[35] Some Haeckelite systems are predicted to be highly metallic (and even superconductors). The calculated Young's modulus of Haeckelite tubes is *ca.* 1.0 TPa.

3. The Future of Carbon Nanostructures: Applications and Emerging Technologies

3.1. *Field emission sources*

When a potential is applied between a carbon nanomaterial surface and an anode, electrons are easily emitted. Using this principle, nanocarbons can be used as efficient field emission sources for the fabrication of multiple electronic devices[36] such as flat panel displays, intense light sources or bright lamps, and X-ray sources. The clear advantages of using carbon nanotubes as electron emission devices are: (1) stable field emission over prolonged time periods; (2) long lifetimes of the components; (3) low emission threshold potentials; (4) high current densities; and (5) absence of the necessity for ultrahigh vacuum. Nanotube-based lamps using MWNTs are relatively cheap to manufacture and have lifetimes > 8000 h and high efficiency for the green (phosphor) color "light bulb". The generation of X-rays can also be achieved if metal targets replace the phosphorus screen and the accelerating voltage is much larger.[36]

3.2. *Scanning probe tips*

It has been possible to attach MWNTs to scanning probe microscope tips so that a better image resolution is achieved when compared to standard nanoprobes.[37] The mechanical strength of carbon nanotubes and the facility to bend and recoil make this material an excellent candidate for the production of long-life microscope tips.

3.3. *Li ion batteries*

For the fabrication of lightweight and efficient batteries, it is possible to intercalate Li ions within graphite-like structures, so that Li^+ can migrate from a graphitic anode to the cathode (usually $LiCoO_2$, $LiNiO_2$, and $LiMn_2O_4$). The charge and discharge phenomena in these batteries are based on the Li^+ intercalation and de-intercalation.[38] These batteries exhibit a superior performance when compared to other batteries such as those with NiCd and NiMH.[38] However, MWNTs and SWNTs could now be used in the fabrication of these types of batteries.

3.4. *Electrochemical devices: Supercapacitors and actuators*

In general, capacitances between 15–200 F/g have been observed for MWNT arrays. The values can result in large quantities of charge being injected when only a few volts are applied (Ref. 36 and references therein). In particular, MWNT supercapacitors are used for applications that require high power capabilities and higher storage capacities (power densities of *ca.* 20 kW/kg at energy densities of *ca.* 7 W-h/kg). On the other hand, carbon nanotube actuators can work at low voltages and temperatures as high as 350°C.[36] For example, the maximum stress observed in SWNTs is *ca.* 26 MPa, a value which is 100 times larger than that observed in natural muscles.[36]

3.5. *Molecular sensors*

Lieber and co-workers[39] were the first to demonstrate, using chemical force microscopy (CFM) techniques, that it is possible to sense functional chemical groups attached at the nanotube ends. Other groups were able to sense ppm of toxic gases such as NO_2 or NH_3,[40] and detect electrical changes due to the different atmospheres (i.e., from vacuum to air; Refs. 41 and 42).

3.6. *Carbon–carbon nanocomposites: Joining and connecting carbon nanotubes*

Various attempts have been made to connect these tubes covalently. Very recently, it was demonstrated[43,44] that controlled electron beam exposure at elevated temperatures is capable of coalescing SWNTs (Fig. 7) and creating "Y", "X", and "T" SWNT molecular junctions (Fig. 8). The authors

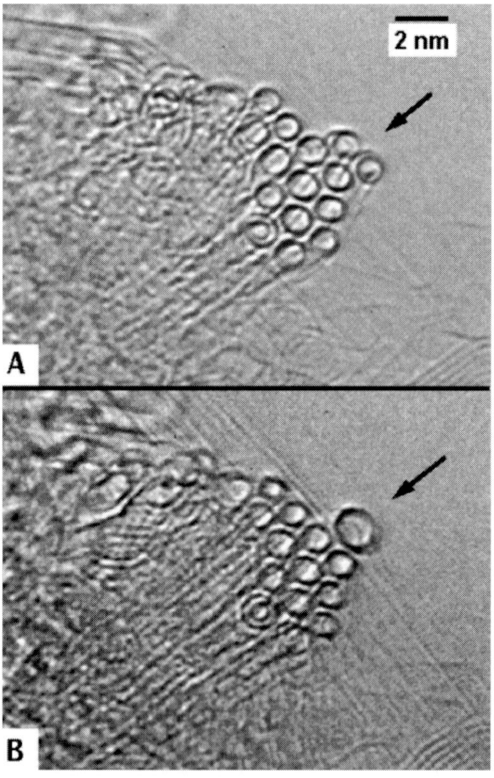

Fig. 7. *In situ* coalescence of SWNTs along the tube axis produced by electron irradiation. One of the nanotubes in (b) is double in diameter (see arrow; Terrones *et al.*, 2000).

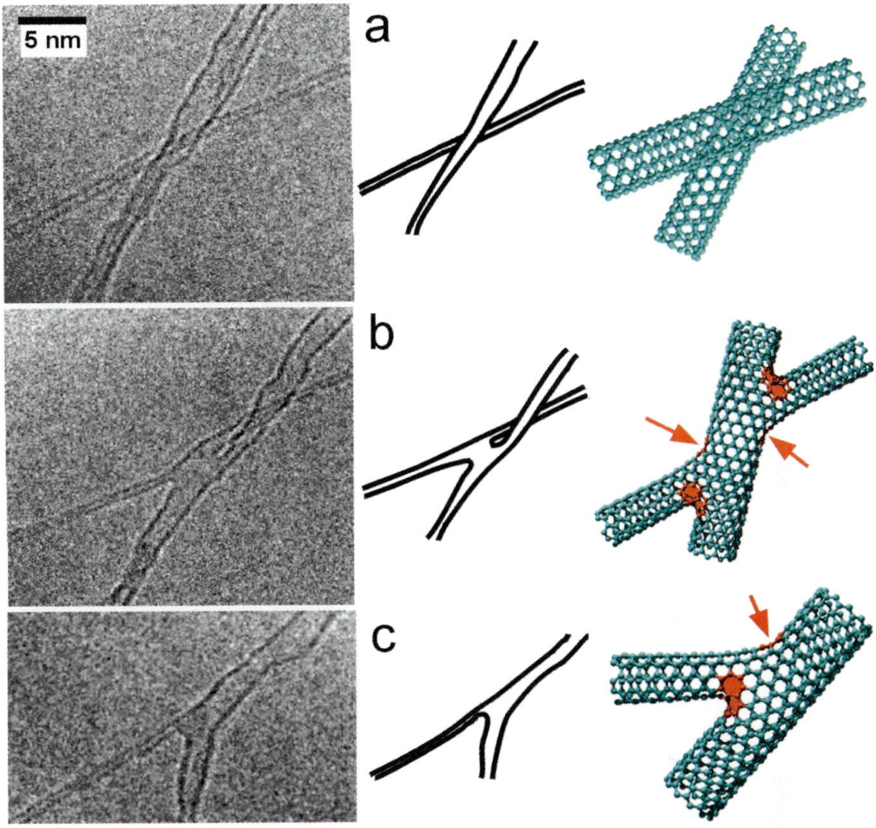

Fig. 8. (a) An SWNT of *ca.* 2.0 nm (running from bottom left diagonally toward top right) crossing with an individual SWNT of *ca.* 0.9 nm; (b) after 60 s, electron irradiation promotes a molecular connection forming an "X" junction (schematics and molecular models are shown for visualization); and (c) subsequent electron irradiation of the structure promotes breakage of the thin extremity, thus resulting in a "Y" junction (Terrones *et al.*, 2002).

concluded that the creation of SWNT junctions should involve: (a) defect formation (e.g., vacancies, interstitials, dangling bonds); (b) surface and atom reconstruction initiated by high electron irradiation; and (c) thermal annealing. The production of these novel carbon–carbon nanocomposites using SWNTs would certainly revolutionize specific areas in electronics (e.g., formation of memory devices, circuits using 2D SWNT matrices, and extra-light and super-robust fabrics using 2D and 3D SWNT networks).

3.7. *Gas and hydrogen storage*

H_2 and Ar can be stored in SWNTs[45,46] and MWNTs,[47] respectively. It is important to emphasize that H_2 storage in carbon nanotubes may be advantageous in the fabrication of fuel cells mainly for powering electric vehicles. Unfortunately, there has been controversy over the high H_2 storage capacity of carbon nanotubes (from 0.1 to 66 wt%).[48] More recent studies have indicated that the hydrogen uptake is lower than 2%, being the highest for HipCo produced SWNTs.[49] Hirscher *et al.*[49] also showed that the influence of impurities such as Ti (coming from the sonication probe) might be responsible for the results reported previously with uptakes up to 7%. At present, it seems that nanotubes may not be the best material for storing hydrogen, but additional experiments and further calculations should be carried out to clarify these results. Nevertheless, porous nanocarbons have shown to have a grater hydrogen intake up to 2.7 wt%.[50]

3.8. *Nanotube electronic devices*

It has been possible to fabricate a three-terminal switchable device based on a semiconducting single nanotube (see Ref. 9). It was also found that metal–metal, metal–semiconductor, or semiconductor–semiconductor nanotube junctions (created by inserting five to seven defects) can indeed be measured.[51] In 2001, two groups demonstrated (see Ref. 36) for the first time the fabrication of field effect transistors, exhibiting a high gain (> 10), a large on-off ratio, and room-temperature operation. Regarding the construction of p–n–p devices, it has been shown that N-doped carbon nanotubes behave as n-type nanowires,[52] whereas B-doped tubes act as p-type conductors.[53] IBM experts envision that in a decade carbon nanotubes could be used in high-performance electronic devices.

3.9. *Biological devices*

There has not been much work on biologic applications of carbon nanotubes. However, it should be possible to inhibit viruses, by attaching them to the surface of nanotubes, similarly to the way in which Au clusters attach to the surface of CN_x nanotubes.[54] In addition, enzymes could be added to the surface of MWNTs so that their activity increases. At present little progress has been carried out along this line, but it is clear that bionanodevices will appear in the near future.

3.10. *Nanotube polymer composites*

The combination of high aspect ratio, small size, strength, stiffness, low density and high conductivity make carbon nanotubes perfect candidates as fillers in polymer composites. It has been demonstrated that the presence of 0.1 wt% of MWNTs in epoxy resins increases the elastic modulus by 20%.[55]

3.11. *Nanotube ceramic composites*

MWNTs coated with SiO_x have revealed a higher oxidation resistance ($< 1200°C$),[56] oxidation being a common drawback of all-carbon materials. Other nanotube composites have been fabricated using alumina and SiC. However, little has been done in this direction.

3.12. *Layered coated nanotubes*

It is also possible to alter the mechanical and transport properties of carbon nanotubes by coating the external tube surface or by inserting metals in the hollow core of the cylinders. These modified tubes can then be used as fillers in specific composites. Ajayan and co-workers were the first to coat MWNTs with V_2O_5.[57] Hsu *et al.* also demonstrated that it is possible to coat carbon nanotubes and nanoparticles with MoS_2 and WS_2.[58] Whitby *et al.*[59] managed to coat individual MWNTs with single layers of WS_2.[59]

4. Conclusions and Future Work

Due to their abundance and great potential, carbon nanomaterials will be taking an important role in the development of emerging technologies in the near future. Carbon nanotubes will be the first of this class to find industrial applications within 1–5 years. However, larger quantities of nanotubes are needed if composite materials are to be launched in various markets (ton/day). This is only the tip of the iceberg since some nanocarbons are ready to be applied in the fabrication of novel devices, and other new carbon structures still need to be synthesized. In the next ten years, conductive paints and plastics, as well as flexible and lightweight magnets containing Fe-filled carbon nanotubes and other nanocarbons will become a reality. The inhibition of some viruses and bacteria using nanoscale carbons will certainly be achieved within the next two years. Some of the predicted structures that are extremely stable (e.g., haeckelites, schwartzites), should also be synthesized very shortly. At present, we are still witnessing novel properties of carbon nanomaterials, such as the ignition of SWNTs when exposed to a standard photographic flash.[60] The nanocarbon field is vast, and various types of nanoscale carbon will undoubtedly become an important part in the development of smart materials in the new millennium. We have to add that the results obtained with carbon have opened new possibilities in other layered materials such as boron nitride, tungsten disulphide,

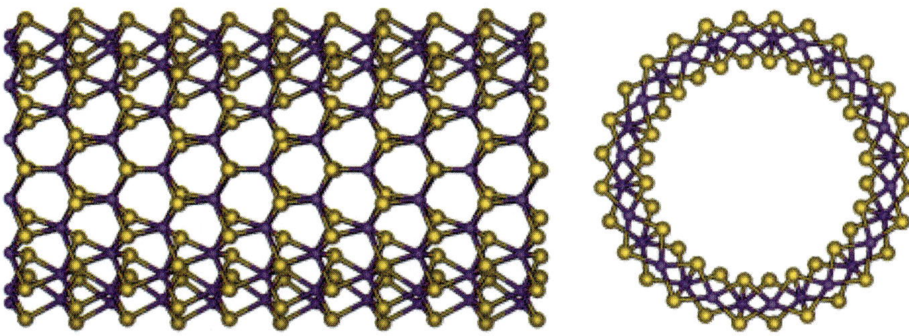

(a)

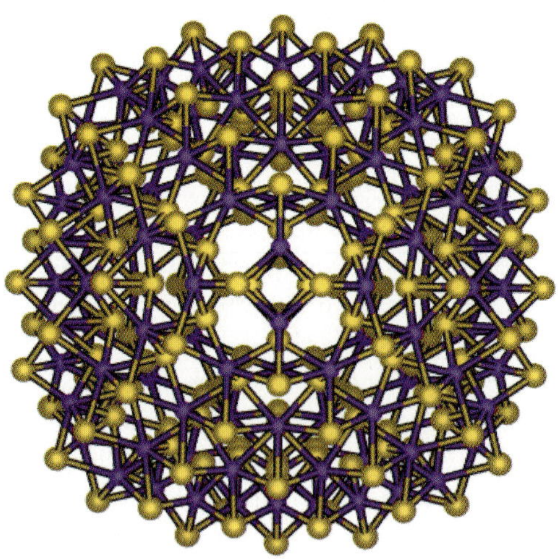

(b)

Fig. 9. (a) Molybdenum sulphide zigzag-type nanotube. (b) Molybdenum sulphide octahedral cage.

molybdenum disulphide, etc., which can acquire curvature to form nanotubes, fullerene-like structures, and other morphologies (see Fig. 9).

Acknowledgments

We are indebted to P. M. Ajayan, A. L. Mackay, H. W. Kroto, F. Banhart, J. C. Charlier, M. Endo, M. Rühle, T. Hayashi, J. A. Rodríguez-Manzo, F. López-Urías and E. Muñoz-Sandoval for stimulating discussions and valuable assistance in some of the work, presented here. We are also grateful to CONACYT-México grants 45762 and 45772, CIAM 41464, CIAM 42428, 2004-01-013/SALUD-CONACYT, 2004-C02-9/Puebla-Fondo-Mixto for financial support.

References

1. H. W. Kroto, C_{60}, buckminsterfullerene: The celestial sphere that fell to earth, *Ang. Chem. Int. Edition* **31**, 111–129 (1992).
2. W. Krätschmer, L. D. Lamb, K. Fostiropoulos and D. R. Huffman, Solid C60: A new form of carbon, *Nature* **347**, 354 (1990).
3. M. S. Dresselhaus, G. Dresselhaus and P. C. Eklund, *Science of Fullerenes and Carbon Nanotubes*, 1st edn. (Academic Press, San Diego, 1996), pp. 1–505.
4. S. H. Friedman, D. L. Decamp, R. P. Sijbesma, G. Srdanov, F. Wudl and G. L. Kenyon, Inhibition of the HIV-1 protease by fullerene derivatives model-building studies and experimental-verification, *JACS* **115**, 6506–6509 (1993).
5. P. J. F. Harris, *Carbon Nanotubes and Related Structures: New Materials for the XXI Century*, 1st edn. (Cambridge University Press, Cambridge, 1999), pp. 1–293.
6. D. Ugarte, Curling and closure of graphitic networks under electron-beam irradiation, *Nature* **359**, 707–709 (1992).
7. M. Terrones and H. Terrones, The role of defects in graphitic structures, *Full. Sci. Tech.* **4**, 517–533 (1996).
8. S. Iijima, Helical microtubules of graphic carbon, *Nature* **354**, 56–58 (1991).
9. M. Terrones, W. K. Hsu, H. W. Kroto and D. R. M. Walton, Nanotubes: A revolution in material science and electronics, *Fullerenes and Related Structures*, Topics in Chemistry Series, Vol. 199, ed. A. Hirsch (Springer-Verlag, 1998), Chapter 6, pp. 189–234.
10. A. Oberlin, M. Endo and T. Koyama, Filamentous growth of carbon through benzene decomposition, *J. Cryst. Growth* **32**, 335–349 (1976).
11. J. C. Charlier and J. P. Issi, Electronic structure and quantum transport in carbon nanotubes, *Appl. Phys. A. Mat. Sci. Proc.* **67**, 79–87 (1998).
12. B. Q. Wei, R. Vajtai, Y. Jung, J. Ward, R. Zhang, G. Ramanath and P. M. Ajayan, Organized assembly of carbon nanotubes — Cunning refinements

help to customize the architecture of nanotube structures, *Nature* **416**, 495–496 (2002).

13. H. J. Dai, A. G. Rinzler, P. Nikolaev, A. Thess, D. T. Colbert and R. E. Smalley, Single-wall nanotubes produced by metal-catalyzed disproportionation of carbon monoxide, *Chem. Phys. Lett.* **260**, 471–475 (1996).
14. H. M. Cheng, F. Li, G. Su, H. Y. Pan, L. L. He, X. Sun and M. S. Dresselhaus, Large-scale and low-cost synthesis of single-walled carbon nanotubes by the catalytic pyrolysis of hydrocarbons, *Appl. Phys. Lett.* **72**, 3282–3284 (1998).
15. P. Nikolaev, M. J. Bronikowski, R. K. Bradley, F. Rohmund, D. T. Colbert, K. A. Smith and R. E. Smalley, Gas-phase catalytic growth of single-walled carbon nanotubes from carbon monoxide, *Chem. Phys. Lett.* **313**, 91–97 (1999).
16. W. K. Maser, E. Munoz, A. M. Benito, M. T. Martinez, G. F. de la Fuente, Y. Maniette, E. Anglaret and J. L. Sauvajol, Production of high-density single-walled nanotube material by a simple laser-ablation method, *Chem. Phys. Lett.* **292**, 587–593 (1998).
17. P. C. Eklund, B. K. Pradhan, U. J. Kim, Q. Xiong, J. E. Fischer, A. D. Friedman, B. C. Holloway, K. Jordan and M. W. Smith, Large-scale production of single-walled carbon nanotubes using ultrafast pulses from a free electron laser, *Nano Lett.* **2**, 561–566 (2002).
18. M. Terrones, W. K. Hsu, J. P. Hare, D. R. M. Walton, H. W. Kroto and H. Terrones, Graphitic structures: From planar to spheres, toroids and helices, *Phil. Trans. Roy. Soc. A* **354**, 2025–2054 (1996).
19. G. Z. Chen, X. D. Fan, A. Luget, M. S. P. Shaffer, D. J. Fray and A. H. Windle, Electrolytic conversion of graphite to carbon nanotubes in fused salts, *J. Electroanal. Chem.* **446**, 1–6 (1998).
20. L. Alvarez, T. Guillard, J. L. Sauvajol, G. Flamant and D. Laplaze, Solar production of single-wall carbon nanotubes: Growth mechanisms studied by electron microscopy and Raman spectroscopy, *Appl. Phys. A* **70**, 169–173 (2000).
21. M. Endo, K. Takeuchi, S. Igarashi, K. Kobori, M. Shiraishi, H. W. Kroto and A. Sarkar, Pyrolytic carbon nanotubes from vapor-grown carbon fibers, *Carbon* **33**, 873–881 (1995).
22. M. Kociak, A. Y. Kasumov, S. Gueron, B. Reulet, I. I. Khodos, Y. B. Gorbatov, V. T. Volkov, L. Vaccarini and H. Bouchiat, Superconductivity in ropes of single-walled carbon nanotubes, *Phys. Rev. Lett.* **86**, 2416–2419 (2001).
23. Z. K. Tang, L. Y. Zhang, N. Wang, X. X. Zhang, G. H. Wen, G. D. Li, J. N. Wang, C. T. Chan and P. Sheng, Superconductivity in 4-anstrom single-walled carbon nanotubes, *Science* **292**, 2462–2465 (2001).
24. P. Kim, L. Shi, A. Majumdar and P. L. McEuen, Thermal transport measurements on individual multiwalled nanotubes, *Phys. Rev. Lett.* **87**, 215502–215505 (2001).
25. J. Hone, B. Batlogg, Z. Benes, A. T. Johnson and J. E. Fischer, Quantized phonon spectrum of single-wall carbon nanotubes, *Science* **289**, 1730–1733 (2000).

26. H. Terrones, T. Hayashi, M. Muñoz-Navia, M. Terrones, Y. A. Kim, N. Grobert, R. Kamalakaran, J. Dorantes-Dávila, R. Escudero, M. S. Dresselhaus and M. Endo, Graphitic cones in palladium catalysed carbon nanofibres, *Chem. Phys. Lett.* **343**, 241–250 (2001).
27. J. C. Charlier and G.-M. Rignanese, Electronic structure of carbon nanocones, *Phys. Rev. Lett.* **86**, 5970–5973 (2001).
28. M. Muñoz-Navia, G. Dorantes-Dávila, M. Terrones, T. Hayashi, Y. A. Kim, M. Endo, M. Dresselhaus and H. Terrones, Synthesis and electronic properties of coalesced graphitic cones, *Chem. Phys. Lett.* **407**, 327–332 (2005).
29. A. L. Mackay and H. Terrones, Diamond from graphite, *Nature* **352**, 762 (1991).
30. S. Iijima, T. Ichihashi and Y. Ando, Pentagons, heptagons and negative curvature in graphite microtubule growth, *Nature* **356**, 776–778 (1992).
31. S. Iijima, P. M. Ajayan and T. Ichihashi, Growth-model for carbon nanotubes, *Phys. Rev. Lett.* **69**, 3100–3103 (1992).
32. R. Martel, H. R. Shea and Ph. Avouris, Rings of single-walled carbon nanotubes, *Nature* **398**, 299 (1999).
33. A. L. Mackay and H. Terrones, Hypothetical graphite structures with negative Gaussian curvature, *Phil. Trans. Roy. Soc. London A* **343**, 113–127 (1993).
34. A. Rodríguez-Manzo, F. López-Urías, M. Terrones and H. Terrones, Magnetism in corrugated carbon nanotori: The importance of symmetry defects, and negative curvature, *Nano Lett.* **4**, 2179–2183 (2004).
35. H. Terrones, M. Terrones, E. Hernández, N. Grobert, J.-C. Charlier and P. M. Ajayan, New metallic allotropes of planar and tubular carbon, *Phys. Rev. Lett.* **84**, 1716–1719 (2000).
36. R. H. Baughman, A. A. Zakhidov and W. A. De Heer, Carbon nanotubes — The route toward applications, *Science* **297**, 787–792 (2002).
37. H. J. Dai, J. H. Hafner, A. G. Rinzler, D. T. Colbert and R. E. Smalley, Nanotubes as nanoprobes in scanning probe microscopy, *Nature* **384**, 147–150 (1996).
38. M. Endo, C. Kim, K. Nishimura, T. Fujino and K. Miyashita, Recent developments of carbon materials for Li-ion batteries, *Carbon* **38**, 183–197 (2000).
39. S. S. Wong, E. Joselevich, A. T. Woolley, C. L. Cheung and C. M. Lieber, Covalently functionalized nanotubes as nanometre-sized probes in chemistry and biology, *Nature* **394**, 52–55 (1998).
40. J. Kong, N. R. Franklin, C. W. Zhou, M. G. Chapline, S. Peng, K. J. Cho and H. J. Dai, Nanotube molecular wires as chemical sensors, *Science* **287**, 622–625 (2000).
41. P. G. Collins, K. Bradley, M. Ishigami and A. Zettl, Extreme oxygen sensitivity of electronic properties of carbon nanotubes, *Science* **287**, 1801–1804 (2000).
42. F. Villalpando-Páez, A. H. Romero, E. Muñoz-Sandoval, L. M. Martínez, H. Terrones and M. Terrones, Fabrication of vapor and gas sensors using films of aligned CN_x nanotubes, *Chem. Phys. Lett.* **336** 137–143 (2004).
43. M. Terrones, H. Terrones, J. C. Charlier, F. Banhart and P. M. Ajayan, Coalescence of single-walled carbon nanotubes, *Science* **288**, 1226–1229 (2000).

44. M. Terrones, F. Banhart, N. Grobert, J. C. Charlier, H. Terrones and P. M. Ajayan, Molecular junctions by joining single-walled carbon nanotubes, *Phys. Rev. Lett.* **89**, 75505 (2002).
45. A. C. Dillon, K. M. Jones, T. A. Bekkedahl, C. H. Kiang, D. S. Bethune and M. J. Heben, Storage of hydrogen in single-walled carbon nanotubes, *Nature* **386**, 377–379 (1997).
46. C. Liu, Y. Y. Fan, M. Liu, H. T. Cong, H. M. Cheng and M. S. Dresselhaus, Hydrogen storage of single-walled carbon nanotubes at room temperature, *Science* **286**, 1127–1129 (1999).
47. M. Terrones, R. Kamalakaran, T. Seeger and M. Rühle, Novel nanoscale gas containers: Encapsulation of N_2 in CN_x nanotubes, *Chem. Commun.* **23**, 2335–2336 (2000).
48. A. Chambers, C. Park, R. T. K. Baker and N. M. Rodriguez, Hydrogen storage in graphite nanofibers, *J. Phys. Chem. B* **102**, 4253–4256 (1998).
49. M. Hirscher, M. Becher, M. Haluska, A. Quintel, V. Skakalova, Y. M. Choi, U. Dettlaff-Weglikowska, S. Roth, I. Stepanek, P. Bernier, A. Leonhardt and J. Fink, Hydrogen storage in carbon nanostructures, *J. Alloy. Compound.* **330**, 654–658 (2002).
50. E. Terres, B. Panella, T. Hayashi, Y. A. Kim, M. Endo, J. M. Domínguez, M. Hirscher, H. Terrones and M. Terrones, Hydrogen storage in spherical nanoporous carbons, *Chem. Phys. Lett.* **403**, 303–366 (2005).
51. Z. Yao, H. W. C. Postma, L. Balents and C. Dekker, Carbon nanotubes intramolecular junctions, *Nature* **402**, 273–276 (1999).
52. R. Czerw, M. Terrones, J.-C. Charlier, X. Blasé, B. Foley, R. Kamalakaran, N. Grobert, H. Terrones, D. Tekleab, P. M. Ajayan, W. Blau, M. Rühle and D. L. Carroll, Identification of electron donor states in N-doped carbon nanotubes, *Nano Lett.* **1**, 457–460 (2001).
53. X. Blase, J.-C. Charlier, A. De Vita, R. Car, P. H. Redlich, M. Terrones, W. K. Hsu, H. Terrones, D. L. Carroll and P. M. Ajayan, Boron-mediated growth of long helicity-selected carbon nanotubes, *Phys. Rev. Lett.* **83**, 5078–5081 (1999).
54. K. Y. Jiang, A. Eitan, L. S. Schadler, P. M. Ajayan, R. W. Siegel, N. Grobert, M. Mayne, M. Reyes-Reyes, H. Terrones and M. Terrones, Selective attachment of gold nanoparticles to nitrogen-doped carbon nanotubes, *Nano Lett.* **3**, 275–277 (2003).
55. X. J. Xu, M. M. Thwe, C. Shearwood and K. Liao, Mechanical properties and interfacial characteristics of carbon-nanotube-reinforced epoxy thin films, *Appl. Phys. Lett.* **81**, 2833–2835 (2002).
56. T. Seeger, Th. Köhler, Th. Frauenheim, N. Grobert, M. Rühle, M. Terrones and G. Seifert, Nanotube composites: Novel SiO_2 coated carbon nanotubes, *Chem. Commun.* **35**, 34–35 (2002).
57. P. M. Ajayan, O. Stephan, P. Redlich and C. Colliex, Carbon nanotubes as removable templates for metal-oxide nanocomposites, *Nature* **375**, 564–567 (1995).
58. W. K. Hsu, Y. Q. Zhu, H. W. Kroto, D. R. M. Walton, R. Kamalakaran and M. Terrones, C-MoS_2 and C-WS_2 nanocomposites, *Appl. Phys. Lett.* **77**, 4130–4132 (2000).

59. R. L. D. Whitby, W. K. Hsu, C. B. Boothroyd, P. K. Fearon, H. W. Kroto and D. R. M. Walton, Tungsten disulphide sheathed carbon nanotubes, *ChemPhysChem.* **2**, 620–623 (2001).
60. P. M. Ajayan, M. Terrones, A. de la Gaurdia, V. Huc, N. Grobert, B. Q. Wei, H. Lezec, G. Ramanath and T. W. Ebbesen, Nanotubes in a flash: Ignition and reconstruction, *Science* **296**, 705 (2002).

Humberto Terrones, born in Mexico City 1962, studied at Iberoamericana University, where he graduated with first class honors in engineering physics in 1987. He obtained his PhD from Birkbeck College, University of London. After a postdoctoral stay at the University of Cambridge, UK, he returned to Mexico to the Institute of Physics, UNAM, in Mexico City, where he worked until 2001, when he was appointed as head of the Advanced Materials Department at a newly formed multidisciplinary institute (IPICYT) in San Luis Potosi, Mexico. In 1997 he was elected a member of the Mexican Academy of Sciences and in 2004 was elected as fellow of the TWAS (Academy of Sciences for the Developing World). His main field of research is the theory and experiment of nanostructured materials and nanotechnology. He has published more than 120 papers in international journals and in the 2000 received the Mexican Academy of Sciences Award to young scientists and the UNAM award to young scientists. He is part of the editorial board of several scientific journals. Other interests include applied mathematics.

Mauricio Terrones, born in Mexico City in 1968, obtained his BSc degree in Engineering Physics with first class honors at Universidad Iberoamericana, and was distinguished as the Best Student of Mexico in Engineering Physics in 1992. In 1994 he started his doctorate degree with Sir Prof. Harold W. Kroto (Nobel Laureate, FRS) and David R. M. Walton (FRSC),

and received his DPhil degree from University of Sussex in 1998. He has co-authored more than 175 publications in international journals, and counts with more than 3500 independent citations to his work. For example, he has published three papers in *Nature*, three in *Science*, seven in *Physical Review Letters*, 16 in *Nano Letters*, etc. In 1999, he was awarded the Alexander von Humboldt Fellowship, and carried out research at the Max-Planck Institut für Metallforschung (Stuttgart, Germany). In 2000, he was recipient of the Mexican National Prize for Chemistry. He also received the Javed Husain Prize and the Albert Einstein medal from UNESCO in 2001, for his contributions to Carbon Nanoscience and Nanotechnology. In 2005, he was awarded the TWAS Prize in Engineering Physics; being Mauricio the youngest scientist ever to receive any TWAS award. He is member of the Mexican Academy of Sciences since 2002. Aged 38, he is currently full Professor at the Instituto Potosino de Investigación Científica y Tecnológica (IPICYT-Mexico), and is leader of the National Laboratory for Nanoscience and Nanotechnology (based at IPICYT). His research now concentrates on the theory, synthesis and characterization of novel nanomaterials and applications.

CHAPTER 2

INORGANIC NANOWIRES

Caterina Ducati

Department of Materials Science and Metallurgy
University of Cambridge
Pembroke Street
Cambridge CB2 3QZ, UK

High aspect ratio inorganic structures with at least one dimension between 1 and 100 nm, such as nanotubes, nanowires, and nanorods, have attracted a keen interest both for fundamental research and for advanced applications because of their unique and fascinating properties, largely contributing to the rise of nanoscience as a new discipline of scientific progress worldwide. Although the synthesis of high aspect ratio nanostructures has been achieved through many different strategies, the exact mechanisms that lead to their formation are not clear and need to be elucidated to control the properties and reliability of these nanoscopic "building blocks". Here I give few examples of growth methods based on chemical and physical processes, and I describe the results of the first characterization of the nanostructures by high-resolution electron microscopy techniques. The analysis of surfaces and interfaces, epitaxial relationships, and defects is expected to advance the understanding of the growth strategies and of the properties at the nanoscale. Far from being a complete review, this report is meant to show the achievements and the difficulties of experimental research in nanoscience, and to introduce the challenges that lay ahead.

Keywords: Semiconductor nanowires, metal oxide nanorods, composite nanowires, synthesis, TEM.

1. Introduction

Nanoscale crystalline materials play a key role as active components in a wide range of advanced technologies, such as nanoelectronics,[1,2] chemical sensing,[3–5] and energy storage.[6,7] They also provide model systems to study how different physical properties can vary, as a function, the number of constituent atoms and shape of the nanostructure: for example, quantum conductance,[8] light emission,[9] and chemical reactivity.[10] Following the discovery of carbon nanotubes,[11] a large worldwide effort has been devoted to the synthesis of high aspect ratio inorganic nanotubes, nanowires, and nanorods with well-defined structural and electronic properties. Layered compound, such as MoS_2, WS_2, and mixed phase metal disulphides, were among the first inorganic materials to be synthesized in nanotube form.[12–14] Semiconductors, metals, and rare earths as pure elements or in compounds have been used to produce nanotubes and nanowires using different methods, from chemical vapor deposition to sol–gel reactions. Among the metal oxides, TiO_2, MoO_3, and V_2O_5 nanotubes and ZnO nanowires have been synthesized recently.[15–17]

"High aspect ratio nanostructure" is a broad term for a class of nanoscale objects. Different names are used to describe the specific morphology: nanowires and nanorods have diameters $>$ 100 or 10–20 times smaller than their length, respectively, with constant cross section; nanotubes have a hollow core; nanobelts are similar to nanowires in terms of aspect ratio but are flat, crystalline strips; nanosaws are nanobelts with at least one toothed edge, etc.

The synthesis of high aspect ratio crystalline nanostructures can be achieved following several approaches that allow some degree of control over dimensions, morphology, and uniformity of the products. In general the so-called "bottom-up" strategies, in which nanostructures are grown from smaller atomic components, have been preferred over their "top-down" counterparts because they achieve smaller dimensions, are more controlled, and create less waste. For the "bottom-up" class of techniques the growth involves two fundamental steps: (i) nucleation and (ii) growth. Both are essential aspects of a phase transformation process that must be understood in detail if we want to control the yield and properties of our nanomaterials. Nucleation in the presence of a seed or catalyst nanoparticle, often referred to as heterogenous nucleation, generally allows control over the position and the diameter of the nanowire. Precipitation of a stable nucleus of the same composition as the nutrient phase (i.e., homogenous nucleation) ensures a cleaner and more uniform product, but is less frequently applied because it is intrinsically less likely to lead to the growth of high aspect ratio nanostructures. As for the proper growth stages, the evolution of the nanostructures is determined by a delicate balance of driving forces and limiting factors. The overall result can be seen as a sum of contributions where each of the partial driving forces (e.g., temperature or concentration gradients) is a function of particle shape and growth rate. The shape of crystals is influenced by equilibrium as well as kinetic effects, that is, by the minimization of the free energy of the system and by the dynamics of layer generation and flow.[18] In the case of nanostructures, where the surface to volume ratio is so high, it is reasonable to assume that the kinetic considerations should prevail. The growth habit has a profound effect on the quality and quantity of the defects incorporated in the nanostructures, as well as on the nature of their surfaces. Defects and interfaces are known to be important in determining the properties of bulk materials, and they are even more crucial for nanoscale materials. For some applications, such as nanoelectronics or nanooptoelectronics, the current approach consists of optimizing the synthesis for the growth of very crystalline nanostructures, ideally free from defects, and to eliminate any unwanted surface termination by post-growth processing.[19] For other applications however, and in particular for catalysis and chemical sensing, controlled defect states and unusual surface reconstruction could have a positive influence on the performance of the devices. In any case, the characterization of defects and surfaces is an essential aspect of the progress in nanoscience and nanotechnology.

Nanostructures of high aspect ratio can be grown (i) exploiting the intrinsic anisotropy of some materials, (ii) via heterogenous nucleation using seeds or catalyst particles, (iii) constraining the growth using capping agents to control the activity of the seed or templates of appropriate shape, (iv) allowing self-assembly of zero-dimensional nanostructures into nanowires or nanorods, and (v) inducing localized concentration and temperature gradients to achieve columnar growth in hydrothermal conditions.

The great variety of recipes and approaches is covered extensively in many review articles.[20–22] In this chapter I am going to present examples of high aspect ratio nanostructures grown with three different approaches, hoping to demonstrate that the synthesis of highly crystalline building blocks can be tailored to the specific needs of tomorrow's devices, both in terms of materials and of morphology. The structural characterization of the nanomaterials was mainly performed using electron microscopy techniques that guarantee accurate determination of shape, crystallinity, and defects of individual nanowires.

2. Synthesis of High Aspect Ratio Inorganic Nanostructures

2.1. *Low-temperature chemical vapor deposition of silicon nanowires*

Among the synthesis strategies, chemical vapor deposition (CVD) is one of the most widely used, both for fundamental and for applied research. The principle behind CVD is the controlled dissociation of reactant gases to give new chemical species that can react with the substrate and form the products. Careful optimization of each stage of the reaction is essential to achieve well-defined products. A basic CVD process consists of the following steps:

(i) a mix of reactant gases and diluent inert gases are introduced at a specified flow rate into the reaction chamber;
(ii) the gas species are dissociated and move to the substrate;
(iii) the gas-phase precursors get adsorbed on the surface of the substrate
(iv) the precursors, then, undergo chemical reactions with the substrate to form the products;
(v) the gaseous by-products of the reactions are desorbed and evacuated from the reaction chamber.

Many designs of CVD reactors have been described in the literature, and they are classified depending on the mechanism used to dissociate the precursor gases (e.g., hot filament, radio frequency, microwave, furnace, plasma assisted, etc.). The specific setup can be optimized for the growth of high aspect ratio nanostructures, and scaled up to produce large quantities (of both films and powders).

In general CVD reactors offer the possibility of adjusting the mixture, pressure, and flux of the gas-phase precursors, as well as the growth temperature. The feature that makes CVD unique for the synthesis of nanotubes and nanowires is that it enables their growth at predetermined locations on the substrate, which is essential for direct integration in electronic devices and sensors. The positional selectivity is achieved through the use of a metal catalyst, patterned or self-assembled before the growth of the nanostructures. CVD synthesis of semiconductor nanowires is often interpreted in terms of the vapor–liquid–solid (VLS) model in which a liquid metal particle catalytically promotes the growth of crystalline nanostructures from gas-phase precursors.[23,24] Careful consideration of the phase diagrams of the elements involved, and in particular of the eutectic temperature of the semiconductor–metal system, guides the choice of experimental parameters enabling a faster optimization of the process. As an example, gallium is known to form a eutectic with small quantities of silicon at 29.8°C and exists as a perfectly mixed liquid over a broad temperature and composition range. The Ga–Si liquidus line indicates that Si precipitation from a Ga–Si alloy is possible at temperatures as low as 100°C. Sunkara and coworkers found that the size of the critical nucleus for silicon crystallization from the Ga–Si melt calculated according to classical nucleation theory is found to correspond to the diameter of the Si nanowires grown at 400°C using a microwave reactor.[25]

The ability to process electronic devices at low temperatures is becoming increasingly important as the dimension of the components is reduced, because of the need to preserve the quality and mechanical stability of the dielectric layers and metal interconnects. If we want to integrate nanowires in state-of-the-art devices, we need to be able to synthesize highly crystalline nanostructures at specific locations using low-temperature processes compatible with the requirements of the semiconductor industry. For this reason we optimized a low-power radio frequency plasma reactor for the growth of silicon nanowires below 400°C, finding the conditions that favor a high growth rate, without decreasing the high crystallinity of the as-grown structures.[26]

In our growth studies we used gold as the catalyst because the bulk Au–Si eutectic temperature is relatively low (363°C) compared to that of other metals, such as Fe and Ti, and at the same time gold has lower diffusivity than gallium and is less likely to contaminate the growth apparatus. The gold catalyst was deposited on passivated Si substrates by evaporation of nanometer-thick films. The substrates were then loaded into the radio frequency parallel-plate CVD system (Plasmalab DP 80), heated to 380°C under vacuum, and allowed to outgas for 3 h. The process gases, silane, and hydrogen were then admitted at different flow rates, maintaining pressures between 0.4 and 2.4 mbar for 15–90 min. The radio frequency power was 13.6 MHz.

From scanning electron microscopy analysis two types of nanostructures are generated in the process: *straight nanowires* with diameters below 20 nm and *worm-like structures* with diameters up to 300 nm. The relative abundance of the two species is determined by the thickness of the evaporated

Au layers, with 5 nm of gold yielding mostly worm-like structures, and subnanometer Au films catalyzing the growth of fine nanowires. The average diameter of the worm-like structures is also controlled by the catalyst film thickness. Transmission electron microscopy (TEM) supports the structural distinction made above, as shown in Fig. 1. For all deposition conditions, the straight nanowires are highly crystalline, whereas the worm-like structures are amorphous, with small isolated crystalline inclusions. A large, spherical gold particle was always found at the tip of the amorphous wires. The fact that most nanowires (70%) do not show a catalyst particle at their tip suggests a base growth mechanism or, at least, indicates that the Au is used up during the synthesis. Some Si nanowires (SiNWs), however, have a spherical gold particle at the tip which seems to determine the diameter of the structure. In particular, the wires with kinks at sharp angles have a spherical Au particle at the tip. Kinking is associated to sudden changes in growth direction, and may be related to growth defects and high growth rates.[27]

The growth direction of many SiNWs is found to be the [110] from analysis of the lattice fringes. However, wires with other growth directions were also observed. The nanowires seem to develop lattice defects, grain

Fig. 1. Bright-field TEM images of silicon nanowires grown by CVD using thin layers of Au as the metal catalyst, at (a) low and ((b) and (c)) intermediate magnification. A gold particle is visible at the tip of the nanowire in (b). A worm-like structure grown from thick Au layers is shown in (c).

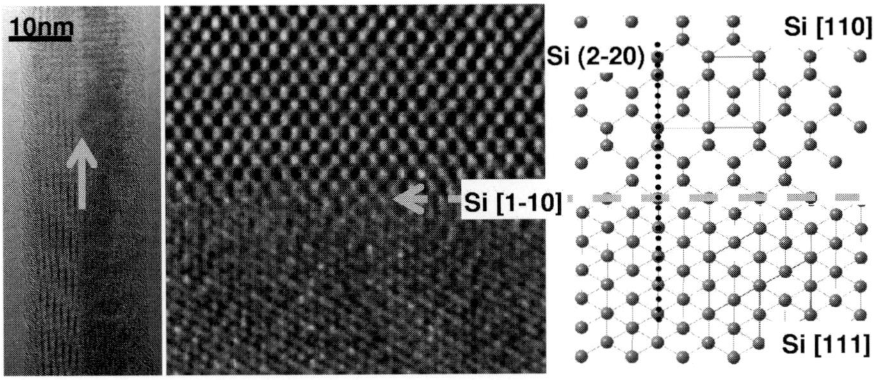

Fig. 2. Bright-field TEM image of a silicon nanowire with a grain boundary running parallel to the wire growth axis, namely Si $[1\bar{1}0]$. High-resolution analysis of the core of the wire (rotated by 90° in the middle panel) shows the reciprocal orientation of the two sides of the wire. A ball-and-stick computer model helps visualizing the interface region.

boundaries, twins, etc., along planes that are parallel to the growth direction. Figure 2 shows the example of a grain boundary running across the center of an Si nanowire (17 nm in diameter), studied by high-resolution TEM. The two halves of the nanowire are seen in the [110] and [111] projection, respectively, aligned so that the $(2\bar{2}0)$ planes run parallel across the grain boundary. Different sets of lattice fringes can be identified, corresponding to the (111) and (220) planes of Si, with interplanar distances of 3.14 and 1.92 Å, respectively. The growth axis for the wire is obviously parallel to the $[1\bar{1}0]$ direction. The ball-and-stick model helps to illustrate the arrangement of the atoms at the interface.

Elemental mapping of small diameter crystalline nanowires confirms that the silicon core is surrounded by SiO_x shell of about 2 nm. The SiO_x layer is most likely the result of surface passivation upon exposure to air after deposition. Energy-dispersive X-ray spectroscopy was performed to test for possible contamination of the as-grown SiNWs. No elements other than Si and O could be traced along the nanowires within the sensitivity of the method. Au could only be found at the base and sometimes at the tip of the SiNWs. It should be emphasized that the purity achieved in this experiment is not common, and accidental impurities are often introduced in the growth process which can act as uncontrolled dopants and affect the electronic properties of the nanowires.[28]

As mentioned before, the vapor–liquid–solid (VLS) process is often used to describe thermal growth of metal-catalyzed semiconductor nanowires.

Although the individual stages of the VLS reaction can be separated, the detailed processes leading to high aspect ratio nanostructures are not well understood. Essentially, the VLS model consists of

(i) catalytic decomposition of the source gas on the surface of the metal particles,
(ii) diffusion of Si from the surface through the catalyst particle or along its surface,
(iii) nucleation of the nanowire and precipitation of Si into the growing structure.

From experimental evidence the liquid intermediate, namely, the metal–Si eutectic melt, is not a necessity, as solid catalyst particles can also lead to SiNW growth.[29] Fast surface and bulk diffusion of Si atoms (and self-diffusion of Au atoms) are however essential for the growth of high aspect ratio nanostructures.

In plasma-enhanced CVD, the effect of the plasma is to pre-ionize and dissociate the precursor gas, as well as to provide local surface heating, which enables an efficient adsorption and diffusion of silicon atoms even at low substrate temperatures. In this study, small diameter SiNWs show a higher growth rate than thicker structures, which suggests proportionality between surface to volume ratio of the Au particle and the Si mass flow "through" the particle. In classical terms, the Gibbs–Thomson equation predicts a decrease of the lower limit for the wire diameter for increasing silane pressure.[30] Here, for a pressure range of 0.4–2.4 mbar, we found no significant change in the SiNW diameter distribution. However, increasing pressure increased the nanowire growth rate. An increased silane pressure appears to enhance the efficiency of the catalytic growth process, reducing the deposition of amorphous material due to self-pyrolysis of silane. Increasing the catalyst film thickness results in the formation of larger gold particles, which nucleate worm-like structures mainly consisting of amorphous silicon. This indicates that diffusive transport of Si through or on the large catalyst particle is fast, supplying Si at a higher rate than the crystallization rate for a given diameter.

Even though the surface of the SiNWs is found to react strongly on exposure to air, a wide range of post-growth modifications is possible with silicon nanowires, such as diameter variation by etching and doping processes. This post-growth tailoring of properties is attractive for many future applications.

2.2. *Synthesis of RuO_2 nanorods in solution*

A less-established technique to produce high aspect ratio nanostructures is the so-called hydrothermal growth, particularly interesting in the case of metal oxide nanorods. Even though the detailed growth mechanism has not yet been established, the possibility to scale up the reaction to industrial quantities makes this approach relevant for commercial applications.

In our study the nanorods are synthesized in aqueous solution using $RuCl_3$ as the Ru precursor, Pt as a growth promoter and Zn as a reducing agent.[31] The diameter distribution of the nanorods ranges from 2 to 30 nm, and peaks around 7 nm, while their length can exceed 100 nm. Most nanorod tips are pointed with an apex of 90–115°, but flat tips can also be observed. The base of the rods is in contact with polycrystalline grains mainly consisting of RuO_2. Figure 3 shows a high-resolution TEM image of a nanorod diameter of 11 nm. The structure of the rod is perfectly crystalline, with the most obvious lattice fringes corresponding to the (110) spacing of RuO_2 ($P4_2/mnm$, rutile structure) viewed parallel to the [001] direction. The rod axis and growth direction are parallel to the $[110]_{RuO_2}$ direction (sketched in the inset of Fig. 3). The left-hand side of the rod apex forms a 45° angle to the longitudinal axis, and the oxide surface consists of (100) planes, ascending and descending in steps of half a cell every

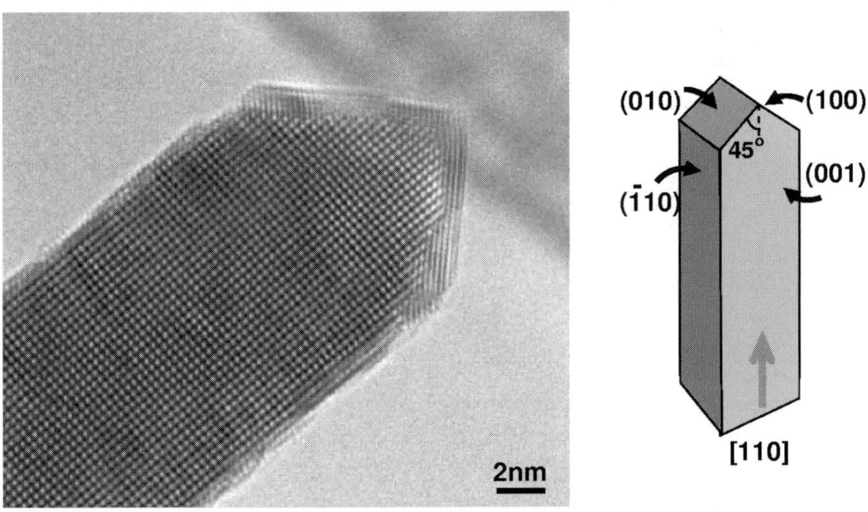

Fig. 3. High-resolution TEM image of a crystalline RuO_2 nanorod, seen along the [001] direction. The 3D sketch shows the low index planes and the rod axis.

four to five cells. The right-hand side of the apex, again at 45° to the rod axis, has a less regular structure with step height ranging from half a cell to two cells. The tip of the rod is decorated with five to six layers of a denser material: the approximately vertical lattice fringes correspond to the (002) planes of the hexagonal close packed Ru metal lattice ($P6_3/mmc$, $a = 2.705$ Å and $c = 4.282$ Å). Although the growth of Ru on RuO_2 is not strictly epitaxial, the planar mismatch is below 5% for most of the observed orientations and can be handled by small lattice distortions. When the (001) or (010) planes of Ru overlap the (100) planes of RuO_2, the planar strain of about 12% is relieved by introducing edge dislocations at the interface.

High-resolution scanning transmission electron microscopy (STEM) was used to further characterize the nanorods (SuperSTEM, Daresbury Laboratory). Images were acquired with a high-angle annular dark-field detector (24 mrad convergence, 70–210 mrad collection semiangle). From electron energy loss spectra the nanoclusters at the surface of the rod in Fig. 4 do not

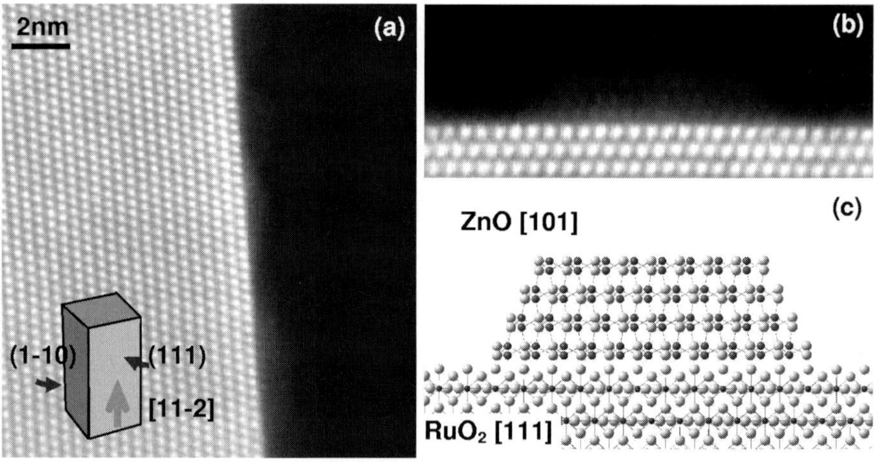

Fig. 4. High-angle annular dark-field (HAADF) images of a nanorod imaged along the [111] direction. The 3D sketch shows the low index planes and the rod axis. Nanoclusters decorate the surface of the rod. A line scan from the center of the rod to the vacuum reveals that the clusters contain Zn and O but not Ru. From analysis of the lattice fringes the ZnO nanocluster in (b) was identified as wurtzite-type ZnO imaged along the [101] direction. In (c) a ball-and-stick computer model suggests the arrangement of the atoms in the two structures (Ru in dark gray, Zn in medium gray, and O in light gray).

seem to contain Ru, but to be made of zinc and oxygen. The ZnO cluster in Fig. 4(b) corresponds to crystalline wurtzite-type ZnO ($a = b = 0.325$ nm and $c = 0.52$ nm) imaged along the [101] direction.

Even though zincblende-type ZnO would give a prefect lattice match to RuO_2 ($a = b = c = 0.449$ nm for ZnO, $a = b = 0.449$ nm and $c = 0.31$ nm for RuO_2) from our observations it appears the wurtzite polymorph is energetically favorable at the nanoscale (at least in UHV conditions).

A mechanism for the growth of the nanorods in solution is based on the theory of columnar (dendritic) growth,[32] with further complexity added by chemical reactions. In the first stages of the synthesis reaction platinum, the most noble metal in solution, is reduced and deposited on the surface of the Zn grains in patches, according to the substrate topography. This causes a depletion of Pt atoms from the solution surrounding the Zn grain. At the same time, there are two competing reactions for the decomposition of $RuCl_3$ (i.e., reduction by Zn to Ru^{3+} and hydrolysis, leading to the formation of $[RuO(H_2O)_4]^{2+}$).

The hydrolysis reaction appears to be more favorable than the reduction to Ru metal, and hence a larger concentration of the oxide precursor is expected. The solution in the proximity of the Zn/Pt grain is now rich in $[RuO(H_2O)_4]^{2+}$, from which RuO_2 starts depositing over the Pt island, releasing the heat of crystallization and warming the solution. Where the topography of the surface is such that the tip of the RuO_2 crystallite emerges from the depleted, warmer region of the solution, there the growth continues in a rod-like shape. Concentration and (inverted) temperature gradients are the driving forces for the growth of the RuO_2 nanorods. A local electric field builds up in the depletion region due to the Pt/Zn galvanic couple potential (standard electrode potential of $+0.76$ V). The gradient of the electric field contributes to the growth of the high aspect ratio nanostructures. Ru metal is co-deposited, but because of its lower concentration, finds a stable configuration by forming nanocrystals at the surface of the oxide rod. This arrangement of the Ru clusters also reduces the surface energy of the system, because of the higher density of the hcp compared to the rutile structure. Sulfuric acid does not remove zinc completely, but some of it must react and form the epitaxial ZnO nanocrystals found at the surface of the rods. Notice that this is an interesting example of self-assembly of quantum dots of ZnO, a high band-gap semiconductor that can be exploited for optoelectronic applications.

2.3. *Physical methods for the synthesis of SiC nanorods and NiS–MoS$_2$ nanowires*

The simplest approach to growing one-dimensional (ID) nanostructures is via vaporization of suitable powder mixture under controlled atmosphere (both pressure and flux are monitored). Using this method a wide variety of materials can be synthesized, and in the following section the case of SiC nanorods are described.

Silicon carbide is a ceramic compound of silicon and carbon. Its high melting point (approximately 2700°C) makes silicon carbide useful for bearings and furnace parts. It is also highly inert. There is currently of much interest in its use as a semiconductor material in electronics, where its high thermal conductivity, high electric field breakdown strength, and high maximum current density make it more promising than silicon for high-powered devices. In addition, it has strong coupling to microwave radiation and that, together with its high melting point permits practical use in heating and casting metals. SiC also has very low thermal expansion coefficient and no phase transitions that would cause discontinuities in thermal expansion. Silicon carbide nanorods have shown excellent mechanical and field-emission properties significantly exceeding those of the bulk SiC crystals.[33,34] Improvement of material characteristics and development of less expensive synthesis methods are of great interest for technological applications.

Our indirect vapor-phase approach to the synthesis of SiC nanorods is characterized by a comparatively low reaction temperature with the use of very simple precursor materials.[35] Finely dispersed silicon and amorphous carbon powders, with a mean grain size of 1 μm and 70 nm, respectively, were used as initial reagents of the reaction. Silicon and carbon were carefully mixed in an equimolar ratio by ball milling. The reaction was carried out in a quartz tube placed horizontally inside an electric furnace. The Si/C powder mixture was put on a fused silica plate in contact with a thermocouple and placed inside the tube. The reaction occurred at temperature of 1200°C over 2 h and under ambient pressure of argon gas containing 0.6 vol% of oxygen. We have observed the formation of SiC nanorods only in the presence of oxygen. Large quantities of randomly arranged SiC nanorods up to 1 μm in length and with a diameter of 10–30 nm were obtained. Most of the nanorods are straight and crystalline, but some show clear evidence of structural defects, for example multiple bends and change of growth axis. In addition to nanorods, SiC nanocrystals with an average size of a few nanometers can be also found. TEM analysis also revealed the presence of large particles of the starting material, probably too stable to react under the processing conditions described above.

The high-resolution TEM analysis reveals that the core of the nanorods consists of crystalline SiC (mainly α-SiC), whereas their surface is covered by a thin layer of an amorphous material, identified as SiO_2, most likely a by-product of the vapor–solid reactions that take place in the furnace. The presence of oxygen is necessary as an intermediate reagent for the gasification of the precursor powders. Since the growth of the nanorods is only observed for temperatures above 1100°C, it is reasonable to suppose that the vapor–solid reactions are only activated at these high temperatures. The formation of SiC nanoparticles together with nanorods can be explained as a result of a solid–solid reaction, occurring in interface regions where silicon and carbon particles are in direct contact with each other. The efficiency of the solid–solid reaction for the formation of nanorods may be lower because the low mobility of reagent atoms does not favor the growth of elongated shapes.[36]

Conceptually simple, but rather difficult to achieve in practice, is the growth of nanowires through heat treatment of a suitable mixture of precursor materials. An effective method of mixing is by creating thin alternating layers of the precursors, and patterning these layers in the shape of pillars of small diameter.[37] High-temperature annealing of pillars of C_{60} and nickel on a molybdenum substrate was found to yield molybdenum–carbon–oxygen.[38] Low-temperature treatment of a similar pillar structure, instead, was found to yield MoS_2 nanotubes filled with metal sulfide nanowires.[39] This surprising, intriguing, and unexpected result, discussed in the following paragraphs, warned us against the effects of hidden contamination.

Thermolysis of pillars made of alternating fullerite powder C_{60}/C_{70}, and transition metal thin film, was found to yield novel structures in the low-temperature regime.[39] In particular, when nickel or cobalt were introduced in the process, we observed the formation of Ni and Co sulfide nanowires at temperatures below 550°C.[40] Pillars of alternating fullerene and metal layers were formed by evaporating C_{60}/C_{70} powder (fullerite, Sigma–Aldrich) and high-purity metal wire (Ni or Co) from tungsten boats are mounted on a four-turret source in a standard evaporator. The films were deposited onto Mo TEM grids (Agar Scientific). The use of TEM grids as substrate material makes the as-grown nanostructures readily accessible for microscopy analysis in their original growth environment, avoiding the destructive processes necessary to disperse and transfer them.

In contrast to most commercially available grids, Mo TEM grids are etched and not electroplated, which makes them suitable for high-temperature annealing experiments, but also prone to contamination from the etching solutions. Indeed we found that sulfur is a major contaminant of standard Mo grids, undesirable for experiments requiring high purity, but nevertheless interesting in this specific case. Patterning of pillars of about $10\,\mu m$ diameter was achieved by using Cu TEM grids as stencil masks. Samples were heat treated at 5–10 mbar N_2 atmosphere at a temperature of 510–550°C for up to 1 h. The structure and composition of the nanowires were analyzed by a combination of electron microscopy techniques.

High aspect ratio nanostructures grown far from the precursor pillars seem to consist of MoO_3, similar to those observed by Durkan and co-workers.[38] Instead, nanowires grown on the pillar sites consist of nickel or cobalt, sulfur and carbon, as determined by energy-dispersive X-ray analysis (EDX) and electron energy loss spectroscopy (EELS). Sulfur was not expected to be incorporated in the wires, and finding its signature was revealing of the profound affinity of Ni and Co for sulfur impurities. Figure 5

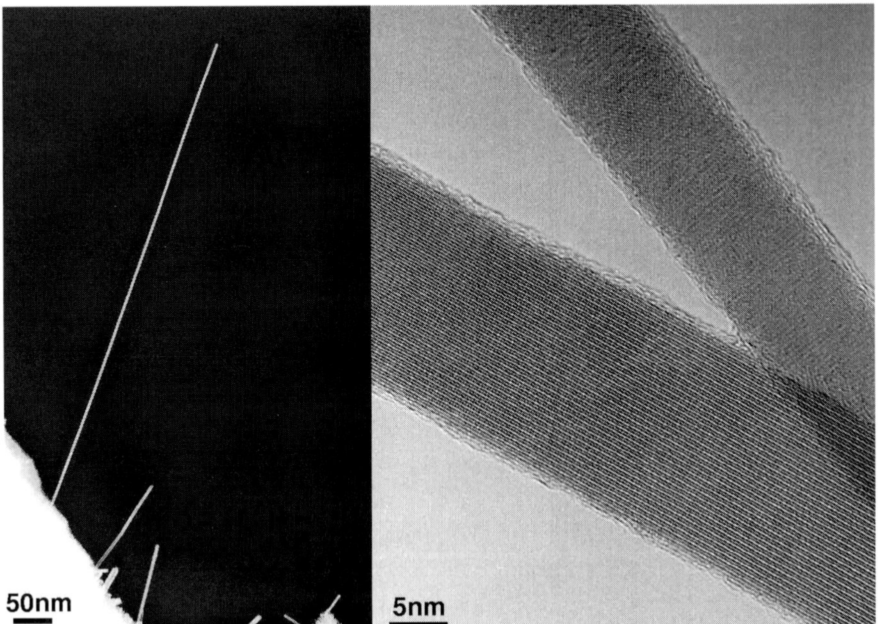

Fig. 5. Nickel sulphide nanorods grown on a Mo TEM grid by thermolysis of pre-patterned pillars.

shows images of the nanowires grown at the edge of the Mo grid. The core of the two wires of Fig. 5(b) is highly crystalline, and surrounded by a thin amorphous layer. It is reasonable to suppose that the carbon signal seen in EDX and EEL spectra originates from this amorphous layer, resulting from decomposition of the fullerenes. Besides Ni and Co sulfide wires, we observed composite structures consisting of a nickel sulfide core coated with a number of tubular layers of MoS_2 ranging between 2 and 15. The most stable form of layered MoS_2 consists of a molybdenum layer sandwiched between two sulfur layers (space group *P*63_*mmc*). Similar to the case of graphite, these triple layers, separated by 0.615 nm, exert van der Waals' interactions on each other.[41,42] Figure 6 shows an example of such composite nanostructure. From fast Fourier transform (FFT) analysis of the structure (Fig. 6(c)) it is possible to measure the frequency of the lattice fringes and the angles between them, and hence determine the crystallographic structure and the relative orientation of the two components. For this composite nanowire the pattern obtained from the nickel sulfide core was identified as a [110] projection of the NiS_2 cubic cell. In this projection the (113)-type planes of NiS_2 are parallel to the MoS_2 layers. The blurred contrast at the interface between the two sulfides may be attributed to a noncomplete separation of the two phases, and possible mixing of the sulfur atoms at the boundary.

The mechanism by which the pillar structure generates nanowires at low temperature must be based on solid-phase transport. The formation of Ni (or Co) and S complexes in the presence of Mo and C may be related to the symmetrical synergism at the basis of hydrotreating catalysis.[43,44] Fullerene appears to be essential in promoting the growth of high aspect metal sulfide nanowires but the specific reaction paths are not clear and must be investigated further.

3. Outlook

The physical and chemical properties of nanoscale materials are different from those of bulk materials because of the higher surface to volume ratio and of the re-organization of electron energy levels determined by the size and the shape of the nanostructures. In general, the properties of nanostructures are thought to be unique, or highly enhanced compared to those of their bulk counterparts, and hence devices or composites based on nanostructured materials receive great attention both from the *academic*

Fig. 6. NiS_2 nanowire encapsulated in a MoS_s multiwall nanotube. In (a) an overall view of the structure, in (b) a magnified portion of the NiS_2 core, in (c) a digital diffraction pattern hinting at an epitaxial relationship between core and walls, in (d) an FFT-filtered version of (b) to enhance visualization of the lattice periodicity. The ball-and-stick model in (e) shows a [100] projection of the MoS_2 lattice and a [110] projection of the NiS_2 lattice reciprocally oriented as in the actual nanowire structure (Mo and Ni are shown as smaller spheres of darker color, S is in lighter gray). The unit cell is also shown for clarity.

community and from *industrial investors*, as witnessed by the rapidly increasing number of publications and patents in this area. In particular high aspect ratio nanostructures with well-controlled composition, diameter, crystal structure, and orientation represent a new class of materials for the investigation of the relationship between structure and properties.

A short list of the most commonly cited properties include: mechanical properties, thermal stability, electric and photoconductivity, linear and nonlinear optical properties, field emission properties, catalytic activity, and magnetic properties; all of which are highly dependent on size, shape, and composition of the nanowires/nanorods. It is easy to realize that the parametric space still to be explored is vast, and might require several years of investigation before nanotechnology can fulfill its promises.

Since the manipulation of nanomaterials is more demanding than that of their bulk counterparts, the experiments to measure their properties must be carefully designed. In some cases it has been necessary to develop new instruments to achieve the resolution and sensitivity required by measurements at the nanoscale. Fundamental issues must be addressed to ensure that the results are not affected by artifacts. Groups around the world have worked for years trying not only to elucidate the physical properties of nanostructures but also to explore different measurement techniques and set a code of practice in the assessment of such properties. The two most prominent cases are those of the Lieber Research Group at Harvard University and Dekker's group at the Kavli Institute of Nanoscience at Delft University. Despite all the effort, the measurement of chemical and physical properties at the nanoscale is far from being standardized, and hence it is important to remain cautious and open minded when dealing with experimental nanometrology.

Besides theoretical and experimental work, our understanding of the nanoworld can be dramatically improved through a computational approach. The complexity of the systems handled by simulations is quickly increasing, coming closer to the description of real nanoscale objects. Calculations generally help determining the minimum energy configuration or the electronic structure of systems up to a few hundreds atoms. An important future development for growth studies would be, for example, to be able to apply fields and gradients to the systems, and study mass transport phenomena on reasonable time scales.

As a final remark, in this field of rapidly evolving concepts and possibilities, I still think that we have not fully explored and understood the growth mechanism of nanostructures, and that if we all want to be more forward with confidence, we must concentrate on these fundamental problems as well as on the fascinating and useful applications that attract most of the funding. This must be evident for the reader that has tried to make sense of the vast body of publications on the subject, but it might not be so obvious for experts of more traditional disciplines where experimental procedures are well established and quality checks have been in place for a number of years.

Acknowledgments

It is a pleasure to acknowledge the support of the Royal Society through a Dorothy Hodgkin Fellowship. For the work done during my PhD, I acknowledge the EC for financial support through the project CARBEN. I would like to thank the laboratory of Prof. John Robertson and the laboratory of Prof. Milani for the synthesis of the nanostructured materials. I am very grateful to Stephan Hofmann for years of fruitful collaboration, and to Paul Midgley for guidance in all aspects of microscopy. I would also like to thank Andrew Blelock and the SuperSTEM team at Daresbury Laboratory, the Isaac Newton Trust, FEI Company and EPRSC for the use of the Tecnai microscope.

References

1. S. J. Tans, R. M. Verschueren and C. Dekker, Room-temperature transistor based on a single carbon nanotube, *Nature* **393**, 49–52 (1998).
2. M. S. Gudiksen, L. J. Lauhon, J. Wang, D. C. Smith and C. M. Lieber, Growth of nanowire superlattice structures for nanoscale photonics and electronics, *Nature* **415**, 617–620 (2002).
3. J. Kong, N. R. Franklin, C. Zhou, M. G. Chapline, S. Peng, K. Cho and H. Dai, Nanotube molecular wires as chemical sensors, *Science* **287**, 622–625 (2000).
4. P. G. Collins, K. Bradley, M. Ishigami and A. Zettl, Extreme oxygen sensitivity of electronic properties of carbon nanotubes, *Science* **287**, 1801–1804 (2000).
5. F. Patolski and C. M. Lieber, Nanowire nanosensors, *Mater. Today* **8**, 20–28 (2005), doi:10.1016/S1369-7021(05)00791-1.
6. L. Schlapbach and A. Züttel, Hydrogen-storage materials for mobile applications, *Nature* **414**, 353–358 (2001).

7. J. Graetz, C. C. Ahn, R. Yazami and B. Fultz, Highly reversible lithium storage in nanostructured silicon, *Electrochem. Solid-State Lett.* **6**, A194–A197 (2003).
8. Y. G. Yoon, P. Delaney and S. G. Louie, Quantum conductance of multiwall carbon nanotubes, *Phys. Rev. B* **66**, 073407 (2002).
9. J. Liu, X.-M. Meng, Y. Jiang, C.-S. Lee, I. Bello and S.-T. Lee, Gallium nitride nanowires doped with silicon, *Appl. Phys. Lett.* **83**, 4241–4243 (2003).
10. E. Lucas, S. Decker, A. Khaleel, A. Seitz, S. Fultz, A. Ponce, W. F. Li, C. Carnes and K. J. Klabunde, Nanocrystalline metal oxides as unique chemical reagents/sorbents. *Chem. Eur. J.* **7**, 2505–2510 (2001).
11. S. Iijima, Helical microtubules of graphitic carbon, *Nature* **354**, 56–58 (1991).
12. L. Margulis, P. Dluzewski, Y. Feldman and R. Tenne, TEM study of chirality in MoS_2 nanotubes, *J. Microsc.* **181**, 68–71 (1996).
13. A. Rothschild, G. L. Frey, M. Homyonfer, R. Tenne and M. Rappaport, Synthesis of bulk WS2 nanotube phases, *Mater. Res. Innov.* **3**, 145–149 (1999).
14. W. K. Hsu, Y. Q. Zhu, S. Firth, M. Terrones, H. Terrones, S. Trasobares, R. J. H. Clark, H. W. Kroto and D. R. M. Walton, $W_xMo_yC_zS_2$ nanotubes, *Carbon* **39**, 1103–1116 (2001).
15. P. M. Ajayan, O. Stephan, Ph. Redlich and C. Colliex, Carbon nanotubes as removable templates for metal-oxide nanocomposites and nanostructures, *Nature* **375**, 564–567 (1995).
16. B. C. Satishkumar, A. Govindaraj, A. M. Vogl, L. Basumallick and C. N. R. Rao, Oxide nanotubes prepared using carbon nanotubes as templates, *J. Mater. Res.* **12**, 604–606 (1997).
17. H. Kim and W. Sigmund, Zinc oxide nanowires on carbon nanotubes, *Appl. Phys. Lett.* **81**, 2085–2087 (2002).
18. W. A. Tiller, *The Science of Crystallization: Microscopic Interfacial Phenomena* (Cambridge University Press, Cambridge, 1991).
19. D. D. D. Ma, C. S. Lee, F. C. K. Au, S. Y. Tong and S. T. Lee, Small-diameter silicon nanowire surfaces, *Science* **299**, 1874–1877 (2003).
20. Y. Xia, P. Yang, Y. Sun, Y. Wu, B. Mayers, B. Gates, Y. Yin, F. Kim and H. Yan, One-dimensional nanostructures: Synthesis, characterization and applications, *Adv. Mater.* **15**, 353–389 (2003).
21. L. Samuelson, C. Thelander, M. T. Bjork, M. Borgstrom, K. Deppert, K. A. Dick, A. E. Hansen, T. Martensson, N. Panev, A. I. Persson, W. Seifert, N. Skold, M. W. Larsson and L. R. Wallenberg, Semiconductor nanowires for 0D and 1D physics and applications, *Physica E* **25**, 313–318 (2004).
22. M. Law, J. Goldberger and P. D. Yang, Semiconductor nanowires and nanotubes, *Ann. Rev. Mater. Res.* **34**, 83–122 (2004).
23. R. S. Wagner and W. C. Ellis, Vapor–liquid–solid mechanism of single crystal growth. *Appl. Phys. Lett.* **4**, 89–90 (1964), doi:10.1063/1.1753975.
24. M. Yazawa, M. Koguchi, A. Muto, M. Ozawa and K. Hiruma, Effect of one monolayer of surface gold atoms on the epitaxial growth of InAs nanowhiskers, *Appl. Phys. Lett.* **61**, 2051–2053 (1992), doi:10.1063/1.108329.

25. M. K. Sunkara, S. Sharma, R. Miranda, G. Lian and E. C. Dickey, Bulk synthesis of silicon nanowires using a low-temperature vapor–liquid–solid method, *Appl. Phys. Lett.* **79**, 1546–1548 (2001), doi:10.1063/1.1401089.
26. S. Hofmann, C. Ducati, R. J. Neill, S. Piscanec, A. C. Ferrari, J. Geng, R. Dunin-Borkowski and J. Robertson, Gold catalyzed growth of silicon nanowires by plasma enhanced chemical vapor deposition, *J. Appl. Phys.* **94**, 6005–6012 (2003).
27. J. Westwater, D. P. Gosain, S. Tomiya, S. Usui and H. Ruda, Growth of silicon nanowires via gold/silane vapor–liquid–solid reaction, *J. Vac. Sci. Technol. B* **15**, 554 (1997).
28. R. J. Barsotti, J. E. Fischer, C. H. Lee, J. Mahmood, C. K. W. Adu and P. C. Eklund, Imaging, structural, and chemical analysis of silicon nanowires, *Appl. Phys. Lett.* **81**, 2866–2868 (2002), doi:10.1063/1.1512827.
29. T. I. Kamins, R. S. Williams, D. P. Basile, T. Hesjedal and J. S. Harris, Ti-catalyzed silicon nanowires by chemical vapor deposition: Microscopy and growth mechanism, *J. Appl. Phys.* **89**, 1008–1016 (2001).
30. A. M. Morales and C. M. Lieber, A laser ablation method for the synthesis of crystalline semiconductor nanowires, *Science* **279**, 208–211 (1998).
31. C. Ducati, D. H. Dawson, J. R. Saffell and P. A. Midgely, Ruthenium coated ruthenium oxide nanorods, *Appl. Phys. Lett.* **85**, 5385–5387 (2004), doi:10.1063/1.1829170.
32. A. Cottrell, *An Introduction to Metallurgy* (IOM, London, 1995).
33. E. W. Wong, P. E. Sheehan and C. M. Lieber, Nanobeam mechanics: Elasticity, strength, and toughness of nanorods and nanotubes, *Science* **277**, 1971 (1997).
34. Z. Pan, H. L. Lai, F. C. K. Au, X. Duan, W. Zhou, W. Shi, N. Wang, C.-S. Lee, N.-B. Wong, S.-T. Lee and S. Xie, Oriented silicon carbide nanowires: Synthesis and field emission properties, *Adv. Mater.* **12**, 1186 (2000).
35. I. N. Kholmanov, A. Kharlamov, E. Barborini, C. Lenardi, A. Li Bassi, C. E. Bottani, C. Ducati, S. Maffi, N. V. Kirillova and P. Milani, A simple method for the synthesis of silicon carbide nanorods, *J. Nanosci. Nanotechnol.* **2**, 453–456 (2002).
36. Y. Zhang, T. Ichihashi, E. Landree, F. Nihey and S. Iijima, Heterostructures of single-walled carbon nanotubes and carbide nanorods, *Science* **285**, 1719–1722 (1999).
37. R. R. Schlittler, J. W. Seo, J. K. Gimzewski, C. Durkan, M. S. M. Saifullah and M. E. Welland, Single crystals of single-walled carbon nanotubes formed by self-assembly, *Science* **292**, 1136–1139 (2001).
38. C. Durkan, A. Ilie, M. S. M. Saifullah and M. E. Welland, Mechanics of nanosprings: Stiffness and Young's modulus of molybdenum-based nanocrystals, *Appl. Phys. Lett.* **80**, 4244–4246 (2002).
39. S. Hofmann, C. Ducati and J. Robertson, Low temperature self-assembly of novel encapsulated compound nanowires, *Adv. Mater.* **14**, 1821–1824 (2002).
40. S. Hofmann, J. Robertson, C. Ducati and R. Dunin-Borkowski, Self-assembly of novel nanowires by thermolysis of fullerene and transition metal thin films, *Nanotechnology* **15**, 601–608 (2004).

41. Y. Feldman, E. Wasserman, D. J. Srolovitz and R. Tenne, High-rate, gas-phase growth of MoS_2 nested inorganic fullerenes and nanotubes, *Science* **267**, 222–225 (1995).
42. G. Seifert, H. Terrones, M. Terrones, G. Jungnickel and T. Frauenheim, Structure and electronic properties of MoS_2 nanotubes, *Phys. Rev. Lett.* **85**, 146–149 (2000).
43. R. R. Chianelli and G. Berhault, Symmetrical synergism and the role of carbon in transition metal sulfide catalytic materials, *Catal. Today* **53**, 357–366 (1999).
44. J. V. Lauritsen, S. Helveg, E. Laegsgaard, I. Stensgaard, B. S. Clausen, H. Topsoe and F. Besenbacher, Atomic scale structure of Co–Mo–S nanoclusters in hydrotreating catalysis, *J. Catal.* **197**, 1–5 (2001).

Caterina Ducati was born in Milan, Italy. She graduated in Physics at Universita' Statale di Milano, with a thesis on the design and realization of a time of flight mass spectrometer for supersonic cluster beams. In 1999, she moved to the Engineering Department of the University of Cambridge, where she worked on nanostructured carbon and carbon nanotubes for field emission and electrochemistry applications, experimental research supported by the European Union under the project CARBEN. After completing her PhD in 2003, she moved to the High Resolution Electron Microscope Group, in the Department of Materials Science, Cambridge, to work on a Knowledge Transfer Partnership Programme sponsored by TTI and Alphasense Limited on the characterization of catalytic materials for electrochemical gas sensors. In 2004, she was awarded a Royal Society Dorothy Hodgkin Fellowship to research on novel nanostructured metal oxides for electrics and catalysis. Caterina is a Junior Research Fellow at Churchill College, Cambridge. Her other interests include exploring parks and playgrounds with her three-year-old son Paolo, and reading.

CHAPTER 3

MULTILAYERED MATERIALS: A PALETTE FOR THE MATERIALS ARTIST

Jon M. Molina-Aldareguia[*,†,‡] and Stephen J. Lloyd[*,§]

[*]*Department of Materials Science and Metallurgy University of Cambridge, Pembroke Street Cambridge, CB2 3QZ, UK*

[†]*Department of Physics, IFM, University of Linköping 58183 Linköping, Sweden*

[‡]*Current address: CEIT and TECNUN, P. Manuel Lardizabal 15 20018 San Sebastian, Spain*
E-mail: jmolina@ceit.es

[§]*Current address: E-mail: Lloyd2000@cantab.net*

Developments in the understanding of how materials behave enable us to design material structures to display specified properties. We introduce multilayered materials as systems in which new properties can emerge, not found in their constituents in bulk form. The importance of transmission electron microscopy to determine structure–property relationships in nanoscale multilayers through characterization of their atomic structure is emphasized. Two examples of technologically useful multilayer systems are considered in more detail: hard coatings made from nitride multilayer films and the new structures and magnetic properties that are found in some metal multilayer systems. Finally, we discuss the future developments that are required to fully exploit the novel properties found in multilayered materials.

Keywords: Multilayer, superlattice, transmission electron microscopy, hardness, magnetic materials, electronic structure.

1. Introduction

William Hume-Rothery, one of the founders of modern metallurgy, once said that "Metallurgy is the oldest art, but the youngest science". He was of course referring to the fact that developments in materials technology have often come about by serendipity and through trial and error rather than via an understanding of the atomic processes involved. For example, even in Roman times blacksmiths knew the surface of iron could be hardened if it was heated in a charcoal bed, but the reasons for this have only been fully explored within the last half-century. This example strikingly illustrates how recently our understanding of even very common materials has been achieved compared with other more basic areas of science. The structure of the atom was known, and famously split by Cockcroft and Walton in 1932, allowing the transmutation of the elements and the fulfillment of the alchemists' dream. However, an understanding of why crystals (consisting of periodic arrays of those atoms) had a much lower mechanical strength than predicted on the basis of the bonding between the atoms only began in 1934 when the existence of line defects (called dislocations) within those crystals was postulated by Taylor and others.[1]

It is only by properly understanding the behavior of materials that we can design their structure to fully exploit their properties, and explore promising new materials combinations. Another historical example makes the point. It was discovered by accident in 1906 that the strength of a freshly made aluminum alloy greatly increased after being left over a "long weekend". But the effect could not be applied to other alloys since no one knew which systems would be likely to show the same behavior. It was only with the understanding (many years later) that the hardening was due to the delayed formation of a new, finely dispersed phase that this phenomenon could be utilized in other alloys.[2]

One of the reasons for the late development of the study of materials as a *science* is their sheer complexity. In 1 cm^3 of a typical metal there are around 10^{23} atoms all "jostling and interacting with one another in the most complicated ways"[3] and, to make any analysis tractable approximations need to be made. Further, material properties are often dependent on defects. (In a phrase often attributed to F. C. Frank, "crystals are like people — they are the defects in them that make them interesting"). And defects (like the dislocations mentioned above) can be more difficult to model than the perfection of a crystal.

The study of materials is also complex because of the near infinite number of ways of combining the 90-odd elements to create new materials. But possibilities do not just arise from the combination of elements. Size also matters. Even in a pure metal its strength depends on the size of its constituent grains.[4] At the nanometer length scale a whole variety of novel properties can appear. In this review we consider the new properties that evolve when size is reduced in just one dimension — to create a thin layer, or combined with other layers periodically to create a multilayer. These synthetic structures provide a platform on which to combine and generate properties — like the colors on an artist's palette — and ultimately to improve on nature's raw materials.

2. Multilayers

Multilayers occur naturally. For example, mollusk shells are made of nacre which comprises layers of aragonite, with a thickness of approximately 250 nm, alternating with very thin layers, of around 10 nm, of protein.[5] On its own aragonite is hard and brittle, but layered in this way the structure becomes both hard and tough — a combination of properties that is usually difficult to achieve. Structures can also be layered at an atomic length scale. High-temperature (T_c) copper oxide (CuO_2) superconductors are natural multilayers with superconducting layers of CuO_2 interleaved with insulating oxide sheets. The goal for us is to create synthetic multilayers to mimic these structures. Through the variation of parameters such as the layer thickness and repeat distance (Λ) (Fig. 1(a)) the properties of the resulting material may be tuned. For example the possible mechanisms of high-T_c superconductivity have been investigated[6] using multilayers of niobium nitride (NbN, low-T_c superconductor) and aluminum nitride (AlN, insulator) (Fig. 1(b)) which are analogs of the high-T_c materials.

As well as being analogs of natural structures, synthetic multilayers have been developed to display unique properties that are not found in their components alone. The properties of the multilayer are not simply some combinations of the properties of the bulk constituents. Perhaps the best known example is the semiconductor quantum well structures developed in the 1970s. Here the periodicity of the multilayer (sometimes referred to as a "superlattice") is superimposed on the periodicity of the atomic lattice, resulting in electronic structure modifications. This allows "band-gap engineering" in which band gaps can be tailored to, for example, modify

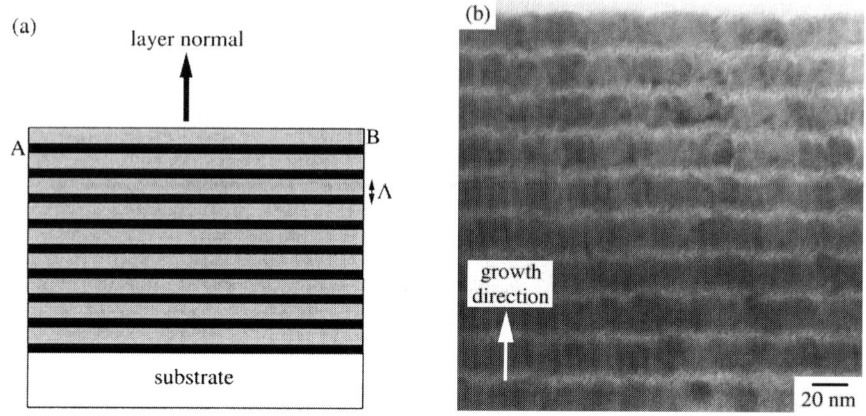

Fig. 1. (a) Schematic of a multilayer made from layers A and B with period, Λ. (b) NbN/AlN superconducting/insulating multilayer ($\Lambda = 21\,\text{nm}$). The lighter layers are AlN (after Lloyd *et al.*, Ref. 7).

the wavelength of light emitted[1] from lasers or light emitting diodes made from these materials.

Multilayers are additionally useful because of the high proportion of interface that they contain, and this can also lead to new properties emerging that are not found in the bulk material. This is particularly important in metallic multilayers, among the first applications of which were in the X-ray optics. Alternating thin layers of materials with very different scattering potential (e.g., silicon and tungsten) produced very high reflectivities, impossible to achieve in conventional materials. The interface atoms can also have different properties that become detectable macroscopically in a multilayer. This phenomenon is used extensively in magnetic devices and memories based on metallic thin films, as well as in naturally occurring systems. Recent simulations by Robinson *et al.*[8] show that the interfaces between two commonly occurring mineral phases (hematite and ilmenite) can be strongly ferromagnetic, even though the two phases are not magnetic on their own. When cooled from a high temperature, this rock system exsolves (dissociates into the hematite and ilmenite phases) on a range of length scales to create many interfaces that give the resulting composite magnetic properties similar to those observed from rocks in both Earth and Mars.

It is not just the layer thickness and Λ that can be varied to control properties. The atomic spacings within the layers can also be modified to a much greater degree than is possible in bulk materials. A different crystal structure may even be stabilized. This arises from a phenomenon known as *epitaxy*. Often very thin films have a well-defined crystallographic relationship with the substrate they are grown on. Thus the film can match the structure and lattice spacings of the substrate rather than those of the bulk material. Spacing changes in the film are known as *coherency* strains due to the coherency between the lattices of the film and substrate (Fig. 2(a)). However above a critical thickness the stored energy in the film becomes too great and the film will transform to a more stable structure with defects introduced at the interface destroying the coherency (Fig. 2(b)).

The constraint of the two-dimensional geometry in multilayers creates a highly anisotropic structure. *Anisotropy* refers to the variation in properties as a function of crystallographic direction and is a general property of crystals. In a synthetic multilayer the anisotropy can be controlled and this is exploited in metal multilayers used in magnetic storage devices, such as hard disk drives.

There is therefore considerable scope for tailoring multilayer properties, but to do so the structure must be grown to high precision. A variety of deposition methods have been developed, but perhaps the most versatile and widely used for the metallic and nitride systems described below is *magnetron sputtering*. This refers to the process of removing an atom in a target through the impact of an energetic ion. The most common method of bombarding the target with energetic ions is to create a glow discharge or plasma from an inert gas, such as argon, introduced into the high vacuum of the deposition chamber. The plasma is confined near the target with a magnetic field. Sputtered atoms ejected from the target then deposit onto substrates, which are often heated to give sufficient mobility to the incoming atoms to ensure high-quality growth. Compound films, such as

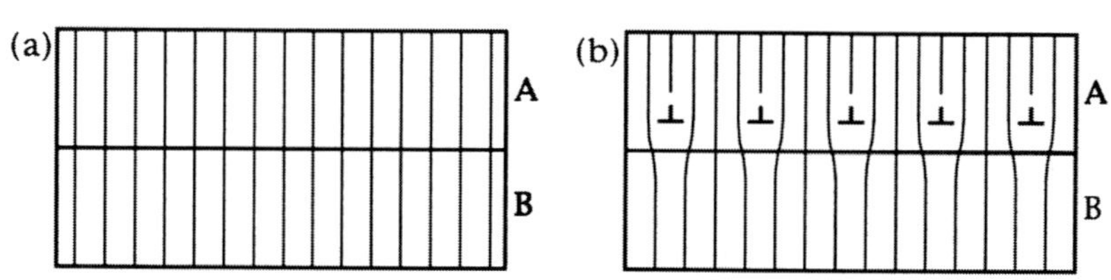

Fig. 2. Interface between two layers, A and B, with the lattice parameter of B greater than that of A. (a) Coherent interface in which lattice spacings of the two layers are equalized perpendicular to the interface. (b) Incoherent interface in which misfit dislocations ($\perp$) are created to relieve strain as the layers relax.

nitrides, can be formed easily by adding nitrogen into the sputtering gas mixture, and commercially used components can be multilayer coated by such a process.[9] To control the properties of the multilayers it is vital to be able to determine the actual structure of the films. We now turn to one major characterization tool that allows structure–property relationships to be determined with confidence: the electron microscope.

3. Electron Microscopy

The electron microscope has been described as the "materials characterization tool of the millennium"[10] and it has been one of the foremost tools in the evolution of the study of materials from an art to a science.[11] Transmission electron microscopy (TEM) employs high-energy electrons (several hundred kiloelectronvolts) to probe the structure of materials. TEM is unrivaled in its ability to image the local structure (including defects), composition and electronic structure at or near atomic resolution through myriad techniques including high-resolution electron microscopy, energy-filtered imaging, and electron holography. Its ability to create *images* at this resolution makes it complementary to nonlocal techniques, such as X-ray diffraction, that measure the structure "averaged" over micrometers.[12] The wide range of material parameters that can be obtained from the same area of specimen in a state-of-the-art TEM today is due to the large number of specimen–beam interactions.[11] However, the strong interaction of the electrons with the specimen under study places a crucial limitation on the technique (the electrons only penetrate a few hundred nanometers of material), hence it is necessary to make a thin foil (*ca.* 100 nm in thickness) before the structure of the material can be examined. The ability to make thin foils from different materials has always been a prerequisite to TEM techniques being applied to new materials problems. One of the early triumphs of TEM was the imaging of moving dislocations in 1956, some 22 years after their existence was first proposed. This achievement was due to the development of new sample-thinning techniques as much as advances in the microscope itself.[13]

In the study of multilayers TEM plays an important role in measuring parameters such as the layer thickness, grain size and orientations in polycrystalline films, waviness and mixing of the layer interfaces, and the breakdown of epitaxial relationships.[14] It can also be used to determine local atomic spacings, composition, and electronic structure changes[15] as well as examining local deformation behavior as described below. In combination with other techniques, TEM provides the information required to understand the complex structure–property relationships in nanoscale synthetic multilayers, essential to give the high degree of control necessary for modern applications. We emphasize the contribution of TEM as we describe multilayers used in two very different applications: mechanically hard nitride coatings and magnetic metallic multilayers.

4. Hard Coatings

Hardness was one of the first material properties to be measured, albeit crudely, by ranking materials in the order of their ability to scratch (Table 1). Today hardness can be measured more quantitatively using indenters that press a hard ball or pyramid (usually diamond) into the surface of a material with a known force. The size of the residual surface impression is then measured. The harder the material, the smaller the impression with the hardness simply defined as a pressure: the force over the projected area of the indent. Pressures under the indent can be enormous, comparable to those obtained in high-pressure cells. In silicon the pressure is sufficient to cause a phase transformation to a ductile metallic phase.

Hard materials, particularly when used as coatings, are vital for technologies as disparate as magnetic recording and high-speed cutting tools.

Table 1. Mohs hardness scale.

Mohs value	Substance	Chemical formula
1	Talc	$3MgO.4SiO_2.H_2O$
2	Gypsum	$CaSO_4.2H_2O$
3	Calcite	$CaCO_3$
4	Fluorite	CaF_2
5	Apatite	$Ca(PO_4)_3F$
6	Orthoclase	$KAlSi_3O_8$
7	Quartz	SiO_2
8	Topaz	$Al_2SiO_4F_2$
9	Corundum	Al_2O_3
10	Diamond	C

This is based on the scratch test: substance A is harder than substance B, if A will scratch B, but B will not scratch A.

Titanium nitride (TiN) is an unusual material: it is an electrical conductor like a metal, but possesses a high hardness more like a ceramic. About 10 years ago it was found that the hardness of multilayers made from TiN and NbN varied with the layer periodicity Λ, reaching a maximum of roughly twice of that of the monolithic components at $\Lambda \sim 5\,\text{nm}$.[16] Other systems showed similar behavior. Metal multilayers can also be much harder than their components, but they are of less interest for many commercial applications since they are less hard and less stable at high temperature than their ceramic counterparts.

The accepted rationale behind creating a hard multilayer is to impede dislocation movement. Dislocation motion (Fig. 3) is a low-energy mechanism to deform a crystal permanently (plastic deformation), akin to the way a carpet can be moved more easily through the passage of a ruck down its length. Figure 4 shows the effect of multiple dislocations traveling along a specific crystal plane (slip plane) to create a step in the layering of a semiconductor multilayer. (This illustrates how useful multilayers are for studying the deformation processes, since the layers act as internal markers that allow us to chart the material flow.) Multilayers can create extra barriers for dislocation motion by, for example, using two layers with different crystal structures (with different planes favored for low-energy slip) or having two layers with different elastic properties so that dislocations are repelled from the stiffer layer. Defects at interfaces and coherency strain can also act as barriers for dislocation motion.[17]

Typically multilayer coatings are less than a few microns in thickness; hence, special techniques must be used to measure their hardness that avoid the substrate affecting the measurement. This is realized through nanoindentation which uses tiny loads to press pointed or spherical diamonds just tens of nanometers into the surface. Thus, the mechanical properties of a

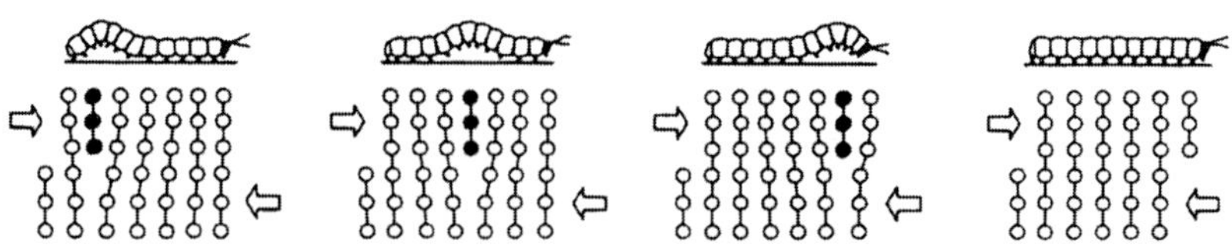

Fig. 3. Schematic illustrating how a crystal lattice can plastically deform through dislocation movement, akin to the segmented motion of a caterpillar. The dislocation is the extra half-plane represented by the filled circles.

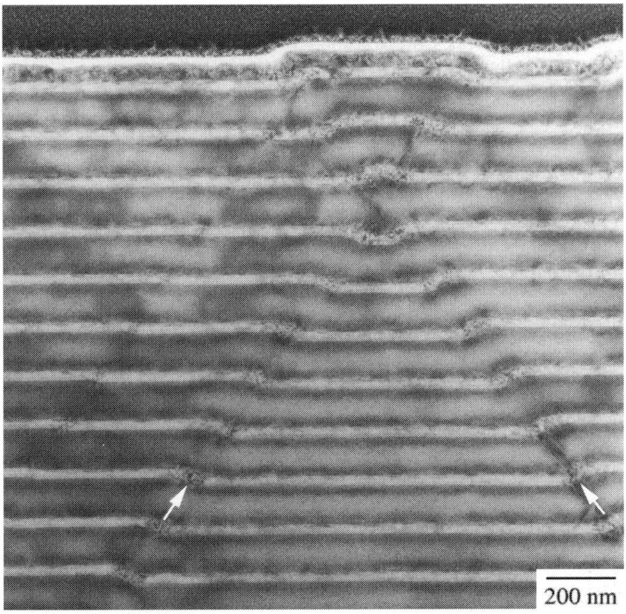

Fig. 4. TEM image of the deformation in a coherently strained $In_xGa_{1-x}As$ multilayer bent in compression at 500° C. The layer contrast observed is predominantly due to the compressive and tensile stresses in the thicker and thinner layers, respectively, resulting from their different indium contents. The ruck in the layering is due to the intersection of two shear bands (shown by arrows), each formed from the passage of around 100 dislocations (after Lloyd *et al.*, Ref. 19).

small volume of material can be measured quickly and conveniently and, because only a small volume is loaded, cracking is usually avoided even in very brittle materials. However, it is one thing to reliably measure hardness, but it is quite another to elucidate the subsurface deformation mechanisms that occur under the indent. The indented surface can be readily examined using, for example, scanning electron microscopy (SEM) and scanning tunneling microscopy. Subsurface damage can also be examined in the SEM in suitably cleaved specimens,[18] but examining deformation processes at nanometer resolution requires TEM. This has been near impossible in the past because of the difficulty in making an electron transparent membrane through a specific site such as an indent. Now this can be done routinely using a focused ion beam (FIB) microscope in which gallium ions mill the cross section containing the indent. The FIB uses the secondary electrons emitted to image the sample as it is being machined (Fig. 5).

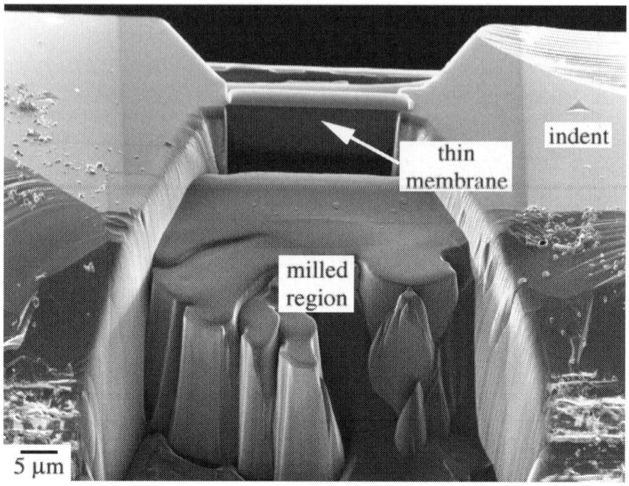

Fig. 5. Secondary electron image of the preparation of a thin membrane through a nanoindent in silicon using a focused ion beam.

Our initial investigations in a range of different materials have revealed a variety of processes that occur around an indent, in addition to dislocation motion. For example, "twins" can form, as seen in Fig. 6(a), showing the deformation under a spherical indent in a coherently strained $In_xGa_{1-x}As$ multilayer. "Twinning" is a very rapid transformation that, in this example, results in the reorientation of the lattice to create a region of crystal that is related to its parent by a rotation of 180°. The shear of the crystal is evident in the high-resolution image (Fig. 6(b)). Due to the low defect density of the crystal, new defects had to be nucleated by the high stress under the indenter. Interestingly twinning was recognized as a deformation mechanism over a century earlier than dislocation flow. At higher temperatures (500°C) dislocations can be activated more readily from preexisting sources, hence the shearing shown in Fig. 4.

The coherency strain between the layers in $In_xGa_{1-x}As$ multilayers can be controlled through the variation of the indium content of each layer through epitaxy. It has been found that the yield stress (i.e., the stress at which material starts to deform plastically) of the system can be varied by a factor of 2 by changing the strain difference between the layers.[20] Our work has shown the complex way coherency stress affects deformation: depending on the loading conditions coherency strains can weaken or strengthen the material.[18] In the following sections, we concentrate on another example:

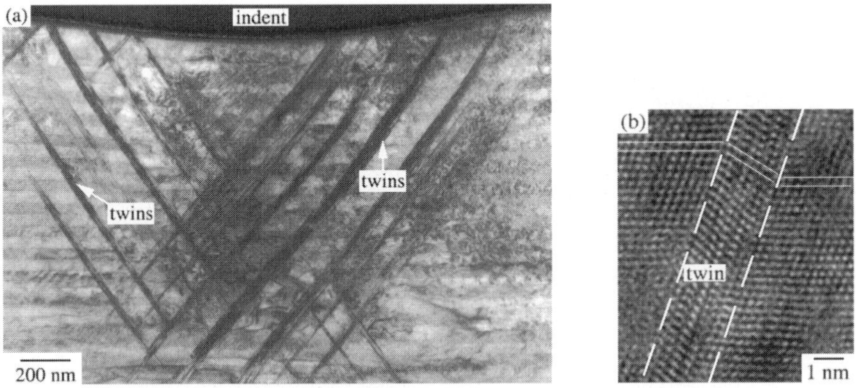

Fig. 6. (a) Deformation under a spherical indent in a coherently strained $In_xGa_{1-x}As$ multilayer observed using TEM. The deformation twins (some of which are indicated by arrows) can be seen cutting across the (horizontal) superlattice layers. Since the twins are in a different crystallographic orientation to the parent crystal, they diffract differently leading to the strong contrast observed. Due to the low defect density of the crystal new defects had to be nucleated by the high stress under the indenter (see Ref. 19 for further details). (b) High-resolution image of a twin in GaAs formed under similar conditions. The atomic planes in the structure can be resolved, as shown by the solid lines, showing the shear of the lattice created by the twin. The dashed lines indicate the boundaries of the twin.

how multilayers can be designed to produce nitride coatings twice as hard as their monolithic counterparts.

4.1. *TiN/NbN multilayers: A case where plastic flow is confined within each layer*

In the case of nitride coatings, we have found evidence that having two layers with different elastic properties can have remarkable effects on the deformation process and the hardness. In monolithic TiN films, slip lines (due to repeated dislocation nucleation and glide on the favored {110} slip planes) are apparent under the indent, and in addition the crystal planes immediately below the indent tip are rotated by as much as 20° (Fig. 7(a)), decreasing rapidly moving away from the surface. The crystal rotations are associated with the strain gradients of the material under the indenter. In contrast, the pattern of rotations seen in a TiN/NbN multilayer ($\Lambda = 14\,\text{nm}$) is different to that seen in the monolithic TiN film (Fig. 7(b)). The rotations are confined to a cone of material defined by the indent impression and a depth below the indent tip that is approximately 0.8 of

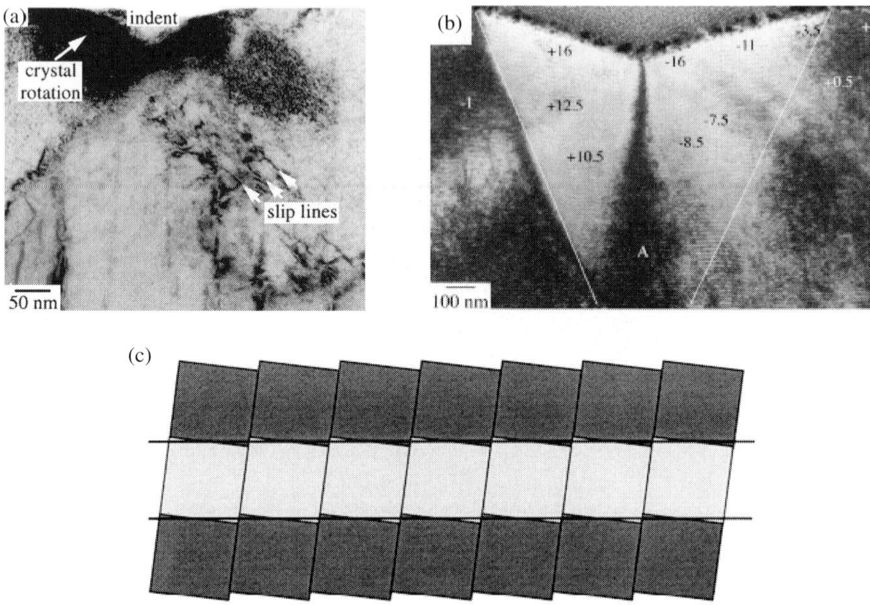

Fig. 7. TEM images of cross sections through nanoindents. (a) A 30-mN indent in a TiN monolithic film showing slip lines and crystal rotation. The region in which the crystal planes have been rotated appears dark here because it is diffracting the electron beam more strongly than the undeformed crystal. (b) A 30-mN indent in TiN/NbN multilayer ($\Lambda = 14\,\text{nm}$), indicating lattice rotations relative to position "A" in degrees. The white lines mark the approximate position of the compressed cone of material. (c) Schematic diagram illustrating how the multilayer may be deforming to give crystal rotations of approximately 10° (indicated by the tilted squares) yet maintaining untilted layer interfaces (indicated by the horizontal black lines).

the width of the indent impression (as approximately indicated by the white lines). Figure 7(b) also shows that the layers become virtually horizontal at approximately 400 nm below the surface, but there is still a crystal rotation of 10°. The observation that the crystal rotation can be greater than the rotation of the layers is at first sight surprising, and illustrates the danger of solely examining the position of the layer interfaces to infer the deformation pattern. Lloyd and co-workers[21] have suggested that this extra rotation can occur if the layers were simultaneously sheared, to maintain untilted layer interfaces on average, as illustrated schematically on Fig. 7(c). Such shearing has been observed elsewhere in W/NbN laminates.[22]

A further indication that plastic flow must occur by the lateral movement of material outward along each layer is given by the fact that no slip

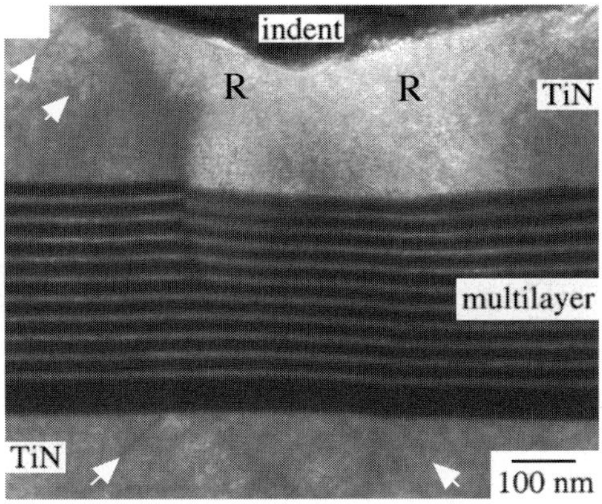

Fig. 8. The 40-mN indent in a film containing a TiN/NbN multilayer sandwiched between "thick" layers of TiN. Slip lines (some of which are indicated by the arrows) in the TiN do not extend into the multilayer, but reappear underneath it. The crystal planes in the regions labeled "R" have undergone significant distortion similar to that seen in Fig. 7(a).

lines are observed across the layers within the TiN/NbN multilayers. This is seen in Fig. 8, a cross section of an indent in a TiN/NbN multilayer ($\Lambda = 20\,\mathrm{nm}$), sandwiched between two "thick" (300 nm) TiN layers which do contain slip lines. Slip lines (some of which are indicated by the arrows) in the TiN do not extend into the multilayer, but reappear underneath it. Hence, it appears that the multilayer interfaces create an extra barrier for dislocation nucleation and propagation that forces the material flow to be confined within each layer. These observations are consistent with the model proposed by Chu and Barnett[17] on the yield stress enhancement of TiN/NbN multilayers, and might explain the large enhancements, up to 100%, in hardness observed with respect to the monolithic components.

4.2. TiN/SiN$_x$ multilayers: A case where columnar growth is interrupted

Our research has also revealed how low levels of porosity can affect the hardness of TiN/NbN multilayers grown by sputter deposition.[23] Sputtered films display often columnar grains: that is, grains elongated in the direction of growth (see for instance, Fig. 10(a)); hence, only a tiny degree of surface roughness can result in pores forming preferentially at the column boundaries due to shadowing (see Fig. 9). Under load, shearing occurs along the columnar boundaries which act as weak links, lowering the hardness

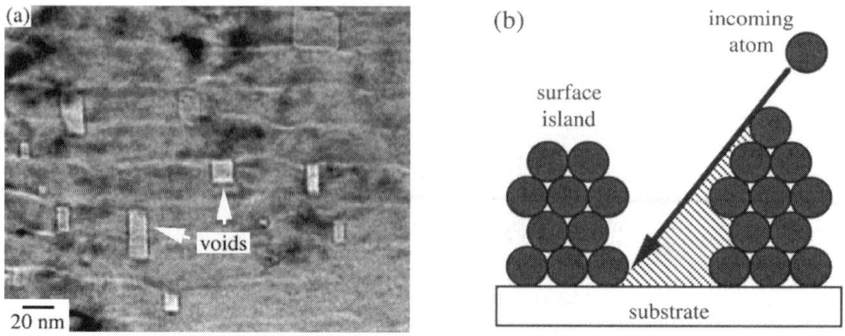

Fig. 9. (a) TEM image showing porosity embedded in a TiN/NbN multilayer. (b) Schematic illustrating the origin of porosity in sputtered films under conditions of limited surface mobility (after Hultman *et al.*, Ref. 25). The shaded region indicates the area shadowed by the deposited islands.

and reducing the benefits of a multilayered structure. In contrast to the single-crystal films discussed above, industrial coatings are often grown on steel and are polycrystalline, hence the undesirable weakening effects associated with porosity at column boundaries are likely to be multiplied. An alternative approach to both suppress the columnar growth and benefit from the hardness associated with small grain sizes is to design a multilayer with alternate crystalline and amorphous layers. This is illustrated in Fig. 10, which compares a TiN polycrystalline, columnar coating and a TiN/SiN_x multilayer, grown under similar conditions, in which the SiN_x layers are amorphous. The amorphous SiN_x layers interrupt the columnar growth of the TiN grains and each TiN layer is forced to renucleate on top of the amorphous SiN_x. A much greater degree of control over the resulting microstructure and roughness is therefore achieved. As a result, the hardness of the TiN/SiN_x multilayer is over 50% greater than the hardness of the monolithic components.[23]

4.3. *TiN/SiN_x multilayers revisited: A case where totally new behavior (not found in the bulk at all) is unraveled when the layers are made extremely thin*

In the examples discussed above the property of the multilayer is more than the "sum of the parts", but the actual component layers are essentially the same as the bulk material. However, the extreme "thinness" of the layers

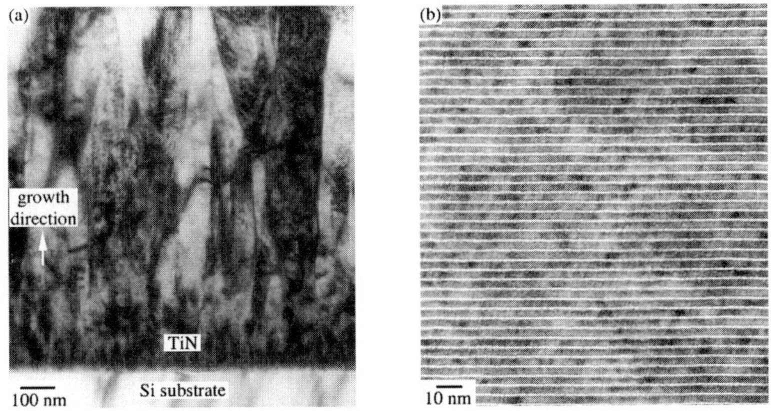

Fig. 10. TEM images showing (a) columnar TiN film grown on Si(001); (b) small region from a TiN/SiN_x multilayer grown on Si(001) in similar conditions. The lighter layers are SiN_x. In (a) the contrast arises from the different crystallographic orientations of the grains. Small, equiaxed grains form near the substrate but, as the film grows, grains in favorable orientations grow preferentially to create a columnar structure. The introduction of amorphous layers in (b) results in a much greater degree of control over the orientation and shape of the grains, as each amorphous layer interrupts grain growth and each TiN layer is forced to renucleate, eliminating the columnar growth and reducing the roughness.

can lead to the evolution of totally new behavior, not found in the bulk at all. This is the case, for instance, with the TiN/SiN_x multilayers shown above when the SiN_x layers are made extremely thin.

Sputtered SiN_x under conventional conditions is grown amorphous, as shown in Fig. 10(b). However, when grown on TiN, the SiN_x adopts the crystalline cubic structure of TiN through epitaxy up to a critical thickness of around 0.3 nm (even though this is not the stable structure at room temperature), because this is more energetically favorable than to introduce defects at the interface (see Fig. 2). For thicker layers, the SiN_x transforms to the conventional amorphous structure. The difference in structure is shown in Fig. 11. Above the critical thickness, the TEM image in Fig. 11(a) shows that the film consists of well-defined TiN layers with equiaxed grains separated by continuous and homogenous amorphous SiN_x interlayers, as shown before. The high-resolution image proves that the growth of a (111) oriented TiN grain (where the atomic planes can be resolved) is interrupted by the amorphous SiN_x layer and that a new TiN grain is renucleated in the subsequent TiN layer. Layer thicknesses are 4.5 and 2.4 nm for TiN and SiN_x, respectively. Below the critical SiN_x layer thickness, the TEM image

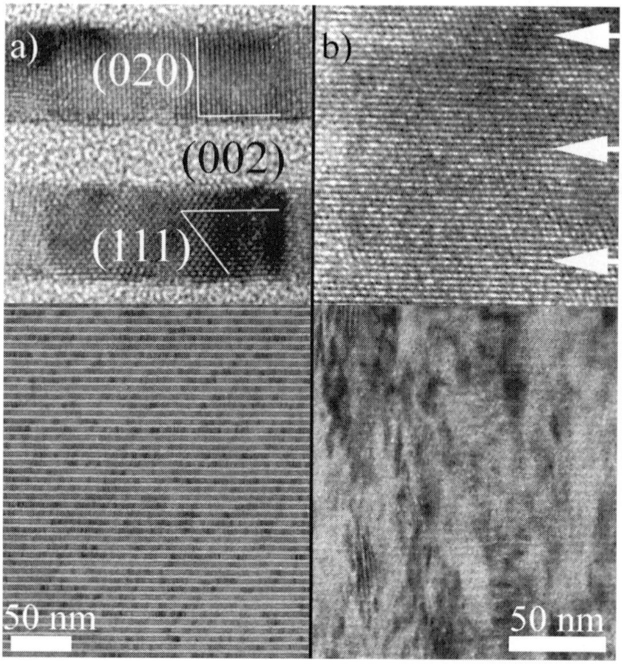

Fig. 11. TEM images (below) and corresponding high-resolution images (above) of TiN/SiN_x multilayers with 4.5-nm-thick TiN layers: (a) 2.4-nm-thick amorphous SiN_x layers and (b) 0.3-nm-thick crystalline SiN_x layers. In (a), the darker layers are TiN, where the atomic columns can be resolved. The lighter layers are amorphous SiN_x and hence, no atomic columns are resolved. The lines indicate the trace of (111) and (002) lattice planes, showing that the amorphous SiN_x interrupts the TiN growth, resulting in equiaxed TiN grains whose size is limited by the layer thickness. In (b), the arrows indicate the position of the SiN_x layers. The atomic columns in the structure can be resolved across all the layers showing that the SiN_x adopts the crystalline structure of the TiN layers, resulting in columnar grains through the entire film thickness, as shown in the image below (after Söderberg *et al.*, Ref. 24).

of Fig. 11(b) shows that the multilayer is composed of columnar grains with retained layering. The corresponding high-resolution image reveals the lack of amorphous SiN_x interlayers for such thin layers and instead the atomic columns can be resolved continuing across all the layers. The latter has remarkable effects on the mechanical properties, with a hardness increase of an additional 20% with respect to multilayers where the SiN_x layers remain amorphous, presumably due to similar hardening effects that those found in TiN/NbN multilayers.

As well as aiding our understanding on the hardness of multilayer films, the results presented here will be increasingly useful in the understanding of local deformation processes such as wear and abrasion, since indentation introduces high plastic strains in a controlled way. Further, if the deformation mechanisms can be understood we will be able to better design materials with specified hardness, and to combine hardness with other important properties such as corrosion resistance. In the next section, we concentrate on the structural variations and correlated magnetic properties that emerge in a metallic multilayer system when the layers are very thin.

5. Metallic Magnetic Multilayers

Multilayers are increasingly used in magnetic memories and sensors because of the novel properties and new phases they can display.[26] For example, as recently as 1988 it was discovered that the electrical resistance across a series of magnetic layers changed enormously depending on whether the magnetization of the layers were parallel or anti-parallel to each other. This was called giant magnetoresistance (GMR) and, due to remarkably fast industrial development, GMR films are already being used as sensors in hard disk drives.

As in the preceding example of hard coatings, new magnetic properties can also arise in the individual thin layers making up a multilayer, combined with a modified crystal structure compared with the bulk material. Here we concentrate on one example: the rich variety of structures and magnetic properties displayed by Fe thin films grown on Cu have very different properties to the iron we usually encounter. At room temperature the stable bulk phase of Fe is body-centered cubic (BCC), and it is ferromagnetic, while at high temperature the nonmagnetic face-centered cubic (FCC) structure is stable (Fig. 12(a)). When grown on Cu, the Fe adopts the FCC structure of the Cu (Figs. 12(b) and 12(c)) up to a critical thickness of around 10 atomic layers (AL), above which the Fe transforms to the bulk BCC structure. However, below the critical thickness the Fe is distorted from cubic symmetry with increased atomic spacings in the direction parallel to the layer normal to give a *tetragonal* structure. The magnitude of the distortion also varies as a function of thickness of the Fe layers, being greatest for the thinnest layers, as summarized in Table 2. Distortions might be expected in thin layers due to coherency strain, as described earlier. However, in this case the bulk FCC Fe structure has smaller atomic spacings than the Cu; hence, coherency strain would increase the Fe spacings in-plane. To

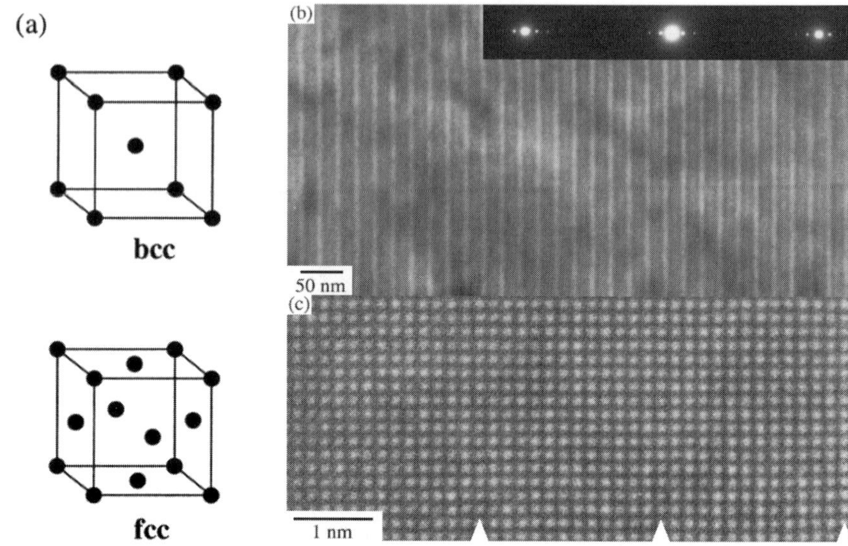

Fig. 12. (a) Unit cells of the BCC and FCC lattices. (b) TEM image of Fe–Cu multilayer ($\Lambda = 2.4\,\text{nm}$) with a diffraction pattern inset. The strong contrast (making the Fe layers appear bright under these imaging conditions) is due to the different mean inner potential of the two materials. The weaker, closely spaced spots on the diffraction pattern arise from the periodicity of the multilayer superimposed on that of the underlying crystal lattice. (c) High-resolution image of the same Fe–Cu multilayer showing the atomic columns in the structure. Arrows indicate the multilayer periods. These are hard to identify here due to the low contrast between the two layers. Using images such as these, and others, the structure of the layers may be determined.

preserve a similar atomic volume, the spacings are expected to decrease perpendicular to the layers (known as Poisson contraction). An expansion is actually observed. It is as if the Fe layers displayed a negative Poisson ratio — unless, as we suggest, the stable "unstrained" structure in this thin-film form is actually tetragonal (Fig. 13). TEM can also provide details of the local electronic structure and these studies also point to modifications in the band structure of Fe, consistent with the tetragonal distortion.[27]

Differing magnetic properties are coupled with these structural distortions (see Table 2). The films are all ferromagnetic (*cf.* nonmagnetic FCC bulk Fe) but with properties that vary with the layer thickness. The moment per atom is greatest for the thinnest layers, and the Curie temperature (at which the magnetization disappears) is greatest for the thickest layers. Finally, the direction of magnetization is also layer-thickness dependent. Usually, thin-films are magnetized in the plane of the film, to reduce the field around the sample, but here a strong magnetic anisotropy forces the

Table 2. Structural and magnetic properties of a series of Fe–Cu multilayers with layer thickness measured in AL.

Fe (AL)	Cu (AL)	% Fe(002) strain	T_c (K)	M_s (μ_B/Fe atom)	Easy axis
2	15	6.8 ± 0.5	~ 300	1.7 ± 0.3	// [001]
4	13	3.6 ± 0.3	~ 300	0.7 ± 0.2	// [001]
6	9	1.8 ± 0.1	~ 400	0.8 ± 0.2	$\perp$ [001]

The strain in the layers is relative to the lattice parameter for FCC Fe (0.3585 nm), T_c is the Curie temperature, M_s the saturation magnetization extrapolated to 0 K.

easy magnetization direction to lie perpendicular to the plane of the film for the thinnest layers. "Perpendicular magnetization" has been observed in a number of multilayer systems and has a very important application in magnetic recording, since the higher stray field above the specimen allows more reliable recording at high densities.

To summarize: the atomic structure, magnetic properties, and electronic structure each vary with the Fe layer thickness and demonstrate that the Fe layers are highly anisotropic. By varying the multilayer design, a whole range of magnetic properties can be displayed — all from pure iron! In creating a multilayer the electronic structure has been modified, stabilizing a new tetragonal structure and properties. It has been suggested[27] that the distortions in the structure could be explained qualitatively in a manner similar to Jahn–Teller distortions in atomic complexes, arising out of the quantum well states created in the thin layers. The differing magnetic properties can be understood from the great sensitivity of the magnetic state of Fe to its atomic volume since the energy changes associated with structural and magnetic transitions are very similar.[28]

In this example the multilayers do not behave analogously to a bulk composite structure. In a bulk composite the properties of the individual layers are essentially unchanged upon incorporation into the composite. Here the multilayers do not simply display a combination of the properties of bulk Fe and Cu but are an entirely new material whose atomic structure and properties vary strongly with multilayer design, pointing the way to how the electronic structure can be engineered. This sort of engineering could be described as a modern (and rather more successful) form of alchemy: changing the properties of elements through modification of their electronic structure. The conventional alchemy required the atomic nucleus to be changed, and ironically, since Fe has the most stable atomic nucleus it is the most resistant to this form of alchemy. Fortunately its electronic structure is more cooperative.

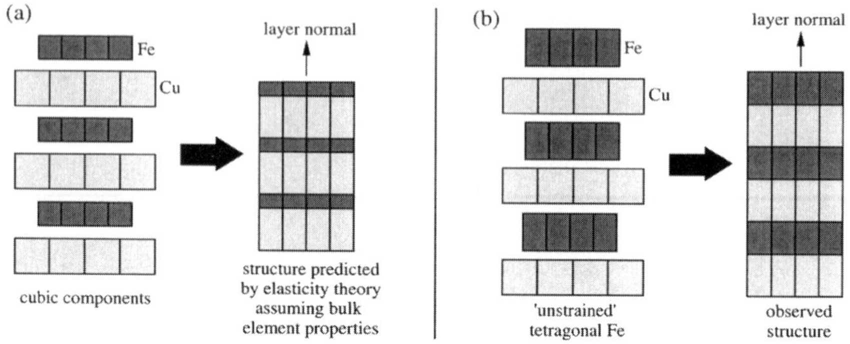

Fig. 13. Schematic illustrating the formation of a multilayer from (a) cubic Fe and Cu, and (b) tetragonal Fe and cubic Cu. Only the latter configuration explains the observed structure.

6. Conclusion and Future Developments

The design of materials with specified properties is the essence of materials science. Yet it is only remarkably recently that we have developed sufficient understanding to control materials' properties at will. Synthetic nanoscale multilayers provide an exciting route to the design of materials with novel magnetic and mechanical properties. To fully exploit the new possibilities that can occur in multilayers, it is vital to elucidate the structure–property relationships that will enable us to design the most promising multilayer structures from the near-infinite number of possibilities.

Future developments will depend on the synergy arising from advances in growth, characterization, and modeling. Growth requirements are stringent as can be seen by the sensitivity of the properties of Fe–Cu multilayers to the Fe layer thickness. In addition, optimum deposition conditions are often a compromise between conflicting factors; for example, high temperatures are favorable for low defect growth but intermixing of the layers is also enhanced. For industrial production, the challenge is to combine precision with reliability and reproducibility. Improvements in ultra-hard coatings may also lie in using the nonequilibrium deposition conditions to synthesize new compounds in the boron–carbon–nitrogen system.[29]

TEM characterization will be made more reliable and quantitative through current developments in instrumentation (e.g., aberration correctors) and specimen preparation allowing surface damage to be removed from thin foils. Improvements in the quantification of atomic and electronic structure characterization are essential to test and refine the modeling of multilayers. *Ab initio* modeling is increasingly sophisticated and can now include magnetic properties, but multilayers still present a challenge in computing time because of the relatively large size of their unit cell. As well as the rigorous *ab initio* approach, simpler, physically perspicuous models are also needed, even if they are quantitatively incorrect, to allow predictions to be made of systems with interesting properties that may then be modeled accurately: models that are more like "cartoons" catch the key features of a person's appearance, as opposed to true portraits.[3]

Materials research is only just beginning to fully utilize and control the vast array of material properties available. Nanoscale multilayers provide new opportunities for materials scientists to display their artistry.

Acknowledgments

We are grateful to the EPSRC, Trinity Hall, Cambridge, the Royal Society, the Basque Government, the Spanish Ministry of Education (Torres Quevedo Program) and the European Social Fund for financial support. We also acknowledge the help and insight of many colleagues, in particular Dr. W. J. Clegg, Prof. L. Hultman, Prof. M. Odén, H. Södeberg and the late Dr. W. M. Stobbs.

References

1. R. W. Cahn, *The Coming of Materials Science* (Pergamon, Amsterdam, 2001).
2. R. F. Mehl, The historical development of physical metallurgy, *Physical Metallurgy*, ed. R. W. Cahn, 2nd edn. (Elsevier, Amsterdam, 1970), pp. 1–31.
3. A. H. Cottrell, The art of simplification in materials science, *Mater. Res. Soc. Bull.* **22**(5), 15–19 (1997).
4. E. Arzt, Size effects in materials due to microstructural and dimensional constraints: A comparative review, *Acta Mater.* **46**, 5611–5626 (1998).
5. G. Mayer, Rigid biological systems as models for synthetic composites, *Science* **310**, 1144–1147 (2005).
6. E. S. Sadki, Z. H. Barber, S. J. Lloyd, M. G. Blamire and A. M. Campbell, Effects of interlayer coupling on the irreversibility lines of NbN/AlN superconducting multilayers, *Phys. Rev. Lett.* **85**, 4168–4171 (2000).
7. S. J. Lloyd, D. M. Tricker, Z. H. Barber and M. G. Blamire, Growth of niobium nitride/aluminium nitride bilayers and multilayers, *Philos. Mag. A* **81**, 2317–2335 (2001).
8. P. Robinson, R. J. Harrison, S. A. McEnroe and R. B. Hargreaves, Lamellar magnetism in the hematite–ilmenite series as an explanation for strong remnant magnetisation, *Nature* **418**, 517–520 (2002).

9. W.-D. Münz, D. B. Lewis, P. Eh. Hovsepian, C. Schönjahn, A. Ehiasarian and I. J. Smith, Industrial scale manufactured superlattice hard PVD coatings, *Surf. Eng.* **17**, 15–27 (2001).
10. D. E. Newbury and D. B. Williams, The electron microscope: The materials characterisation tool of the millennium, *Acta Mater.* **48**, 323–346 (2000).
11. P. B. Hirsch, Electron microscopy in materials science, *The Electron. Proceedings of the International Centennial Symposium on the Electron,* eds. A. Kirkland and P. D. Brown (Institute of Materials, London, 1998), pp. 464–469.
12. S. J. Lloyd, J. M. Molina-Aldareguia and W. J. Clegg, Structural characterisation of TiN/NbN multilayers: X-ray diffraction, energy-filtered TEM and Fresnel contrast techniques compared, *J. Microsc.* **217**(3), 241–259 (2005).
13. M. J. Whelan, The early observations of defects in metals by transmission electron microscopy, *Understanding Materials*, ed. C. J. Humphreys (Institute of Materials, London, 2002), pp. 17–36.
14. J. M. Molina-Aldareguia, S. J. Lloyd, Z. H. Barber, M. J. Blamire and W. J. Clegg, The microstructure and nanoindentation behaviour of TiN/NbN multilayers, *Mater. Res. Symp. Proc.* **594**, 9–14 (2000).
15. S. J. Lloyd and R. E. Dunin-Borkowski, Electron structure of face centred tetragonal Fe in ferromagnetic Fe–Cu multilayers, *Phys. Rev. B* **59**, 2352–2362 (1999).
16. M. Shin, L. Hultman and S. A. Barnett, Growth, microstructure and microhardness of epitaxial TiN/NbN superlattices, *J. Mater. Res.* **7**(4), 901–911 (1992).
17. X. Chu and S. A. Barnett, Model of superlattice yield stress and hardness enhancements, *J. Appl. Phys.* **77**(9), 4403–4411 (1995).
18. M. R. Castell, A. Howie, D. D. Perovic, D. A. Ritchie, A. C. Churchill and G. A. C. Jones, Plastic deformation under microindentations in GaAs/AlAs superlattices, *Philos. Mag. Lett.* **67**, 89–93 (1993).
19. S. J. Lloyd, K. M. Y. P'ng, A. J. Bushby, D. J. Dunstan and W. J. Clegg, Effect of coherency strain on the deformation of $In_xGa_{1-x}As$ superlattices under bending and nanoindentation, *Philos. Mag.* **85**(22), 2469–2490 (2005).
20. N. B. Jayaweera, A. J. Bushby, P. Kidd, A. Kelly and D. J. Dunstan, Control of plasticity with coherency strain, *Philos. Mag. Lett.* **79**, 343–349 (1999).
21. S. J. Lloyd, A. Castellero, F. Giuliani, Y. Long, K. K. McLaughlin, J. M. Molina-Aldareguia, N. A. Stelmashenko, L. Vandeperre and W. J. Clegg, Observations of nanoindents via cross-sectional transmission electron microscopy: A survey of deformation mechanisms, *Proc. Royal Soc. A* **461**, 2521–2543 (2005).
22. D. E. Kramer, M. F. Savage, A. Lin and T. Foecke, Novel method for TEM characterization of deformation under nanoindents in nanolayered materials, *Script. Mater.* **50**, 745–749 (2004).
23. J. M. Molina-Aldareguia, S. J. Lloyd, M. Odén, T. Joelsson, L. Hultman and W. J. Clegg, Deformation structures under indentations in TiN/NbN single-crystal multilayers deposited by magnetron sputtering at different bombarding ion energies, *Philos. Mag. A* **82**, 1983–1992 (2002).

24. H. Söderberg, M. Odén, J. M. Molina-Aldareguia and L. Hultman, Nanostructure formation during deposition of TiN/SiN_x nanomultilayer films by reactive dual magnetron sputtering, *J. Appl. Phys.* **97**, 114327 (2005).
25. L. Hultman, L. R. Wallenberg, M. Shinn and S. A. Barnett, Formation of polyhedral voids at surface cusps during growth of epitaxial TiN/NbN superlattice and allow films, *J. Vac. Sci. Technol. A* **10**, 1618–1624 (1992).
26. P. Grünberg, Layered magnetic structures in research and application, *Acta Mater.* **48**, 239–251 (2000).
27. S. J. Lloyd, The electron structure of ferromagnetic Fe–Cu multilayers examined using transmission electron microscopy, *The Electron. Proceedings of the International Centennial Symposium on the Electron*, eds. A. Kirkland and P. D. Brown (Institute of Materials, London, 1998), pp. 464–469.
28. G. Prinz, Metastability in epitaxial magnetic thin films, *J. Magn. Magn. Mater.* **100**, 469–480 (1991).
29. W. Kulisch, Deposition of diamond-like superhard materials in the B/C/N system, *Phys. Stat. Sol. (A)* **177**, 63–79 (2000).

Jon M. Molina-Aldareguia Born in San Sebastian, Basque Country, Spain, Jon Molina studied at TECNUN, the Engineering School of the University of Navarra, where he graduated with first class honors in Materials Engineering in 1998. He then moved to the University of Cambridge, where he obtained his PhD in 2002. During 2002 he was a postdoctoral researcher investigating the physics of thin films at the University of Linköping, Sweden. He has worked on the processing and the mechanical and microstructural characterization of multilayered and nanostructured thin films, including TiN/NbN and TiN/SiN_x multilayers and CN_x films. Aged 30, he is currently a Researcher at the Centro de Estudios e Investigaciones Técnicas de Gipuzkoa (CEIT) and Associate Professor of Materials Science and Engineering at TECNUN.

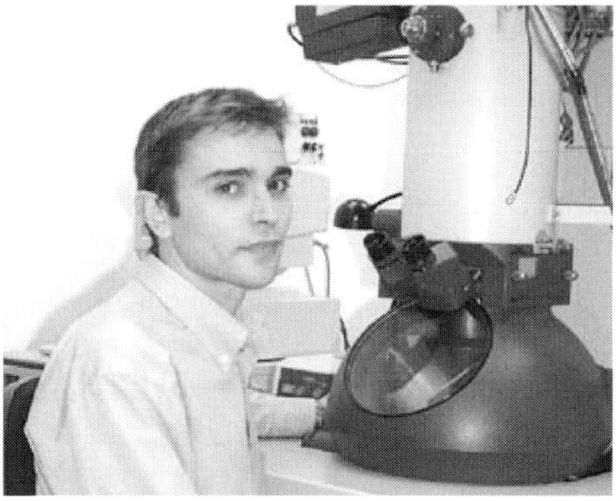

Stephen J. Lloyd Stephen J. Lloyd is 34. Born in Croydon, Surrey, he studied Natural Sciences at the University of Cambridge where he graduated with the top first in Materials Science and Metallurgy in 1993. Continuing study in Cambridge he obtained his PhD, and was appointed to a Junior Research Fellowship at Trinity Hall in 1996. From 2000 to 2004 he has held a Royal Society University Research Fellowship in Cambridge. His research interests have focused on the development of electron microscopy techniques to study structure–property relationships in multilayered materials, and most recently, the characterization of deformation associated with indentation. He has published over 50 papers. Married with two daughters, he enjoys being a husband and father.

CHAPTER 4

NATURE AS CHIEF ENGINEER

Simon R. Hall

Centre for Organized Matter Chemistry
School of Chemistry, University of Bristol
Bristol BS8 1TS, UK

With the benefit of millions of years of natural selection, it is no surprise that when it comes to fitting form to function, Nature is second to none in engineering the right structure for the right job. The creation of architectural elegance, despite using the minimal amount of material has long fascinated man, who has endeavored to understand how intricate construction can be accomplished merely through the flow of inorganic ions and strategically placed macromolecules. When the demands of the environment are the blueprints of the construction, structures are produced with the utmost efficiency. This was first stated explicitly by D'Arcy Thompson, in his seminal work *On Growth and Form* (1917).[1] However, even as far back as the 16th century, scientist and astronomer Johannes Kepler noted that "Nature uses as little as possible of anything". Both these and many other luminaries held Nature in the highest esteem as an engineer par excellence.

In the 19th century, people used these tenets and took their inspiration for daring and innovative architectural and engineering projects from Nature. In the 21st century therefore, when advanced nanoscale materials are required, it is only right that man should once again turn to Nature for solutions to their creation.

Keywords: Biomimetic, template, synthesis, inorganic, superconductor.

1. Nature Inspires Engineering

"... To hold, as 'twere, the mirror up to nature ..."
Hamlet Act 3, Scene II

As the Victorians ushered in the golden age of engineering with such marvels as the Forth Rail Bridge and the Eiffel Tower, it was a time of daring experimentation. With new materials and ever larger projects under construction, it was to Nature that men began to turn to provide the architecture required; an architecture that had been proven through many millennia to provide the most propitious form for the job in hand, in the most efficient manner possible.

One of the first, and most famous "bioinspired" architectural project was the construction of the Crystal Palace for the Great Exhibition of 1851. Joseph Paxton conceived the Crystal Palace largely as a result of the work he did as Head Gardener to the Duke of Devonshire at Chatsworth House, Derbyshire. While at Chatsworth, Paxton built the largest conservatory in the world at that time, utilizing glass and iron for strength and durability. In 1837, the arrival of a lily from Guyana demanded a custom-built heated pool which Paxton designed. He became fascinated by the huge leaves of the plant which he dubbed "a natural feat of engineering" and tested their strength by floating his daughter on one of them. A closer examination of the underside of the leaves revealed to Paxton the secret of their mechanical stability; an array of radiating ribs connected with flexible cross-ribs (Fig. 1). Continuous experimentation over the following years enabled Paxton to improve on his glass and iron structures, culminating in the incorporation of the waterlily's structural features in his design for the Crystal Palace (Fig. 2).

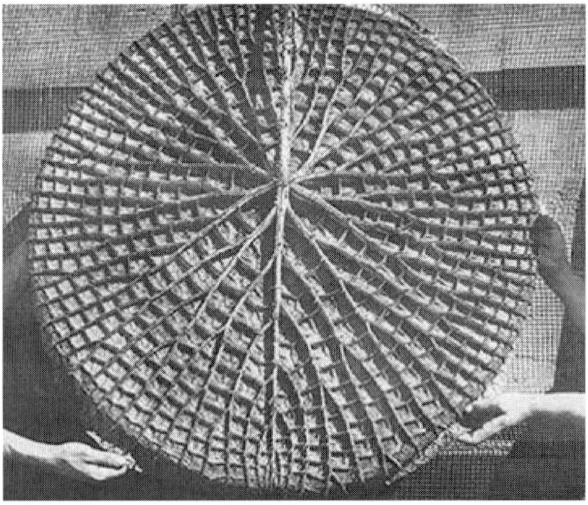

Fig. 1. Underside of a waterlily leaf.

Fig. 2. The façade of the Crystal Palace.

Another striking example of engineering inspired by Nature can be seen in Isambard Kingdom Brunel's Royal Albert Bridge near Plymouth. The bridge is a clever combination of arch and suspension bridge. An arch bridge produces a net outward thrust at the abutments, whereas a suspension bridge pulls the abutments inward. By combining the two concepts in one bridge, the overall force at the abutments is almost zero. In Brunel's bridge, the arches consist of iron tubes with an oval cross section, which produce the outward thrust to balance the inward pull of the draped chains. The minimal force carried by the abutments allows for a lighter and more importantly, cheaper construction. Inspiration for this may have come from the observation that this method of force balancing is one which every four-legged animal adopts. For example, in an elephant, the legs are the abutments, the belly the chains, and the spine the arch of the bridge. Although it is not known whether Brunel (in the manner of Paxton) first considered Nature before embarking on his design, it is likely that such a natural analog would not have been far from his mind.

In the 20th century, scientists began to take an active interest in the architectural constructs of the biological world, particularly keen to understand the procedures flora and fauna use in the production of inorganic structural elements. The Father of this approach was R. J. P. (Bob)

Williams of Oxford University, who instigated a study of the detailed functional use of inorganic elements in biological systems.[2] By applying principles from inorganic chemistry such as the complex-ion formation and redox potential, to biological systems, he was able to deduce many hitherto unknown mechanisms in biomineralization. Among the discoveries from this time were the elucidation of the special inorganic chemistry of unusual metal-binding sites in Nature, and the role and mode of action of calcium in the formation of calcified structures (summarized in Ref. 3).

One of the students of the Williams "Oxford School" was Stephen Mann. Mann realized that by understanding the processes of biomineralization in terms of the movement and precipitation of inorganic elements within a "biological environment" it should be possible to replicate or mimic them under laboratory conditions.[4] Mann supposed that as mineralization usually takes place due to constraint within an organism, then by replicating those constraining factors synthetically, either *physically* in the form of for example, vesicles, or *chemically* by control of localized supersaturation, bioanalogs could be created. These experiments in "biomimetics" (e.g., Refs. 5 and 6) yielded complex and often strikingly "lifelike" inorganic materials, by following closely (but not exactly) the protocols used in the natural analog.

Currently, research worldwide into biomimetic control of mineralization is strong, producing many diverse and often industrially valuable materials (see e.g., Refs. 7 and 8). All syntheses however still rely either directly or indirectly on the "boundary organized biomineralization" concepts introduced by Williams, Mann *et al.*

2. Nature Becomes Engineering

"Have no fear of perfection — you'll never reach it"
Salvador Felipe Jacinto Dali Y Domenech (1904–1989).

The complexity of biological structures and the complex systems which give rise to them are not easily replicated in the laboratory. Even the most advanced and succinct synthetic protocols can only ever offer a poor imitation of the natural analog. This has convinced many scientists to "cut out the middle man" and directly utilize naturally occurring materials as part of their synthetic procedures. The advantage of this approach is clear; by using a preformed, often hierarchically complex material, the scientist aims to transfer the physical properties of the original, to that of the synthetic analog. Many intact biosystems have been used to date, such as bacterial and fungal colonies,[9–11] viroid particles,[12–15] insect wings,[16] and diatoms,[17] to produce materials with potential applications in catalysis, separations science, electronics, and photonics. One particularly striking example of this approach was the use of spider silk as a template to produce organic/inorganic fiber composites.[18]

Spider silk is a biopolymer with an extraordinarily high tensile strength, far exceeding high tensile steel and Kevlar on a weight-for-weight basis. The impetus for this research was the attempt to incorporate the incredible tensile strength of spider silk in an organic–inorganic composite of a composition which could be tailored to the desired application. One advantage of using silk in a composite material is its inherent biocompatibility. This would allow for potential biomedical applications where strength is required, such as artificial tendons or nonallergenic sutures. Spider silk was able to be successfully coated with a wide variety of inorganic materials among which, nanoparticulate magnetite and cadmium sulfide were of the most interest, as they imbued the composite with magnetic and semiconducting properties, respectively. Fibers were shown to retain their flexibility and strength even after mineralization, and thus would be able to find applications as smart, structural fabrics.

The most highly prized property of the majority of bioconstructs however, is that of a high surface area. In the fields of catalysis, chromatography and drug delivery, a high surface area ($> 300\,\mathrm{m}^2/\mathrm{g}$) greatly improves performance. Micro- and nanoscale features provide greater numbers of reactive sites and points of nucleation, enabling a high surface area material to outperform a chemically identical, low surface area one.

Zeolites are inorganic high surface area materials, with a microporous ($< 5\,\mathrm{nm}$) structure. In addition to the 48 naturally occurring mineral examples, more than 150 structurally distinct zeolite types have been synthesized in laboratories. Zeolites are highly prized for their ion-exchange capabilities and have found uses in water purification, molecular sieving, and as high surface area catalyst supports. Traditionally, in a laboratory synthesis, a structure-directing agent such as a bulky organic cation, for example, tetrapropylammonium bromide will be used, with an aluminosilicate gel condensing around it to form the zeolite. These materials have a monodisperse (a uniform size) pore structure, the size of which is dependent on the bulky organic cation. Extension of the pore size would allow for a greater range of reactions/applications for zeolites and much research has been devoted to the exploration of different templating agents. An ideal solution would be to use a porous naturally occurring material with pores at a larger length scale on which to "grow" a zeolite.

This has been achieved by Greil *et al.*[19] utilizing wood as a sacrificial biotemplate. Using rattan palm as a precursor material (Fig. 3(a)), simple pyrolysis to carbon replicas followed by an infiltration with liquid silicon at 1550°C led to a biomorphous interconnected SiC skeleton, with no change in the overall morphology of the carbonized wood template. A small degree of shrinkage was observed in the initial carbonization step, but the main structural features and porosity was preserved. Due to their highly porous morphologies and chemical composition, these templates were finally used as supports for zeolite coatings, via a support self-transformation method. The large content of Si in the Si–SiC ceramic replicas allowed zeolite to be crystallized on the inner channel walls by a hydrothermal route. The Si–SiC bioreplicas acted as a carrier for the zeolite layer and also as the unique source of silicon in the subsequent zeolite crystallization (Fig. 3(b)).

The successful formation of a crystalline zeolitic phase was confirmed by X-ray diffraction (Fig. 4) which showed peaks due to crystalline Si and SiC in the replicas, and the presence of an MFI-type (10-membered oxygen ring system) zeolite material.

The novel combination of biotemplating of a wood structure and zeolite crystal growth by self-transformation in the support enables the creation of a hierarchical porous material which would otherwise be difficult to synthesize in the laboratory. The final composites exhibit a high thermomechanical strength and a bimodal micro/macroporous architecture which makes them

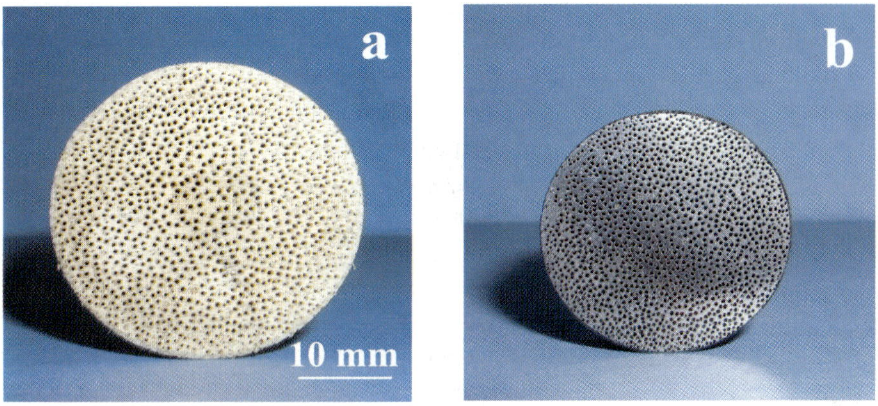

Fig. 3. Images of (a) rattan wood and (b) rattan-derived Si–SiC zeolitic bioceramic.

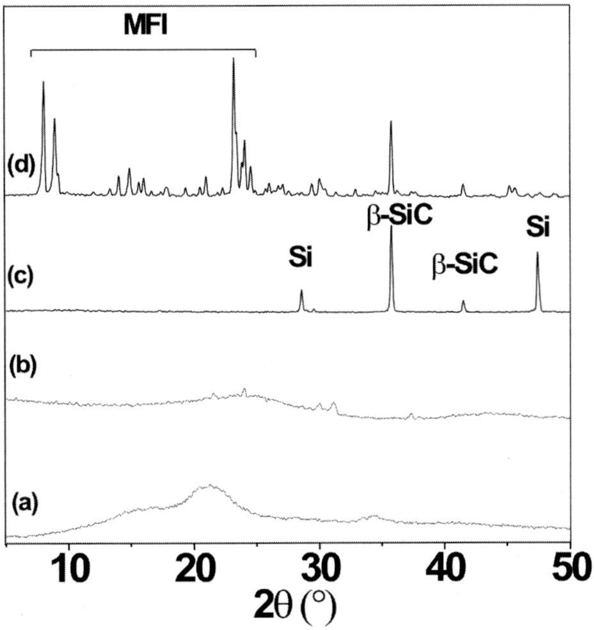

Fig. 4. X-ray diffraction patterns of (a) rattan wood, (b) carbonized rattan wood, (c) rattan-derived Si–SiC, and (d) Si–SiC zeolite composites.

eminently suitable for applications such as molecular sieving and catalytic activity.

High surface area silica materials are able to be synthesized in a number of length scales, due to the ability of silica to form open, highly porous networks. Even amorphous silica can have a surface area over $300\,m^2/g$.[20] With the use of surfactants, amorphous silica has been shown to condense around organic micelles to form materials well above $1000\,m^2/g$.[21] The high porosity in the final material is therefore a result of the fine structural feature of the template. When looking for a natural material which can be used to create high surface area inorganic materials, one must look for structural stability and intricate morphologic features.

One such material that lends itself well to inorganic replication is pollen. A pollen grain, most types of which are in the size range 25–35 μm, consists of the living cell in the center surrounded by a layer of cellulose known as the *intine* and a waxy, resistant coat, the *exine*, which consists of a polymeric substance called "sporopollenin". The function of the exine is to protect the pollen grain from desiccation and microbial attack, which accounts for its durability.

Pollen is ubiquitous and is easy to collect and incredibly stable when stored. Pollen grains have been found in the tombs of Egyptian pyramids which have been identifiable from their unchanged morphology even after almost 7000 years of quiescent storage. The majority of species have a high surface area and an intricate design, so can be selectively chosen for specific applications. As a template for mineralization, however, pollen has only recently been investigated.[22] The replication was achieved by soaking freeze-dried, mixed-flower pollen in various metastable solutions followed by thermal removal of the biological template to produce silica, calcium phosphate, or calcium carbonate facsimiles.

Figure 5 shows representative scanning electron microscope (SEM) images of native and mineralized pollen grains from mustard plants (*Brassica*). The native pollen grains were approximately 25 μm in length and exhibited a characteristic ellipsoidal morphology consisting of four longitudinal segments with foam-like surface structure. These features were

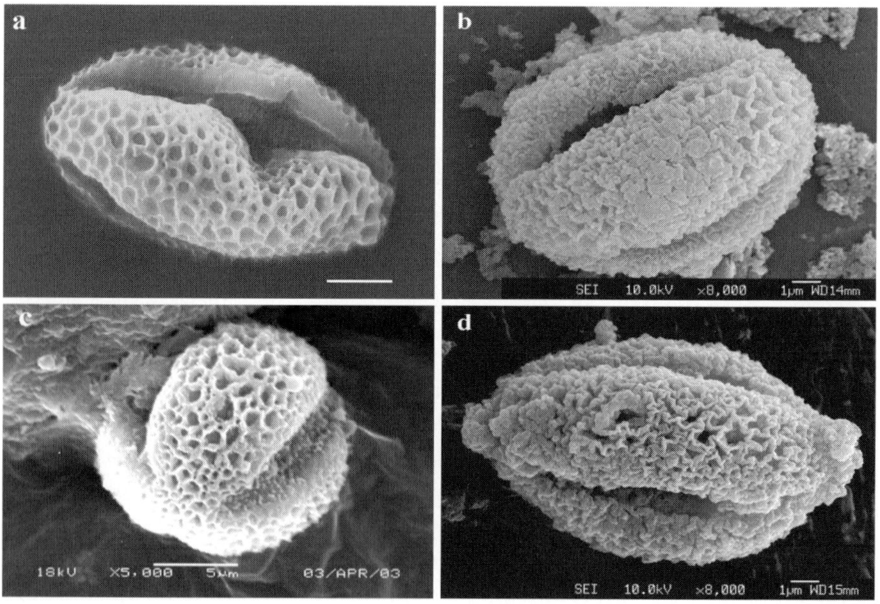

Fig. 5. SEM images of (a) uncoated pollen grain; ((b)–(d)) inorganic replicas consisting of (b) calcium phosphate, (c) calcium carbonate, and (d) silica. Scale bars: (a) and (c) 5 μm, (b) and (d) 1 μm.

also displayed in the hollow inorganic replicas, along with an additional macroporosity associated with removal of the biological template, and reduction in the size of the particles to 10–15 μm due to the thermal processing. Energy-dispersive X-ray mapping of individual grains confirmed that the calcined replicas consisted of silica, calcium carbonate, or calcium phosphate with negligible organic content.

Thermogravimetric studies indicated a typical organic weight loss during heating of approximately 70%. ^{29}Si NMR spectra of the silica replicas showed Q^4, Q^3, and Q^2 ($Q^n = \mathrm{Si(OSi)}_n\mathrm{(OH)}_{4-n}, n = 2$–4) compositions of approximately 75%, 9% and 16%, respectively, indicating that even after calcination at 600°C approximately 25% of the siloxane centers were not fully condensed and potentially reactive to further functionalization. This is an interesting area for further exploration. The attachment of reactive functional groups to the surfaces of these replicas could extend their range of applications considerably.

Powder X-ray diffraction indicated the formation of a highly crystalline calcite ($CaCO_3$) replica, whereas the calcium phosphate replicas consisted of a mixture of poorly crystalline brushite ($CaHPO_4{\cdot}2H_2O$) and monetite ($CaHPO_4$). Nitrogen adsorption–desorption measurements on the silica replicas showed that they were highly porous with relatively high surface areas ranging from 300 to as high as 817 m^2/g. The data suggest that the high surface area and mesoporosity arise as a consequence of the complex foam-like surface morphology of the native pollen grains and outgassing of organic components during thermal degradation. Similar results were not obtained however for the calcium carbonate and calcium phosphate replicas, which showed only weak gas–solid interactions and surface areas of 2 and 10 m^2/g, respectively.

Porous silica is used as a support in column liquid chromatography. For this the pore structure and specific surface area of the material must be controlled and tailored to the type of chromatography. Liquid–solid chromatography (LSC) requires the silica to be mesoporous (2–50 nm) with a high specific surface area whereas liquid–liquid chromatography (LLC) utilizes macroporous ($>$ 50 nm) silica with a low specific surface area. By this classification the silica pollen replicas produced in this work may have a potential application as a support in LSC. These morphologically complex particles could also have important applications in controlled storage–release processes and heterogenous catalysis. As proof of concept, the calcined silica replicas were functionalized by impregnation of the complex structures with silver or magnetite (Fe_3O_4) nanoparticles to produce metallic or magnetic derivatives, respectively. In the former, *in situ* photoreduction of adsorbed Ag(I) ions resulted in a homogenous dispersion throughout the silica matrix of silver nanoparticles with mean size of 11 nm as confirmed by transmission electron microscopy (TEM) studies on fractured samples (Fig. 6). Similar data were obtained when the silica particles were immersed in a sol of magnetite nanoparticles, which

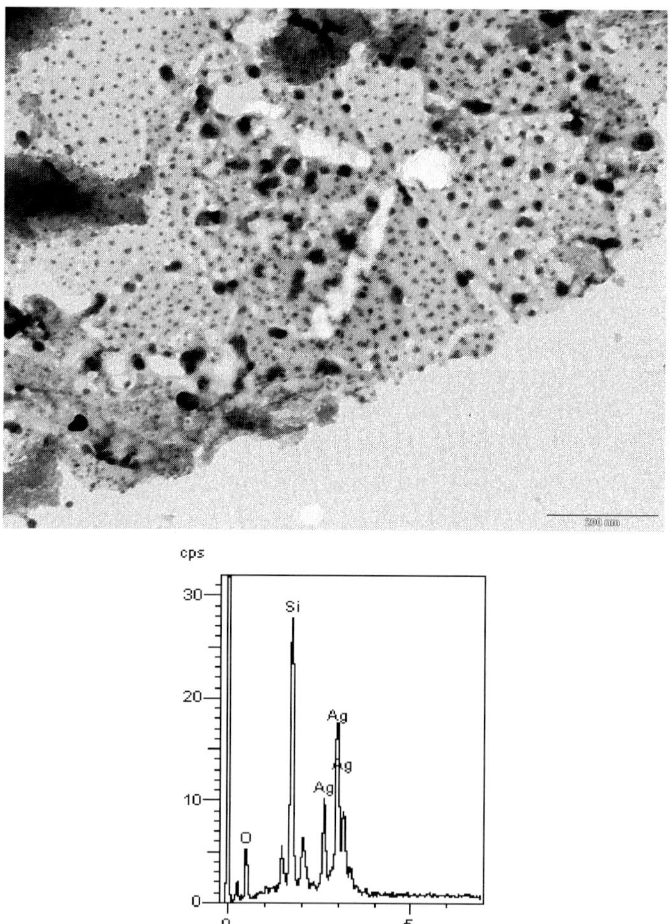

Fig. 6. TEM image showing high surface area silica pollen replica with silver nanoparticles grown within the nanoporous structure. The graph shows the energy-dispersive X-ray analysis (EDXA) signal from the silica–silver composite material. Scale bar is 200 nm.

resulted in the sequestration of nanoparticles with mean size of 7.6 nm, presumably by capillary uptake into the macro- and mesopores of the silica matrix.

Another possible application of the porous silica material produced in this work is as a drug-delivery system. Inorganic materials produced by the sol–gel route are used extensively in medicine, for example for implantation into the human body to augment bone repair. Utilizing the ability to encapsulate biologically active molecules into the material while retaining their activity has led to research into the use of silica as drug-delivery systems. This potential application was explored by a simple soaking of the pollen replicas in solutions of the nonsteroidal anti-inflammatory drug (ibuprofen) and the antihistamine drug (chlorpheniramine) followed by monitoring the corresponding release profiles in a simulated body fluid (SBF) (Fig. 7). Both the drugs have an indicative peak in the UV region, and the release of these drugs is therefore able to be monitored as an increase in the absorbance over time.

The high surface area silica replicas adsorbed significantly increased levels of the drug compared with the calcium-containing minerals. However, in each case, the pollen-templated replicas released approximately 50% of the entrapped drug within the first 10 min of immersion, after which the rate of release decreased significantly, such that 70–80% of the drug was transferred into the SBF after 3 h (Fig. 7), suggesting that with their much

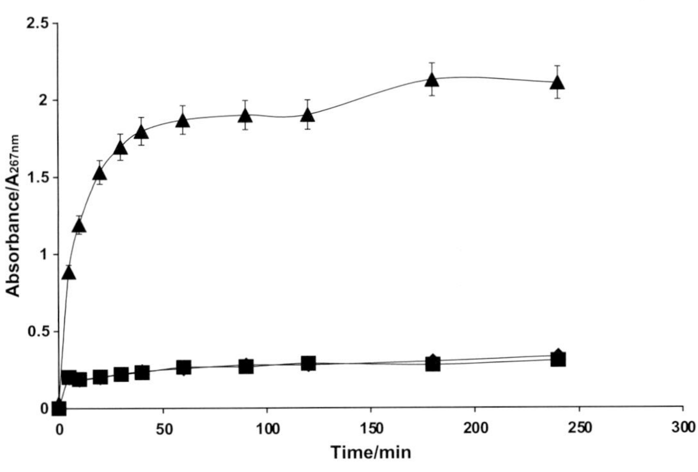

Fig. 7. Controlled release of drugs from silica (▲), calcium carbonate (♦), and calcium phosphate (■) pollen replicas.

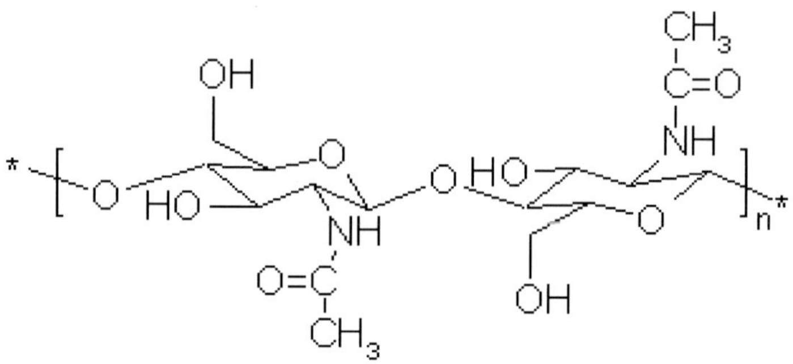

Fig. 8. The molecular structure of chitin.

higher surface area, the silica replicas in particular, could be useful in controlled release applications.

The inner layers of pollen are comprised of cellulose, which is the most abundant biopolymer in the biosphere. Second in abundance is chitin (Fig. 8).

Chitin is one of the main components in the cell walls of fungi, and insect and arthropod exoskeletons. It is a polysaccharide composed of *N*-acetyl-*D*-glucos-2-amine subunits, joined together by β-1, 4-glycosidic bonds. The acetylamino side groups provide the chitin molecule with an increased potential for hydrogen bonding when compared to cellulose, thereby giving structures composed of chitin an increased mechanical strength. As with cellulose, Nature uses chitin to form complex architectural structures in many phyla, from tough crab and lobster shells, to strong yet flexible mushroom gills. In cuttlefish, it is chitin which directs the morphology of the cuttlebone, a device which enables the cuttlefish to control its buoyancy and also survive the extreme pressures when diving to great depths. Chitin is also capable of producing stunning optical effects. In the marine worm *Aphrodite*, bundles of chitin fibers, of sizes comparable to the wavelength of light, are aligned to form a spine which appears red when viewed in normal light. However, when light falls perpendicularly to the spines, they have the appearance of opal, with blues, greens and yellow reflected from the regular chitinous arrays. In butterflies too, it is the regular array of chitin and not pigment molecules which gives rise to the wonderful iridescent colors (Fig. 9).

Even though they are molded by Nature into these fantastic three-dimensional (3D) morphologic complex structures, chitin and its derivatives have a remarkable structure all of their own. Chitin molecules adopt a

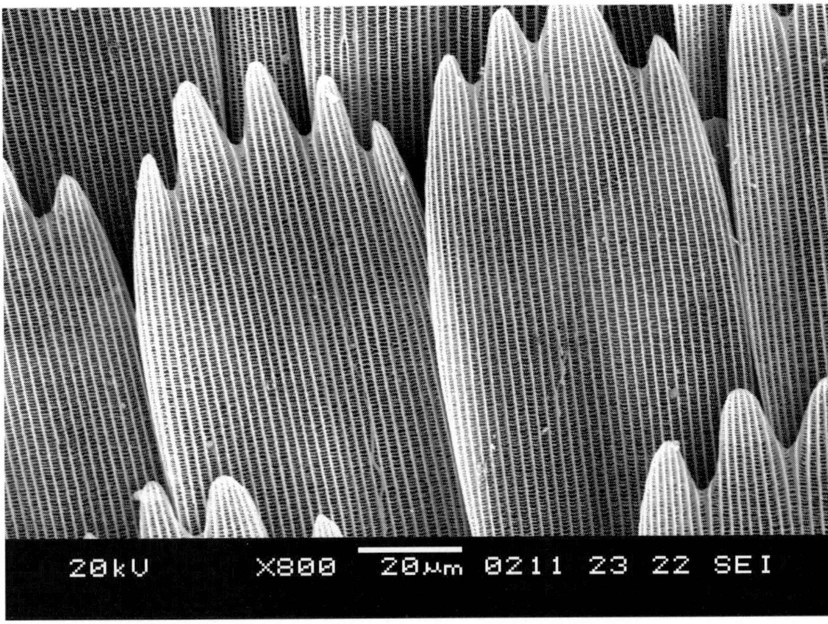

Fig. 9. SEM image of iridescent wing scales from a peacock butterfly.

helical conformation, due to the presence of chiral carbon centers formed by glycosidic linkages, and these helices can then pack together to form fibers and bundles of fibers.

Perhaps the most important derivative of chitin is chitosan. Chitosan (2-amino-2-deoxy-$\beta(1 \rightarrow 4)$-D-glucan) is formed by the deacetylation of chitin, and has found applications in areas as diverse as medicine,[23] cosmetics,[24] and food technology[25] due to its low cost and environmentally benign nature. Of particular relevance to mineralization studies is the fact that chitosan has the ability to preferentially sequester transition and post-transition metal ions from aqueous solutions.[26]

Chitosan is a remarkably stable biomaterial, able to withstand temperatures of up to 160°C without undergoing any significant molecular change. This stability is conferred by the formation of chain-like, helical structures on crystallization, which only begin to undergo thermal oxidation and degradation from approximately 200°C onward. The fibrous nature of the biopolymer is retained however, and it is this stability that has been used successfully to produce carbon fibers, simply through calcination.[27]

This stability makes it ideal as a template in which to perform hydrothermal syntheses without appreciable loss of template structure before inorganic morphology is determined. There are many inorganic materials which benefit from the morphologic constraint offered by a chitinous matrix; control over crystal morphology is the ultimate goal of the modern engineer when designing new materials, as fundamental properties of the material are determined from the way the material is constructed on the molecular scale.

There is one very important field that has never felt the touch of biomimetic control over crystallization however, and that is the field of superconductor research. Historically, metal oxide superconductors have been made by repeatedly grinding together metal and/or metal oxide powders, followed by sintering. More recently, advances in the understanding of materials chemistry have led to the use of sol–gel methods to produce finer and more homogenous superconducting particles (see Ref. 28). By producing nanosized superconducting particles, important fundamental properties such as the critical current density (J_c) and the superconducting critical temperature (T_c) can be greatly improved.[29,30] Despite this advance, researchers are still left with the problem of grinding and pelletizing the resulting oxide, in an attempt to control the bulk morphology, or by producing anisotropic "tapes" of superconducting material. These weakly self-organizing tapes are an attempt to extend the length over which these materials are able to operate. This self-organization is important, as it is believed that grain boundaries act as weak links in limiting the critical current density of bulk high-T_c superconductors. The elimination or minimization of large-angle grain boundaries by morphologic control is therefore a very desirable target. For applications in computer circuitry, superconducting nanodevices and low-dissipation power transmission, a pure crystal of a morphologically favorable geometery is required.

By being able to control the morphology of the superconductor, from the individual component nanoparticles to the way these are ordered in 3D space, will provide a more satisfactory method of producing these materials, rather than the more haphazard current method of "heating" and "beating". The ideal morphology for these applications is the nanowire. A superconducting nanowire would allow for direct incorporation into circuitry and would have the additional benefit of minimizing the irregular grain boundaries which limit the performance.

In our research group at the University of Bristol, synthesis of $YBa_2Cu_4O_8$ (Y124) nanowires are carried out in the presence of chitosan.

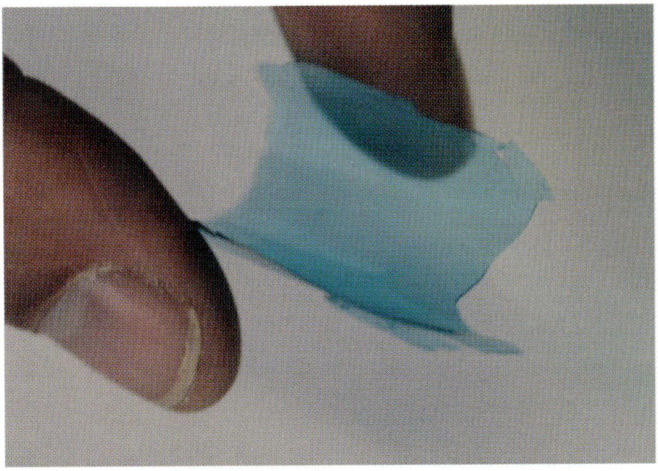

Fig. 10. A thin, clear blue film of chitosan/Y124 superconductor precursor.

Composite materials are made by addition of 1 ml of Y124 precursor solution to 10 ml of chitosan solution, followed by vigorous mixing to ensure a homogenous, light blue sol. The sol is then cast into Petri dishes, and allowed to gel at 35°C overnight, to form a thin and flexible, clear blue film (Fig. 10). This film is then calcined to form crystalline superconductor nanoparticles and nanowires, depending on the rate of heating.[31]

A transmission electron microscope (TEM) image of the control Y124 (no chitosan) sample is shown in Fig. 11. By allowing the sol to gel, and then calcining, large and irregular shaped particles are produced. Confirmation of the correct crystalline phase was by powder X-ray diffraction (PXRD), which showed sharp peaks indexable to the Y124 phase. Superconducting quantum interference device (SQUID) magentometry shows that the T_c of this material is 89 K.

With chitosan-templated Y124, the morphology of the crystals produced are intimately linked to the rate of heating in the calcination step. With a heating rate of 1°C/min, discrete nanoparticles begin to nucleate and grow in the chitosan matrix from around 300°C. This temperature corresponds to the formation of copper oxide from copper acetate. Sintering of the nanoparticles which nucleate are prevented by the presence of the slowly degrading chitosan, and they thus remain the discrete centers for subsequent outgrowth of Y124 nanowires.

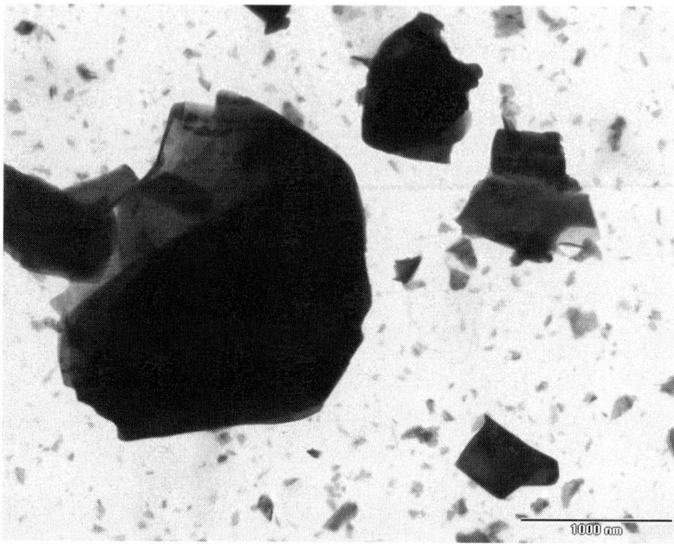

Fig. 11. TEM image showing large, irregular particles of Y124 synthesized via the sol–gel method. Scale bar is 1000 nm.

Nanowires are able to grow from these "seed" nanoparticles from around 350°C onward. TEM images of the calcination quenched at 300°C show nanowires associated with nanoparticles, at a temperature when the templating influence of the chitosan is still being felt. Nanowires are then able to undergo growth along the crystallographic *c*-axis by way of a matrix-mediated growth mechanism. This leads to formation of nanowires, all around 50 ± 5 nm in the short axis and from 600 to over 1000 nm in length (Fig. 12). Superimposition of the diffraction pattern onto the nanowire image showed that in each case, nanowire growth is always along the crystallographic *c*-axis (Fig. 13). These were all single crystals of Y124, as evidenced from selected area electron diffraction (SAED) which gave an unvarying diffraction pattern along the whole length of the nanowire.

SQUID magnetometry of the nanowires reveals that T_c is 85 K (Fig. 14). This is within 4 K of the control sample, and proves that morphology can be finely controlled without significant loss of superconducting performance.

With their preferential crystallographic-oriented growth, these nanowires will provide a better way of studying and understanding superconductivity in anisotropic systems.

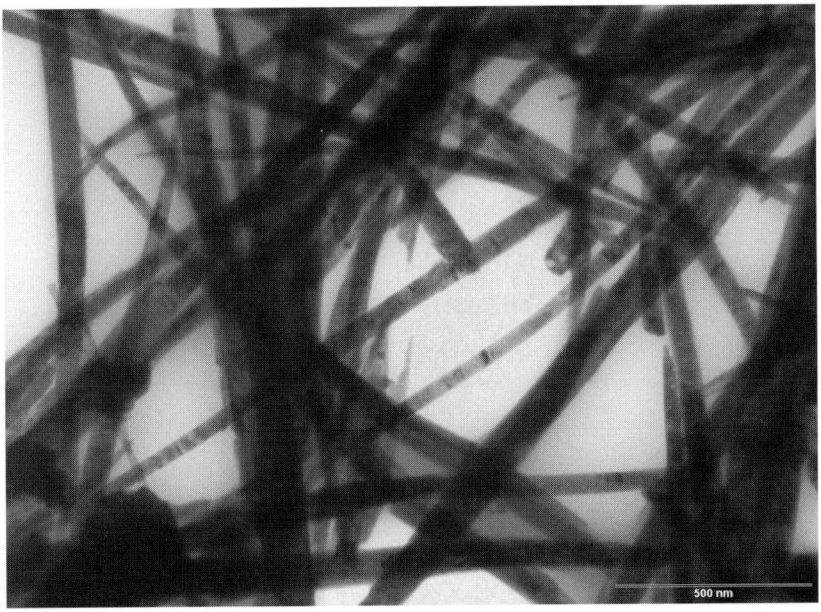

Fig. 12. TEM image showing Y124 nanowires grown in a chitosan matrix. Scale bar is 500 nm.

With a heating rate of 10°C/min, there is no opportunity for *c*-axis growth of nanowires, owing to the relative rapidity of degradation of the chitosan template. A small degree of control over the size of the nanoparticles can be achieved by adjusting the amount of precursor Y124 sol that is added to the chitosan solution. With 0.5 ml of Y124 sol added to 10 ml chitosan solution, the nanoparticles formed are 10 ± 7 nm. Increasing the amount of Y124 sol to 1 ml produces nanoparticles that are more monodisperse and have a diameter of 28.5 ± 3 nm (Fig. 15).

Confirmation of the Y124 phase of these nanoparticles is given by SAED on the TEM (Fig. 16). The polycrystalline ring pattern can be indexed to the Y124 phase. The absence of nanowires in materials calcined at 10°C/min suggest that the extended presence of the chitosan template at lower calcination heating rates is supporting vapor–liquid–solid-oriented growth.

As a template for the nanoengineer, the materials which Nature provides are second to none. By providing a wide range of complex structural motifs and in combination with a resistivity to chemical processing, we are able to pick natural templates which are almost perfectly suited for the

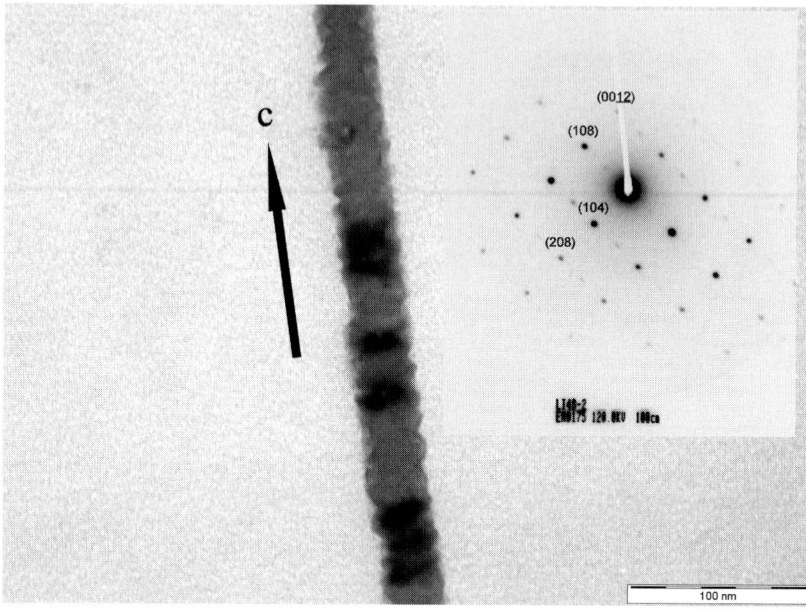

Fig. 13. Y124 nanowire with electron diffraction pattern (inset). The pattern confirms nanowire growth occurs along the crystallographic c-axis. Scale bar is 100 nm.

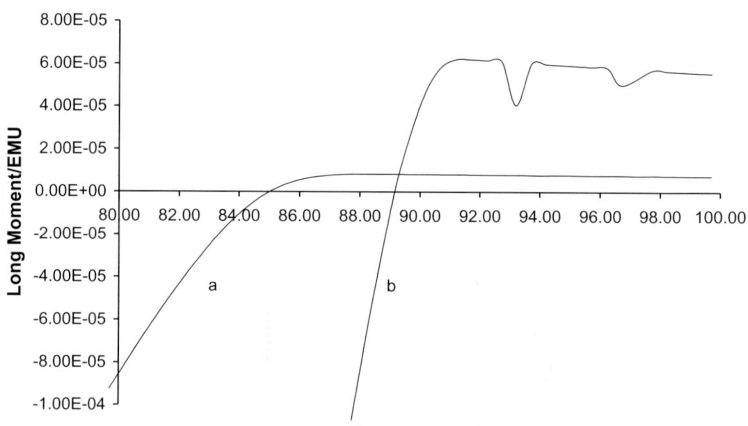

Fig. 14. SQUID magnetometry of (a) nanowire and (b) control Y124 superconductors. The superconducting critical temperature is where the curve crosses the temperature axis.

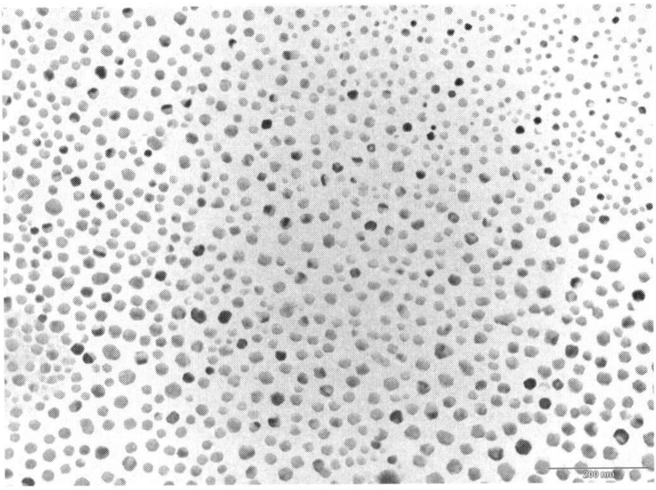

Fig. 15. TEM image of nanoparticles of Y124 grown in a chitosan matrix. Calcination was 10°C/min. Scale bar is 200 nm.

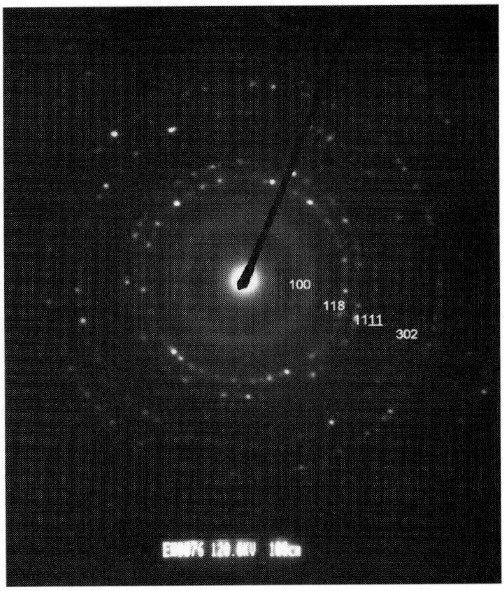

Fig. 16. Electron diffraction pattern of Y124 nanoparticles in Fig. 15.

nanoengineering to be performed. "Almost" is not a tenable state when engineering is concerned however, and it did not take long before scientists began to wonder if they could play an active role in modifying the natural template to fit their needs.

3. Engineering Nature

3.1. *The future*

Man is never satisfied with technological stagnation. Even with the bewildering array of biological templates available, it is only a small leap of imagination to imagine actively tailoring the biotemplate to fulfill the needs of a particular application. There may be many tens of thousands of different types of pollen, all with different morphologies, suited to slightly different applications, but to be able to request a pollen morphology "made to measure" for your application would provide a technological advantage many would gladly pay for. It is possible that the template already exists somewhere out there in the biosphere, but has yet to be discovered. In "Life, the Universe and Everything", the author Douglas Adams wrote:

> "Very few things actually get manufactured these days, because in an infinitely large Universe such as, for instance, the one in which we live, most things one could possibly imagine, and a lot of things one would rather not, grow somewhere."

Unfortunately, most scientists and engineers are not prepared to wait for the ideal template to be discovered and have begun to take it upon themselves to create the biotemplate they require. It is only recently though with the advent of genetic engineering, that the active construction of a biotemplate is a very real possibility.

With viruses (Fig. 17), the simplest of biologic entities, it is possible to alter the protein sequences which provide the main structural impetus for inorganic ion-binding affinity. Douglas *et al.*[12] took cowpea chlorotic mottle virus and by genetically engineering the N-terminus of the protein, the inner surface of the virus is changed from having cationic to anionic character.

This increases markedly, the electrostatic attraction of the inner surface of the capsid (protein shell) to inorganic cations, while not affecting

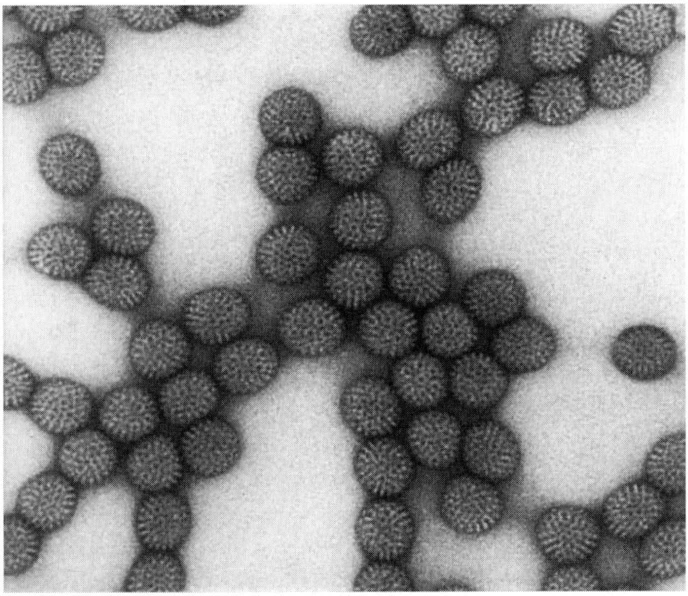

Fig. 17. TEM image of viral capsids.

the morphology of the virus. This enabled the group to produce mineralization of iron oxide nanoparticles, using the inner cavity of the mutant virus as a preferential site of nucleation and constrained growth. Similarly, another capsid, the small heat shock protein cage from the extremophile *Methanococcus jannaschii* has been genetically engineered to increase the amount of thiol (−SH) groups on the inner surface of the protein.[32] Using a polymerase-chain-reaction-mediated site-directed mutagenesis, glycine residues on the inner surface of the protein were replaced with thiol-containing cysteine residues. After expression of the mutated gene in *Escherichia coli*, the mutated thiolated proteins could then be harvested and purified. As before, the mutation did not produce any change in the protein-folding characteristics of the capsid, and the overall morphology was retained when the mutated protein was allowed to self-assemble. Given the affinity of many metals for the thiol group, such a mutation will further enable the preferential sequestration and mineralization of the inner cavity of the capsid, leading to monodisperse metallic nanoparticles.

It will surely not be long before the genetic alteration of the sequences in biomineralizing organisms themselves will enable scientists to direct organisms to grow inorganic materials to order. This will signal a change in the

Table 1. The controlling force in templated mineralization.

Protocol	Assembly of template	Mineralization of template
Bioinspired	Man	Man
Biotemplate	Nature	Man
Engineered passive template	Nature/man	Man
Engineered biomineralizing template	Nature/man	Nature

use of biologic systems from "passive" to "active" templates, thereby obviating the need for the engineer to "do chemistry" on the template. Having the biological entity acting both as tailor-made template and performing the inorganic replication is the next logical step. It is clear that the hand of man is being successively removed as a controlling force in the engineering of these materials (Table 1).

We may still be a long way from the fully independent, self-replicating genetically engineered entities which are capable of building structures to our blueprints, but as long as Nature can provide the complex precursors, man will continue to improve the procedures involved in creating nano- and microengineering marvels.

As inspiration and template over all length scales, Nature will act as Chief Engineer serving the needs of science and engineering for many years to come.

The wonderful thing about science is that it's *alive.*

R. P. Feynman (1918–1988)

References

1. T. D'Arcy, *On Growth and Form*, 1st edn. (Cambridge University Press, 1917).
2. R. J. P. Williams, Metal ions in biological systems, *Biol. Rev.* **28**, 381–415 (1953).
3. R. J. P. Williams, Inorganic biochemistry and myself, *J. Inorg. Biochem.* **28**, 81–84 (1986), doi: 10.1016/0162-0134(86)80072-1.
4. S. Mann, On the Nature of boundary-organized biomineralization (BOB), *J. Inorg. Biochem.* **28**, 363–371 (1986), doi: 10.1016/0162-0134(86)80101-5.

5. S. Mann, *Biomimetic Materials Chemistry*, 1st edn. (VCH Publishers Inc., 1996).
6. D. Walsh and S. Mann, Fabrication of hollow porous shells of calcium-carbonate from self-organizing media, *Nature* **377**, 320–323 (1995), doi: 10.1038/377320a0.
7. S. Mann *et al.*, Bacterial templating of zeolite fibres with hierarchical structure, *Chem. Commun.*, 781–782 (2000), doi: 10.1039/b001528h.
8. S. J. Bryant, Biomaterials: Where we have been and where we are going, *Ann. Rev. Biomed. Eng.* **6**, 41–75 (2004), doi: 10.1146/annurev.bioeng.6.040803.140027.
9. S. Mann *et al.*, Bacterial templating of ordered macrostructures in silica and silica-surfactant mesophases, *Nature* **385**, 420–423 (1997), doi: 10.1038/385420a0.
10. S. Mann, Crystal tectonics: Chemical construction and self-organization beyond the unit cell, *J. Chem. Soc. Dalton Trans.* **21**, 3753–3763 (2000), doi: 10.1039/b004066p.
11. C. A. Mirkin *et al.*, Living templates for the hierarchical assembly of gold nanoparticles, *Angew. Chem. Int. Ed.* **42**, 2306–2309 (2003), doi: 10.1002/anie.200351231.
12. T. Douglas and M. Young, Host–guest encapsulation of materials by assembled virus protein cages, *Nature* **393**, 152–155 (1998), doi: 10.1038/30211.
13. T. Douglas and M. Young, Virus particles as templates for materials synthesis, *Adv. Mater.* **11**, 679–681 (1999), doi: 10.1002/(SICI)1521-4095(199906)11:8<679::AID-ADMA679>3.0.CO;2-J.
14. S. Mann *et al.*, Inorganic–organic nanotube composites from template mineralization of tobacco mosaic virus, *Adv. Mater.* **11**, 253–256 (1999), doi: 10.1002/(SICI)1521-4095(199903)11:3<253::AID-ADMA253>3.0.CO;2-7.
15. S. Mann *et al.*, Organization of metallic nanoparticles using tobacco mosaic virus templates, *Nano Lett.* **3**, 413–417 (2003), doi: 10.1021/nl034004o.
16. G. Cook, P. L. Timms and C. Goltner, Exact replication of biological structures by chemical vapor deposition of silica, *Angew. Chem. Int. Ed.* **42**, 557–559 (2003), doi: 10.1002/anie.200390160.
17. K. H. Sandhage, Novel, bioclastic route to self-assembled, 3D, chemically tailored meso/nanostructures: Shape-preserving reactive conversion of biosilica (diatom) microshells, *Adv. Mater.* **14**, 429–433 (2002), doi: 10.1002/1521-4095(20020318)14:6<429::AID-ADMA429>3.0.CO;2-C.
18. E. L. Mayes, F. Vollrath and S. Mann, Fabrication of magnetic spider silk and other silk-fiber composites using inorganic nanoparticles, *Adv. Mater.* **10**, 801–805 (1998), doi: 10.1002/(SICI)1521-4095(199807)10:10<801::AID-ADMA 801>3.0.CO;2-I.
19. P. Greil *et al.*, Biomorphic cellular SiSiC/Zeolite ceramic composites: From rattan palm to bioinspired structured monoliths for catalysis and sorption, *Adv. Mater.* **17**, 344–349 (2005), doi: 10.1002/adma.200400672.
20. R. K. Iler, *The Chemistry of Silica: Solubility, Polymerization, Colloid and Surface Properties and Biochemistry of Silica* (John Wiley & Sons Ltd., Canada, 1979).

21. J. S. Beck *et al.*, Ordered mesoporous molecular sieves synthesized by a liquid-crystal template mechanism, *Nature* **359**, 710–712 (1992), doi: 10.1038/359710a0.
22. S. R. Hall, H. Bolger and S. Mann, Morphosynthesis of complex inorganic forms using pollen grain templates, *Chem. Commun.*, 2784–2785 (2003), doi: 10.1039/b309877j.
23. M. Hasegawa, K. Yagi, S. Iwakawa and M. Hirai, Chitosan induces apoptosis via caspase-3 activation in bladder tumor cells, *Jpn. J. Cancer Res.* **92**, 459–466 (2001).
24. P. Perugini, I. Genta, F. Pavanetto, B. Conti, S. Scalia and A. Baruffini, Study on glycolic acid delivery by liposomes and microspheres, *Int. J. Pharm.* **196**, 51–61 (2000), doi: 10.1016/S0378-5173(99)00439-1.
25. S. S. Koide, Chitin-chitosan: Properties, benefits and risks, *Nutrition Res.* **18**, 1091–1101 (1998), doi: 10.1016/S0271-5317(98)00091-8.
26. K. Ogawa, K. Oka and T. Yui, X-ray study of chitosan-transition metal complexes, *Chem. Mater.* **5**, 726–728 (1993), doi: 10.1021/cm00029a026.
27. M. Bengisu and E. Yilmaz, Oxidation and pyrolysis of chitosan as a route for carbon fiber derivation, *Carbohyd. Polym.* **50**, 165–175 (2002), doi: 10.1016/S0144-8617(02)00018-8.
28. A. Manthiram and J. N. Goodenough, Synthesis of the high-T_c superconductor $YBa_2Cu_3O_{7-\delta}$ in small particle size, *Nature* **329**, 701–703 (1987), doi: 10.1038/329701a0.
29. C. T. Chang *et al.*, A sol–gel route to prepare Tl–Ca–Ba–Cu–O superconductor with T_c above 120 K, *Physica C* **162**, 113–114 (1989), doi: 10.1016/0921-4534(89)90943-X.
30. K. Fossheim, $Bi_2Sr_2CaCu_2O_x$ bulk superconductor with MgO particles embedded, *J. Alloys Comp.* **340**, 1–5 (2002), doi: 10.1016/S0925-8388(01)02021-7.
31. S. R. Hall, Biomimetic synthesis of high-T_c, type-II superconductor nanowires, *Adv. Mater.* **18**, 487–490 (2006), doi: 10.1002/adma.200501971.
32. T. Douglas *et al.*, The small heat shock protein cage from *Methanococcus jannaschii* is a versatile nanoscale platform for genetic and chemical modification, *Nano Lett.* **3**, 1573–1576 (2003), doi: 10.1021/nl0347861.

Simon R. Hall Born in Stratford-Upon-Avon, Simon Hall read chemistry with geology at the Open University. Upon graduation, he joined the laboratories of Prof. Stephen Mann at the University of Bath, moving with him to the University of Bristol where he obtained his PhD on template control of the structure of minerals in 2000. He remained there for the next three years in a postdoctoral position, during which he collaborated with Toyota on the synthesis of advanced materials from biopolymers and with Vectura and Spinox Ltd. in the creation of pharmaceutical excipients and biomedical implants, using naturally occurring materials as complex templates. In 2004, he was made a Royal Society University Research Fellow, where his research has focused on biomimetic routes to advanced superconducting materials. His other interests include enjoying the works of Richard Feynman, Salvador Dali, and Jim Morrison and the taste of a good Islay single malt whisky. He has a black belt in the samurai sword art of Iaido.

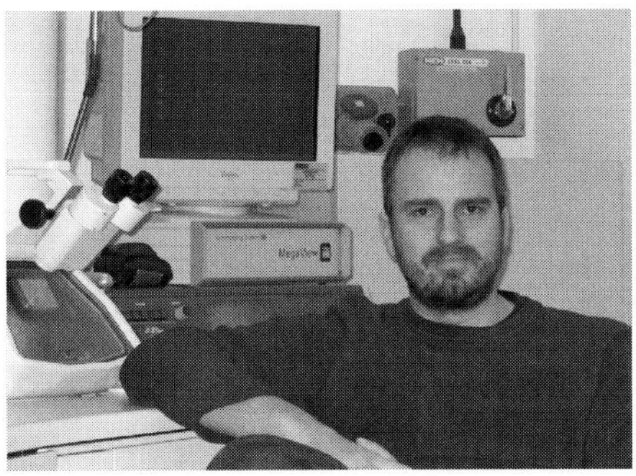

CHAPTER 5

SUPRAMOLECULAR CHEMISTRY: THE "BOTTOM-UP" APPROACH TO NANOSCALE SYSTEMS

Philip A. Gale

School of Chemistry, University of Southampton
Southampton, SO17 1BJ, UK
E-mail: philip.gale@soton.ac.uk

Supramolecular chemistry is the chemistry of "designed noncovalent intermolecular interactions". In this chapter we will trace the beginnings of supramolecular chemistry from molecular recognition through the more recent advances in the design of prototypical molecular machines.

Keywords: Chemistry, structure, design, molecular recognition, molecular machines.

1. Introduction

Molecular chemistry is concerned with making and breaking covalent bonds to form new molecules. Supramolecular chemistry is different because it deals with synthetic molecular systems that are held together by weaker *noncovalent* interactions such as electrostatic forces, hydrogen bonds, π–π stacking interactions, van der Waals' forces, or hydrophobic effects. These noncovalent interactions taken individually are weak, but when several are used in concert, very stable molecular ensembles or *complexes* may result.* Jean-Marie Lehn encapsulated these ideas succinctly when he described supramolecular chemistry as "chemistry beyond the molecule".

*The strength of interaction between molecular components of complexes can be expressed as stability constants (K) which are the equilibrium constants for the formation of the complexes from their component parts — so the larger the magnitude of the stability constant, the stronger the complex. In the expression below a term shown

Early work concentrated on the *recognition* of metal ions, which is the selective binding of a particular metal ion (sometimes referred to as a guest) by a receptor molecule (or host) using noncovalent interactions. This can be achieved by careful design and synthesis of the receptor molecule so that it is complementary to the desired guest. Borrowing ideas introduced in enzyme chemistry by Nobel Laureate Emil Fischer,[1] the receptor could be regarded as a "lock" and the guest as a "key" that perfectly fits the lock.

Some interactions, such as hydrogen bonds and metal–ligand interations, are directional. Chemists have learned how to use these interactions to "program" information into molecular subunits so that they *self-assemble* into potentially useful supramolecular superstructures. This self-assembly process, when combined with traditional synthetic covalent modification, gives us access to new molecules (e.g., molecular knots and chains) that previously could not be made, in addition to larger scale assemblies on the nanoscale.

This chapter will look back to the birth of modern supramolecular chemistry with the formation of simple crown ether–alkali metal complexes through to more recent examples of supramolecular approaches to nanoscale molecular machinery.[2]

2. Molecular Recognition

The birth of supramolecular chemistry can be traced back to the pioneering work of Charles Pedersen, an industrial chemist who worked for DuPont. Pedersen was interested in finding ways to prevent the oxidative degradation of petroleum products and rubber, which is caused by trace amounts of metal ion impurities, such as copper and vanadium. To do this, he developed a series of compounds known as "metal deactivators". These molecules bind the metal ions, converting them into inactive complexes and so suppressing their catalytic activity. In 1960, while attempting to synthesize one of these compounds he isolated some white crystals in very low yield.

---

in square brackets refers to a concentration:

$$\mathbf{host} + \mathbf{guest} \overset{K}{\rightleftharpoons} \mathbf{complex} \quad K = \frac{[\mathbf{complex}]}{[\mathbf{host}][\mathbf{guest}]}.$$

They are frequently quoted as log K values, which are proportional to the free energy change for the complexation process. Solvent molecules play an important role in this process and consequently, stability constants will vary from solvent to solvent for the same complex. Therefore, whenever a stability constant is quoted, the solvent in which it was measured is also mentioned.

His curiosity was piqued by the fibrous, silky quality of the crystals. Even more interestingly, they were soluble in alcohols when sodium cations were present but insoluble in their absence. Elemental analysis and mass spectrometry showed that the crystals were a cyclic polyether **1** (Fig. 1(a)) that had formed due to the presence of a small amount of catechol (Fig. 1(b)) impurity in his reaction mixture.[3]

Pedersen observed from a space-filling model that a sodium ion is held in the cavity of this macrocycle by attractive electrostatic ion–dipole interactions between the cation and the six oxygen atoms in the polyether ring. This binding mode accounted for the interesting solubility properties of the

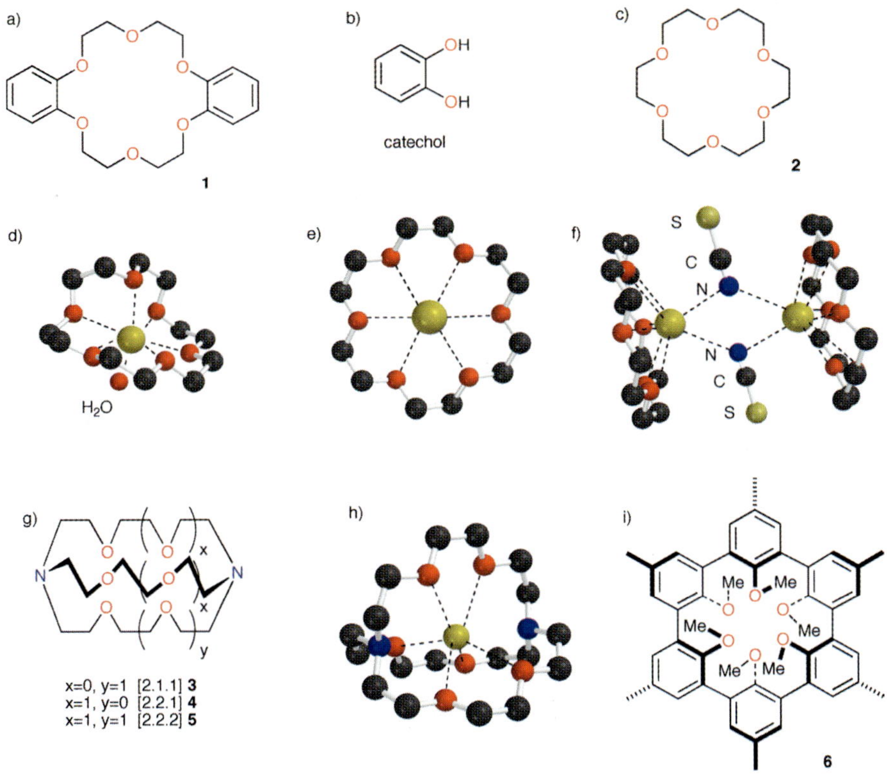

Fig. 1. The chemical structures of the cation-binding agents: (a) dibenzo[18]crown-6 **1**, (b) catechol, and (c) 18-crown-6 **2** with the crystal structures of the sodium (d), potassium (e), and cesium (f) cation complexes of [18]crown-6 (in the latter case two cesium complexes are shown with associated thiocyanate anions). The chemical stucture of [2.1.1], [2.2.1], and [2.2.2]cryptands (**3**, **4**, and **5**, respectively) is shown in (g) with the crystal structure of the potassium complex of [2.2.2]cryptand (h), and (i) spherand **6**. Cations are yellow, oxygen is red and nitrogen is blue.

compound. It was later found that the metal ion acts as a template during the formation of the macrocycle by causing the reactants to wrap around it and orienting them in a favorable way to form the 18-membered ring, **1**.

As the model of compound **1** looked like a crown, Pedersen called this class of compound the "crown ethers". This particular compound was named dibenzo[18]crown-6. The term "[18]" refers to the number of atoms in the macrocycle and the numeral 6 to the number of oxygen atoms in the ring. This is much easier to remember than the systematic name of compound **1** which is 2, 3, 11, 12-dibenzo-1, 4, 7, 10, 13, 16-hexaoxacyclooctadecane.

Many different sizes of crown ether have now been synthesized and coordination studies have shown that a relationship exists between the cavity size, cationic radius, and stability of the resulting complex. In short, the better the fit of the cation into the crown, the stronger the complex formed. This phenomenon is referred to as *optimal spatial fit.* For example, [18]crown-6 (**2**, Fig. 1(c)) forms complexes with sodium, potassium, and cesium cations in methanol with stability constant ($\log K$) values of 4.32, 6.10, and 4.62, respectively.[4] The crystal structure of the potassium complex of [18]crown-6[5] (Fig. 1(e)) shows that the cation fits snugly in the middle of the crown ether. Smaller cations, such as sodium, cause the crown ether to distort[6] (Fig. 1(d)), by wrapping itself around the metal to maximize the electrostatic interactions which increases the strain in the molecule. Larger cations such as cesium must perch over one face of the macrocycle because they are too large to fit into the cavity (Fig. 1(f)).[7]

Although Pedersen's discovery was serendipitous,[8] it demonstrated that selectivity could be introduced into synthetic receptors by making them complementary to the desired guest.

In 1969, improving on the cation-binding ability of the crown ethers, Jean-Marie Lehn and his co-workers at the Université Louis Pasteur in Strasbourg reported a new class of "three-dimensional" cation receptors called *cryptands.*[9–11] These materials are cage-like bicyclic molecules that contain three polyether strands strung between two nitrogen bridgehead atoms (Fig. 1(g)). The cryptands have been found to complex group 1 and 2 metal cations with stability constants much higher than those of analogous crown ethers. For example, in methanol [2.2.2]cryptand (**5**) is selective for potassium cations (diameter = 2.66 Å), binding them with a stability constant ($\log K$) of 10.4 which is over four orders of magnitude higher than [18]crown-6. The crystal structure of this complex is shown in Fig. 1(h). By changing the length and number of oxygen atoms in each polyether strand, Lehn found that he could tune the receptor to be selective for smaller cations. For example, [2.1.1]cryptand (**3**) is selective for lithium cations (diameter = 1.36 Å) and [2.2.1]cryptand (**4**) is selective for sodium (diameter = 1.94 Å).

Both the crown ethers and cryptands require some degree of rearrangement to form a complex with a metal cation. Recognizing this, Donald Cram reasoned that rigid receptors with binding sites fixed in an octahedral arrangement around an enforced cavity would show enhanced binding over flexible receptors. With the help of molecular models he designed a new class of receptor, the spherands, that contain an enforced spherical cavity (compound **6** in Fig. 1(i)). Spherands bind sodium and lithium cations very strongly (compound **6** forms complexes with sodium with a stability constant $K = 1.2 \times 10^{14}\,M^{-1}$ in chloroform-*d* saturated with water). Cram formalized these observations with the principle of *preorganization* that states "the more highly hosts and guests are organized for binding and low solvation prior to their complexation, the more stable will be their complexes".[12]

The pioneering work of Pedersen,[8] Lehn,[13] and Cram[12] ignited interest in supramolecular chemistry and was recognized with the award of the 1987 Nobel Prize for Chemistry. Since then, the field of molecular recognition has grown with the development of receptors for a wide range of cationic, anionic, and neutral guests as well as more complex biomolecules.

In the years between this pioneering work and the present day, molecular recognition has been extended to be anionic[14] and neutral guest species,[15–17] and more recently to ion-pairs.[18] There are different challenges in each of these areas and the construction of these types of receptor is still an active research area for many groups around the world due to the applications of these systems in many areas including drug delivery, sensor technology, and environmental remediation.

Molecular recognition has advanced to the point where the complexation of complex biomolecules is now possible. In 1997, Hamilton has attached four peptide loops (peptides are chains of amino acid molecules — the constituent parts of proteins) to the upper rim of a calix[4]arene (a bucket-shaped molecule that, in this case, is being used as a piece of molecular scaffolding holding the four peptide loops close together in space).[19] Similar peptide loops are present in fragment antigen-binding regions of antibodies. X-ray analysis of antibody–protein complexes has shown that binding occurs by the formation of a large and open interfacial surface composed of residues capable of noncovalent-binding interactions with the antibody. In the majority of cases, the antibody contacts the antigen via at least four peptide loops. In Hamilton's synthetic antibody **7** (Fig. 2(a)), the rigid cone-like calixarene acts as a molecular scaffold holding the peptide loops in an arrangement that defines a peptide-binding domain.

The peptide loops present in **7** contain the negatively charged Gly–Asp–Gly–Asp sequence (a chain of four amino acids containing negatively charged carboxylate groups). Hamilton decided to study the interaction of **7** with cytochrome *c* (cyt *c*), a protein that carries a number of positively charged groups on its surface. This was achieved using a technique called *affinity chromatography*. It was shown that **7** binds cyt *c* strongly, with a binding constant similar to that of the natural cytochrome *c* oxidase–cyt *c* complex. Nuclear magnetic resonance studies indicate that it is likely that the calixarene binds to cyt *c* in a region close to the heme edge. This is a part of the protein that contains an iron(III) metal ion bound in a heme group (a circular molecule involved in electron transfer, shown in yellow in Fig. 2). Figure 2 shows a model of the calixarene docked with the crystal structure of the protein. This demonstrates that the receptor can interact with four of the five available positively charged lysine residues and cover a large portion of the protein surface. It was found that the rate of reduction (gain of electrons) of the heme group in the calixarene–cyt *c* complex in the presence of excess ascorbate (a chemical which acts as a reducing agent, i.e., a donator of electrons to other molecules) was considerably slower than that observed for the free protein. This is further evidence that the calixarene binds at the heme edge of the protein and thus hinders the approach of the reducing agent to the heme group.

3. Self-Assembly

The formation of the deoxyribonucleic acid (DNA) double helix from two complementary DNA strands is perhaps the clearest example of a self-assembling molecular system in biology. The thermodynamically stable double-helical structure forms spontaneously and reversibly as the strands are mixed together under the right conditions, and hydrogen bonds form between complementary base pairs. The rapid reversibility of the process ensures that any error that may have occurred during assembly can be corrected.

Chemists can use self-assembly to access new noncovalently linked molecular architectures by combining appropriately designed and yet simple subunits. These molecular components contain within them the information

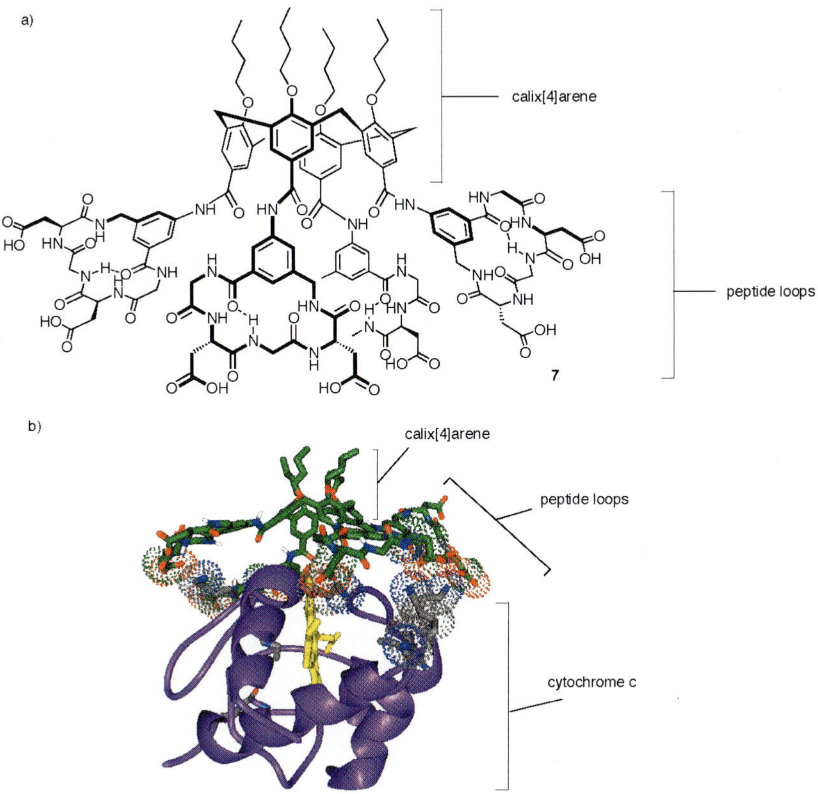

Fig. 2. The chemical structure of receptor **7** (a) and its proposed complex with cytochrome *c* (heme group shown in yellow) (b). Reproduced with permission from *Angew. Chem. Int. Ed. Engl.* **36**, 2680 (1997). Copyright (1997) Wiley-VCH.

required to construct the self-assembled architecture in terms of the position and directionality of their binding sites, the distribution of electron density over their surfaces, or their oxidation states. This information is accessed when the components are mixed together and "read out" as the self-assembled structure.

Hydrogen bonds have been used to direct the assembly of many different types of noncovalently linked molecular architectures. One early example is the formation of a "rosette-like" insoluble complex **8** between melamine and cyanuric acid discovered by Whitesides and co-workers.[20] Melamine can be regarded as having three faces, each of which can donate two hydrogen bonds from the amine groups and accept one to the aromatic nitrogen atom. In contrast, cyanuric acid can accept two hydrogen bonds and donate one from each of its "faces". When mixed together the two compounds form an insoluble hexagonal network resembling a rosette as shown in Fig. 3(a).

Lehn and co-workers adapted this hydrogen-bonding motif by blocking one face of each of the subunits with alkyl chains.[21] Steric interactions cause these compounds to form a linear molecular ribbon **9**, which is a noncovalently linked polymer (Fig. 3(b)).

Julius Rebek, Jr. and co-workers have synthesized a series of self-assembled "tennis balls" from curved molecular subunits.[22] Compound **10** (Fig. 4) contains hydrogen bond donating groups (blue) and hydrogen bond accepting groups (red) that cause two of these molecules to wrap around each other and form a spherical complex with a large central cavity. This cavity contains solvent molecules that are released on guest binding making guest binding an entropically favorable process. This particular tennis ball is capable of encapsulating *p*-quinone and cyclohexadiene simultaneously and, by increasing their effective concentration, accelerating the rate of the reaction between them by 200 times (Fig. 4).

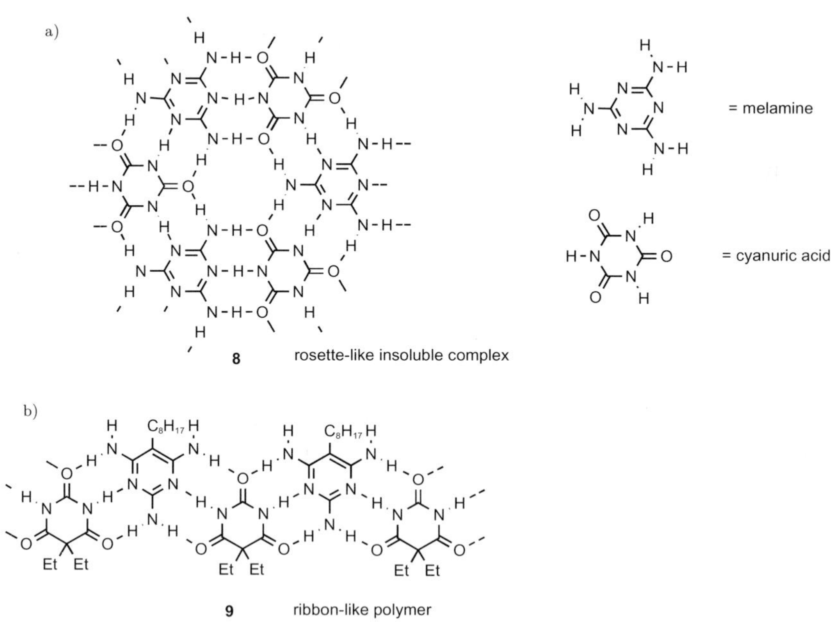

Fig. 3. Melamine and cyanuric acid form an insoluble "rosette-like" complex **8**, whereas sterically hindered analogs form a molecular ribbon **9**.

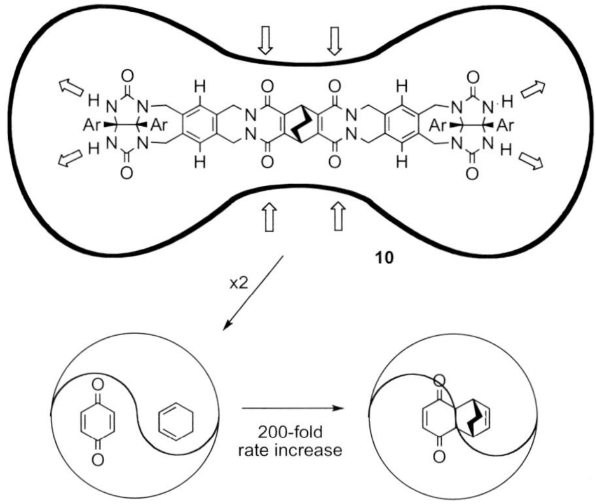

Fig. 4. The curved molecule **10** dimerizes to form a molecular tennis ball. This self-assembled sphere can be used as a molecular reaction vessel, accelerating the rate of Diels–Alder reactions by up to 200 times.

In cell membranes, proteins, such as gramicidin A, form channels that allow the passage of protons and alkali metal cations across the membrane (the membrane is composed of two layers of compounds called lipids, and is consequently also known as a lipid bilayer). They do this by forming a α-helix (this type of helix contains a central channel) in which amide (N–H) groups from the peptide backbone alternatively point up and down, forming hydrogen bonds with amide carbonyl oxygen atoms. Two of these helices, assembled in a head-to-head arrangement, are required to span the membrane. A supramolecular approach to the formation of synthetic ion channels has been pioneered by M. Reza Ghadiri and his research group at the Scripps Research Institute in California.[23] These researchers use cyclic α-peptides and cyclic D, L-α-peptides to form self-assembled peptide nanotubes. The cyclic peptides adopt conformations wherein the peptide NH and CO bonds point up and down in a similar fashion to the hydrogen bonds of the gramicidin A helix. This allows the formation of multiple hydrogen bonds between cyclic peptides and drives the formation of the peptide nanotube. One example is shown in Fig. 5. In this case the cyclic α-peptides **11** have self-assembled into a nanotube within a lipid bilayer.[24] The opening and closing of this channel can be monitored by measuring the flow of potassium ions through the channel, which reveals sharp opening

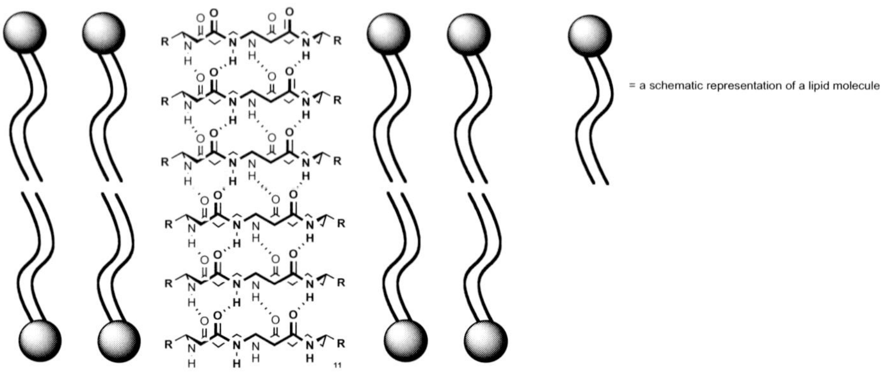

Fig. 5. A schematic representation of a cyclic peptide-based, self-assembled transmembrane channel structure **11** embedded in a lipid bilayer membrane.

and closing events corresponding to either conformational changes in the assembly or to a dynamic assembly–disassembly processes occurring in the membrane. Larger versions of these channels have been shown to transport bigger molecules, such as glucose, across membranes.[25]

Other approaches to synthetic transmembrane channel systems have been pioneered by Matile and co-workers in Geneva. Matile has developed new synthetic supramolecular α-barrels that are constructed from molecular units containing a rigid octiphenyl stave from which short peptide chains are attached (Fig. 6). The octiphenyl stave is a rigid rod composed of eight phenyl rings linked together. The rings can rotate around the axis of the rod. Matile has used this molecule as a piece of molecular scaffolding from which he has appened eight chains of amino acids. When chains on adjacent octiphenyl groups interdigitate they form antiparallel α-sheets held together by multiple hydrogen bonds. Twists in the stave due to the arene–arene torsion angles and steric crowding of the N- and C-terminal amino acid residues cause the self-assembled structures to curve so allowing the formation of rigid-rod β-barrel structures. The most important feature of this approach to synthetic channel formation is its versatility in that it is possible to choose the identity of the amino acid residues that coat the exterior and interior surfaces of the barrel and hence tune the barrel's solubility (exterior) and transport ability (interior). Matile calls this property the *functional plasticity*,[26] as the barrels are adaptable for use in different applications without making changes to the underlying structure that constitutes the barrel. For example, barrel **12** shown in Fig. 6 has leucine

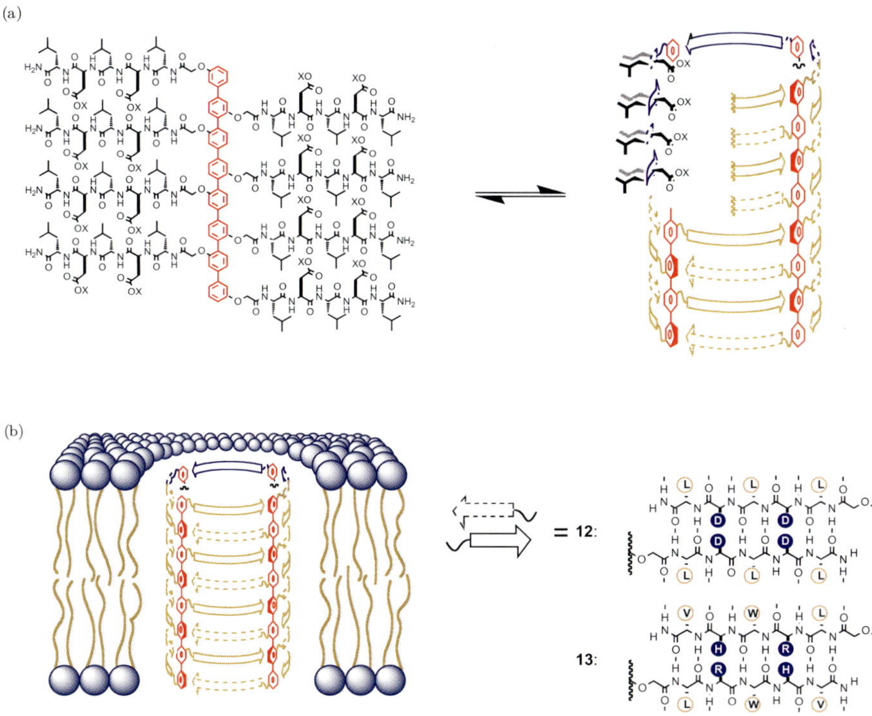

Fig. 6. (a) Self-assembly of peptide functionalized *p*-octiphenyl monomeric rods. The octiphenyl staves are highlighted in red while the exterior surface of the barrel is brown and the interior surface blue. (b) The α-barrel in a lipid bilayer. Amino acids located on the outer surface are black on white; internal ones are white on blue. D: aspartate; H: histidine; L: leucine; R: arginine; V: valine; and W: tryptophan. Reproduced with permission from *ChemBiochem.* **4**, 1299 (2003). Copyright 2003, Wiley-VCH.

groups (amino acids containing hydrophobic groups) on its exterior that confer solubility in a lipophilic (nonpolar "organic solvent-like") environments (such as within a cell membrane) and a polar interior lined with aspartate groups while both barrels **12** and **13** have been used as a unique type of sensor for enzyme activity.[27]

Transition metals can also be used to direct the assembly of noncovalently linked molecular ensembles. Although metal–ligand bonds are not noncovalent interactions, they are commonly used by supramolecular chemists because although they are thermodynamically strong interactions, they have varying degrees of lability. This means that the bonds may persist for a long time or may be short lived, continuously breaking and remaking in a reversible, dynamic process. This reversibility allows a type of "error-checking" in the assembly process that will consistently lead to the thermodynamically most stable assembly. Additionally, due to ligand field effects, transition metal ions often have very specific coordination geometry requirements, which afford very precise control of the structure of the molecular assembly. By changing the oxidation state of the transition metal in an assembly, it may be possible to alter its preferred coordination geometry and so electrochemically switch the assembly between two different states. The electrochemical control of assemblies will be discussed in the final section of this chapter.

4. Self-Assembly with Covalent Modification

When a strategy of self-assembly is used in concert with covalent modifications, previously inaccessible molecular topologies may be obtained. J. Fraser Stoddart and his co-workers have employed electron-rich and electron-poor aromatic components that self-assemble primarily via a charge transfer interaction to produce a wide variety of self-assembled structures, including catenanes (molecular chains). This work stemmed from Stoddart's discovery that bisparaphenylene[34]crown-10 (an electron-rich crown ether) will bind the positively charged and electron-poor paraquat dication (a positively charged rod-like molecule carrying a 2+ positive charge) to form the charge transfer complex **14** (Fig. 7(a)). Catenanes can be synthesized from precursors similar to this complex. One example, the [5]catenane **15** ([5] designates the number of interlocked rings) is shown in

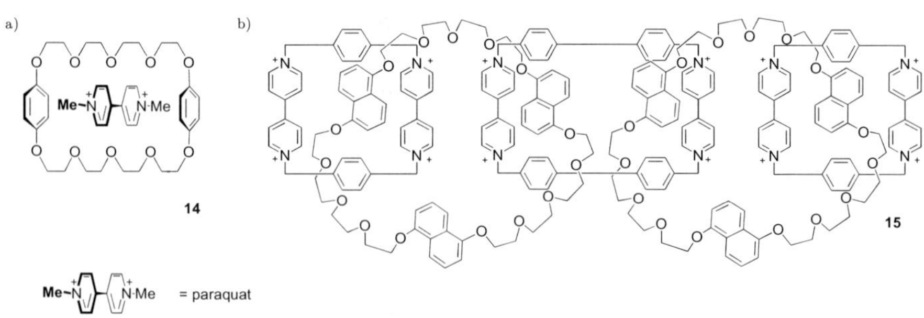

Fig. 7. (a) A paraquat dication threaded through a bisparaphenylene-[34]crown-8 forming a charge transfer complex **14** and (b) the chemical structure of the [5]catenane, olympiadane **15**.

Fig. 7(b). This compound is called "Olympiadane" because of its similarity to the five inter-linked Olympic rings.[28,29]

Among the most ambitious chemical goals that has been reached so far, using self-assembly techniques, is the synthesis of a molecular knot. Jean-Pierre Sauvage and his research team at the Université Louis Pasteur, Strasbourg are pioneers in this area of chemistry.[30] Sauvage used a transition metal (in this case copper(I), an ion that prefers to adopt a tetrahedral coordination environment) to direct the assembly of the double-helical complex **16**. The ends of the helix were linked together covalently to form the metallated trefoil knot **17** and the metal ions removed to afford the metal-free knot **18** (Fig. 8). The knot is one continuous strand that crosses itself three times. Again, it should be emphasized that the syntheses of this remarkable material would be practically impossible without self-assembly strategies.

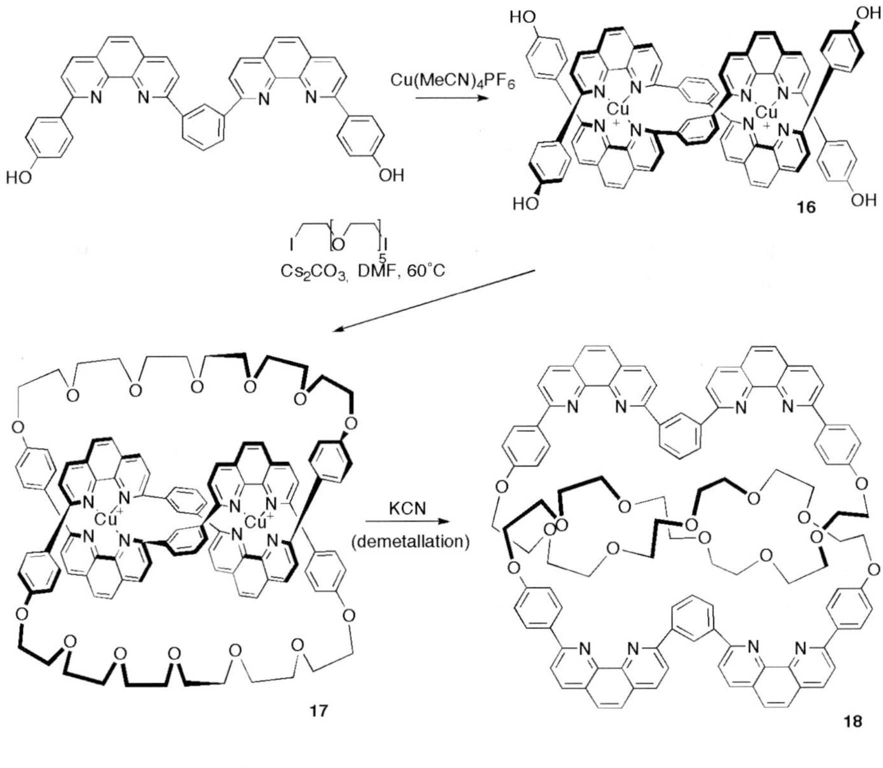

Fig. 8. The metal-directed assembly of a trefoil knot **18**.

5. Supramolecular Approaches to Molecular Machines

Recently, the examples of supramolecular systems that could be classed as "supramolecular machinery" have appeared. Many of these systems contain catenane or rotaxane components. Rotaxanes are composed of a cyclic molecule (macrocycle) through which an axle is threaded. The ends of the axle are blocked by bulky groups that prevent the macrocycle from slipping off. These systems also have potential uses for the storage of information. In 1997, Sauvage and co-workers reported that the rotaxane **19** (Fig. 9) can act as an abacus-like molecular shuttle.[31] The complex was produced using transition metal-directed self-assembly to thread the macrocycle onto the axle with subsequent covalent modification of the axle by the introduction of blocking groups. The axle contains a 1, 10-phenanthroline group containing two nitrogen atoms and a terpyridine group containing three nitrogen atoms. These groups are referred to as "stations" where the macrocyclic component of the rotaxane can bind. The macrocycle also contains a 1, 10-phenanthroline group and, in the presence of copper(I), the metal ion coordinates to the phenanthroline station (complex **19** in Fig. 9) locking the macrocycle to this position. Copper(II) requires a higher number of coordinated groups than copper(I). When the copper ion is oxidized to + 2 oxidation state it jumps to the three-coordinate terpyridine station so increasing its coordination sphere to 5, forming complex **20** (Fig. 9) and in doing so moving the macrocycle to the terpyridine station. The position of the macrocycle on the thread is controlled by the oxidation state of the copper ion allowing us to exert control over a "microscopic" event (the movement of the macrocycle) by controlling a "macroscopic" factor (redox potential).

Dave Leigh and co-workers at the University of Edinburgh have synthesized a molecular motor **21** (Fig. 10) consisting of a [2]catenene consisting of one large and one small macrocycle.[32] The large ring has three stations for binding the smaller macrocycle and by switching the binding of the smaller ring to these stations on and off, it is possible to drive the smaller ring around the larger one as shown in Fig. 11. Here the larger macrocycle contains three stations, **A**, **B**, and **C**, each with different binding affinities for the smaller ring (stability constants, K_a) such that $K_a(\mathbf{A}) > K_a(\mathbf{B}) > K_a(\mathbf{C})$. In State I, the smaller macrocycle binds to station **A**. The binding affinities of the stations can be modulated by light, heat, or chemical stimuli that changes the shape of the stations and so changes their ability to accept hydrogen bonds. If station **A** is converted to **A′**, a group with

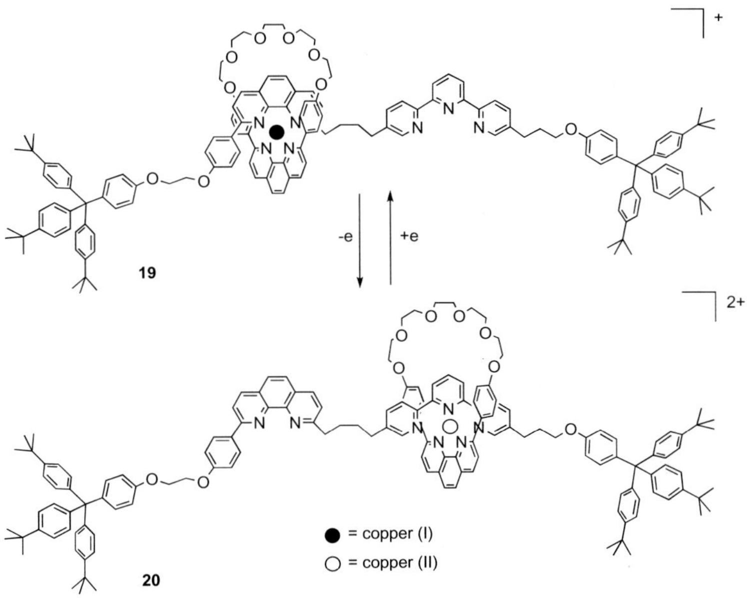

Fig. 9. An electrochemically controlled molecular shuttle **19/20**.

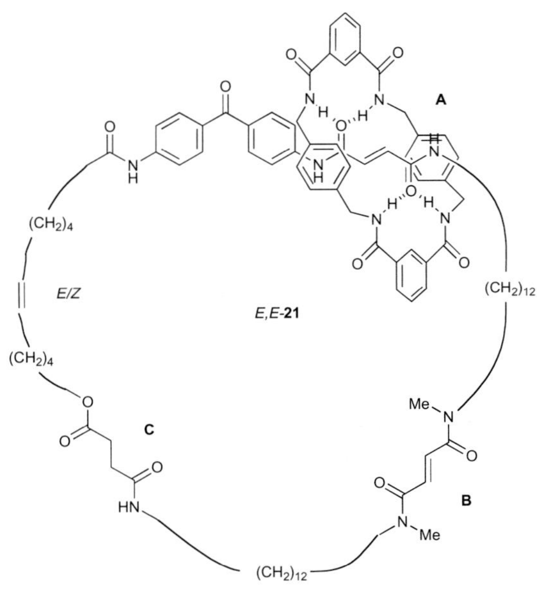

Fig. 10. The structure of molecular motor **21**.

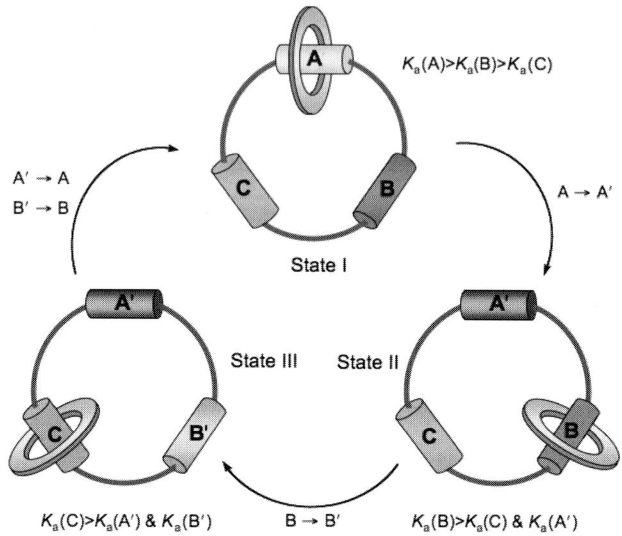

Fig. 11. Stimuli-induced sequential movement of a macrocycle between three different binding sites in [2]catenane **21**. Reproduced with permission from *Nature* **424**, 174 (2003). Copyright 2003, Macmillan Magazines Ltd.

a lower binding affinity for the macrocycle than **B** or **C**, then the smaller macrocycle will move via biased Brownian motion to site **B** (State II). If **B** is then changed into a station **B′** such that $K_a(\mathbf{B}') < K_a(\mathbf{C})$, then the macrocycle will move to site **C** (State III). Changing **A′** back to **A** and **B′** to **B** resets the motor and the smaller ring returns to its original position.

Credi and co-workers have continued their studies in two-station [2]rotaxanes and prepared a molecular machine that behaves like a nanoscale molecular elevator.[28,33] The authors synthesized a "platform" **22**, consisting of three dibenzo[24]crown-8 rings fused together by a triphenylene core. This molecule is able to form a 1:1 complex with the tripod-shaped molecule **23** which contains both dialkylammonium and pyridinium groups (Fig. 12) with each leg of the tripod threading through one of the crown ether groups (Fig. 13). By attaching a bulky 3, 5-di-*tert*-butylbenzyl group to the end of each leg of the tripod, the crown ether platform is prevented from slipping off. The crown ether groups bind preferentially to the ammonium ($-NH_2^+-$) groups on the tripod. However, when the elevator is exposed to a base such as t-BuNP(NMe_2)$_3$ the ammonium groups lose a proton, forming amine ($-NH-$) groups. The crown ether platform has a much lower affinity for amine than ammonium, and will then move to bind

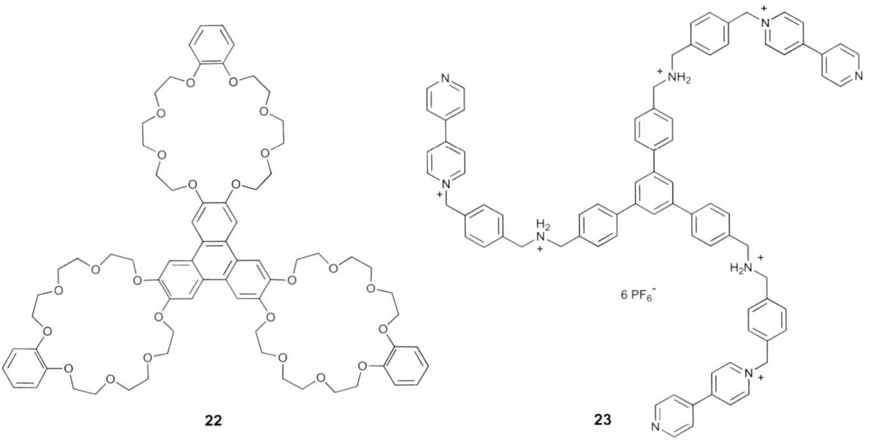

22 **23**

Fig. 12. The components of the molecular elevator: platform **22** and tripod **23**.

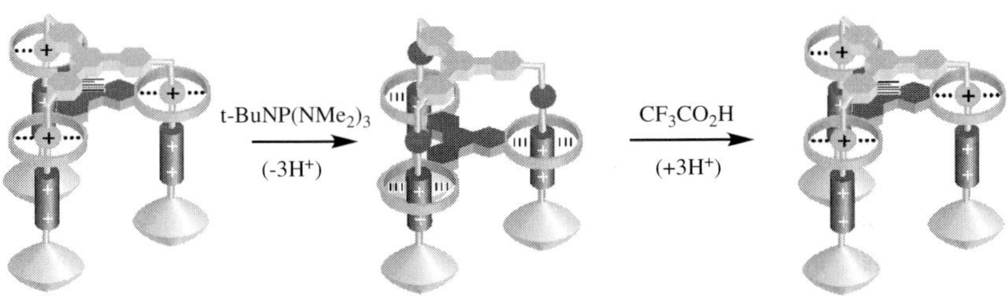

Fig. 13. Crown ether platform **22**, tripod **23**, and schematic representation of the acid-base-induced operation of the molecular elevator. Reproduced with permission from *Science* **303**, 1845 (2004). Copyright 2004, AAAS.

the pyridinium groups (the elevator moves down). Subsequent addition of acid (CF_3CO_2H) reprotonates the amine groups back to ammonium and the platform moves back up to the top of the tripod (Fig. 13).

The synthesis of these prototypical systems and others like them may come to be viewed as the genesis of supramolecular machinery: that is, the generation of molecular-sized devices such as gears, switches, and motors built by chemical and supramolecular chemical techniques from the ground up. The alternative approach of miniaturizing our macroscopic world to produce machinery in the micrometer-sized range can be achieved through the use of silicon micromechanics.[34] However this technique produces machines that have a limited lifespan and are orders of magnitude larger than the molecular devices that are currently emerging from supramolecular chemistry.

6. Conclusion

This short chapter has surveyed supramolecular chemistry from birth of the area with the development of simple molecular recognition systems for cations through to the construction of functional molecular machines. As well as spanning the divide between the "chemical" and "nano" worlds, as we have seen, supramolecular systems can also bridge between the "chemical" and "biological" arenas. There can be no doubt that more esoteric supramolecular complexes, with uses that can now only be dreamt of in the pages of science fiction novels, will emerge in the coming years.

Acknowledgment

I would like to thank the Royal Society for a University Research Fellowship.

References

1. E. Fischer, Einfluss der configuration auf die wirkung derenzyme, *Ber. Dt. Chem. Ges.* **27**, 2985–2993 (1894).
2. P. A. Gale, Supramolecular chemistry: From complexes to complexity, *Phil. Trans. R. Soc. Lond. A* **358**, 431–453 (2000).
3. C. J. Pedersen, Cyclic polyethers and their complexes with metal salts, *J. Am. Chem. Soc.* **89**, 2495–2496 (1967).
4. H. K. Frensdorff, Stability constants of cyclic polyether complexes with univalent cations, *J. Am. Chem. Soc.* **93**, 600–606 (1971).
5. P. Seiler, M. Dobler and J. D. Dunitz, Potassium thiocyanate complex of 1, 4, 7, 10, 13, 16-hexaoxacyclooctadeca, *Acta Cryst. B* **30**, 2744–2745 (1974).
6. M. Dobler, J. D. Dunitz and P. Seiler, Hydrated sodium thiocyanate complex of 1, 4, 7, 10, 13, 16-hexaoxacyclooctadecane, *Acta Cryst. B* **30**, 2741–2743 (1974).
7. M. Dobler and R. P. Phizackerley, Caesium thiocyanate complex of 1, 4, 7, 10, 13, 16-hexaoxacyclooctadecane, *Acta Cryst. B* **30**, 2748–2750 (1974).
8. C. J. Pedersen, The Discovery of crown ethers, *Angew. Chem. Int. Ed. Engl.* **27**, 1021–1027 (1988).

9. B. Dietrich, J.-M. Lehn and J.-P. Sauvage, Diaza-polyoxa-macrocycles et macrobicycles, *Tetrahedron Lett.* **10**, 2885–2888 (1969).
10. B. Dietrich, J. M. Lehn and J. P. Sauvage, Cryptates XI: Complexes macrobicycliques, formation, structure, proprietes, *Tetrahedron* **29**, 1647–1658 (1973).
11. B. Dietrich, J. M. Lehn, J. P. Sauvage and J. Blanzat, Cryptates X: Syntheses et proprietes physiques de systemes diaza-polyoxa-macrobicycliques, *Tetrahedron* **29**, 1629–1645 (1973).
12. D. J. Cram, The design of molecular hosts, guests, and their complexes (Nobel lecture), *Angew. Chem. Int. Ed. Engl.* **27**, 1009–1020 (1988).
13. J.-M. Lehn, Supramolecular chemistry — Scope and perspectives molecules, supermolecules, and molecular devices (Nobel lecture), *Angew. Chem. Int. Ed. Engl.* **27**, 89–112 (1988).
14. J. L. Sessler, P. A. Gale and W. S. Cho, *Anion Receptor Chemistry*, 1st edn. (Royal Society of Chemistry, Cambridge, 2006).
15. S. K. Chang and A. D. Hamilton, Molecular recognition of biologically interesting substrates — Synthesis of an artificial receptor for barbiturates employing 6 hydrogen-bonds, *J. Am. Chem. Soc.* **110**, 1318–1319 (1988).
16. A. D. Hamilton, *J. Chem. Ed.* **67**, 821–828 (1990).
17. K. Odashima, A. Itai, Y. Iitaka and K. Koga, Host–guest complex formation between a water-soluble polyparacyclophane and a hydrophobic guest molecule, *J. Am. Chem. Soc.* **102**, 2504–2505 (1980).
18. G. J. Kirkovits, J. A. Shriver, P. A. Gale and J. L. Sessler, Synthetic ditopic receptors, *J. Inc. Phenom. Macrocyc. Chem.* **41**, 69–75 (2001).
19. Y. Hamuro, M. C. Calama, H. S. Park and A. D. Hamilton, A calixarene with four peptide loops: An antibody mimic for recognition of protein surfaces, *Angew. Chem. Int. Ed. Engl.* **36**, 2680–2683 (1997).
20. J. P. Mathias, E. E. Simanek, J. A. Zerkowski, C. T. Seto and G. M. Whitesides, Structural preferences of hydrogen-bonded networks in organic solution — The cyclic $CA_3{\cdot}M_3$ "rosette", *J. Am. Chem. Soc.* **116**, 4316–4325 (1994).
21. J.-M. Lehn, M. Mascal, A. De Cian and J. Fischer, Molecular recognition directed self-assembly of ordered supramolecular strands by co-crystallization of complementary molecular components, *J. Chem. Soc. Chem. Commun.*, 479–481 (1990).
22. J. de Mendoza, Self-assembling cavities: Present and future, *Chem. Eur. J.* **4**, 1373–1377 (1998).
23. J. D. Hartgerink, T. D. Clark and M. R. Ghadiri, Peptide nanotubes and beyond, *Chem. Eur. J.* **4**, 1367–1372 (1998).
24. T. D. Clark, L. K. Buehler and M. R. Ghadiri, Self-assembling cyclic beta(3)-peptide nanotubes as artificial transmembrane ion channels, *J. Am. Chem. Soc.* **120**, 651–656 (1998).
25. J. R. Granja and M. R. Ghadiri, Channel-mediated transport of glucose across lipid bilayers, *J. Am. Chem. Soc.* **116**, 10785–10786 (1994).
26. S. Matile, En route to supramolecular functional plasticity: Artificial-barrels, the barrel-stave motif, and related approaches, *Chem. Soc. Rev.* **30**, 158–167 (2001).

27. P. A. Gale, A "Holey" supramolecular approach to the detection of enzyme activity, *ChemBiochem.* **4**, 1299–1302 (2003).
28. D. B. Amabilino, P. R. Ashton, A. S. Reder, N. Spencer and J. F. Stoddart, Olympiadane, *Angew. Chem. Int. Ed. Engl.* **33**, 1286–1290 (1994).
29. D. B. Amabilino, P. R. Ashton, V. Balzani, S. E. Boyd, A. Credi, J. Y. Lee, S. Menzer, J. F. Stoddart, M. Venturi and D. J. Williams, Oligocatenanes made to order, *J. Am. Chem. Soc.* **120**, 4295–4307 (1998).
30. C. O. Dietrich-Buchecker, J.-P. Sauvage, A. De Cian and J. Fischer, High-yield synthesis of a dicopper(I) trefoil knot containing 1,3-phenylene groups as bridges between the chelate units, *J. Chem. Soc. Chem. Commun.*, 2231–2232 (1994).
31. P. Gavina and J. P. Sauvage, Transition-metal template synthesis of a rotaxane incorporating two different coordinating units in its thread, *Tetrahedron Lett.* **38**, 3521–3524 (1997).
32. J. V. Hernández, E. R. Kay and D. A. Leigh, A reversible synthetic rotary molecular motor, *Science* **306**, 1532–1537 (2004).
33. J. D. Badjić, V. Balzani, A. Credi, S. Silvi and J. F. Stoddart, A molecular elevator, *Science* **303**, 1845–1849 (2004).
34. R. T. Howe, R. S. Muller, K. J. Gabriel and W. S. N. Trimmer, Silicon micromechanics: Sensors and actuators on a chip, *IEEE Spectrum* **27**, 29–35 (1990).

Philip A. Gale was born in Liverpool in 1969. He graduated with a BA(Hons.) in chemistry from Wadham College, Oxford in 1992 (MA (Oxon. 1995)) and a DPhil (for work conducted in Prof. Paul Beer's Research Group) from Linacre College, Oxford in 1995. In that year he was awarded a Fulbright Post-Doctoral Fellowship which was taken up at the University of Texas at Austin working under the supervision of Prof. Jonathan L. Sessler. In 1997 he was awarded a Royal Society University Research Fellowship and started his independent career at the Inorganic Chemistry Laboratory, Oxford. He moved to the School of Chemistry at the University of Southampton in September 1999 as Royal Society URF and Lecturer and was promoted as Senior Lecturer in 2002 and Reader in 2005. His interests in supramolecular chemistry are focused on anionic species and specifically on synthetic anion receptor chemistry and the roles anions can play in self-assembly and in crystal engineering. He was awarded a Society of Porphyrins and Phthalocyanines Young Investigator Award in July 2004, the 2004 Bob Hay Lectureship by the RSC UK Macrocycles and Supramolecular Chemistry Group and a 2005 Corday-Morgan medal and prize by the Royal Society of Chemistry. He is the Co-editor of the journal *Supramolecular Chemistry* and a Member of the Editorial Board of *Chemical Society Reviews*, and the International Editorial Advisory Boards of *Coordination Chemistry Reviews*, the *Encyclopaedia of Supramolecular Chemistry*, *Chemistry World*, and *Chemical Communications*. In 2004 he joined the International Scientific Committee of the *International Symposium on Macrocyclic and Supramolecular Chemistry* (ISMSC).

CHAPTER 6

MOLECULAR SELF-ASSEMBLY: A TOOLKIT FOR ENGINEERING AT THE NANOMETER SCALE

Christoph Wälti

Institute for Microwaves and Photonics
School of Electronic and Electrical
Engineering, University of Leeds
Leeds LS2 9JT, UK

Among the many challenges facing the development of a molecular-based nanotechnology, the directed assembly of discrete molecular objects, and their controlled integration into macroscopic structures, is fundamental. The selective self-assembly or self-organizing characteristic inherent to certain molecules, for example DNA, is a property that could be exploited to address these challenges. This integration problem can be separated into more fundamental tasks: attaching molecular anchors to the macroscopic structures with high spatial resolution; assembling the discrete molecular objects; and positioning and attaching the molecular objects onto the macroscopic structures.

Here, several aspects of these tasks will be discussed, for example using short DNA molecules as molecular anchors and their attachment to electrodes separated by a few ten nanometers; the generation of branched DNA complexes by molecular self-organization; and using AC electric fields for the manipulation, orientation, and positioning of DNA molecules.

Keywords: Bionanotechnology, self-assembly, DNA, bioelectronics, AC electrokinetics.

1. Introduction

Microelectronics has transformed our world in less than 50 years, liberating and dramatically extending the capabilities of individuals and communities. It can be argued that this revolutionary explosion started with the invention of the transistor. In 1945, Mervin Kelly, the Director of Bell Telephone Laboratories, assembled a team to research on fundamental solid-state physics with the vision of replacing vacuum tubes and electromagnetic switches with small, low-power, solid-state devices. The team was led by William Shockley and included Walter Brattain and John Bardeen, who, in 1947, developed the first working solid-state amplifier, the point-contact transistor. A detailed discussion of the physics of the transistor is given in Ref. 1. Although individual transistors were now available, they still had to be connected individually to other electronic components such as resistors and capacitors to build useful electronic circuits. It did not take long before the limits of this circuit fabrication technique were reached and circuits assembled from individual transistors and other components became too large and difficult to build. The transition from the discrete circuitry of transistors, and the inherent difficulties in individually assembling such components into circuits, required the consolidation of all components into a single device, the integrated circuit (Fig. 1). In 1958, Jack Kilby at Texas Instruments, and virtually at the same time, Robert Noyce at Fairchild Semiconductors, found a way of fabricating all necessary electronic components from a single material and integrating them into a single device. Shortly afterward Lionel Kattner and Isy Haas, also at Fairchild

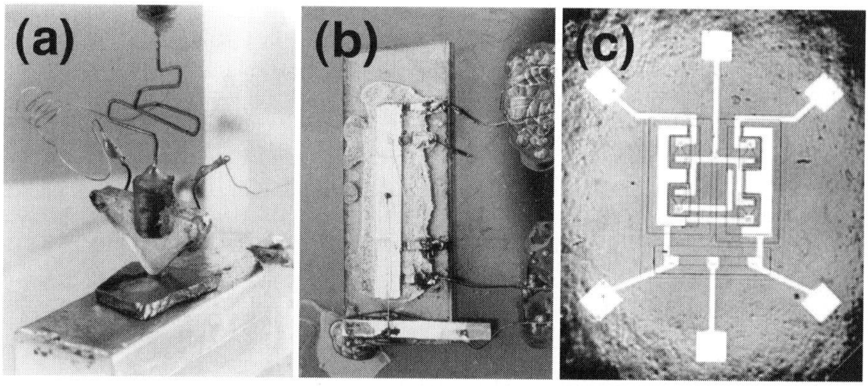

Fig. 1. The transistor and the integrated circuit, the inventions that arguably triggered the revolutionary explosion of microelectronics. (a) The first transistor invented at Bell Telephone Laboratories in 1947 by Brattain and Bardeen. Reprinted with permission of Lucent Technologies Inc./Bell Labs. (b) The first integrated circuit, invented by Jack Kilby at Texas Instruments in 1958 and virtually the same time by Robert Noyce at Fairchild Semiconductors. Courtesy of Texas Instruments. (c) The first planar integrated circuit, designed and built by Lionel Kattner and Isy Haas at Fairchild Semiconductor in 1961. Courtesy of Fairchild Semiconductor.

Semiconductors, fabricated the first planar integrated circuit using a single multi-step process. It was these inventions that led to the revolutionary explosion in high-speed switching, logic, and memory circuitry that has had such an impact on the modern world.

Nanotechnology has arguably reached a similar crossroads. Although a wide range of nanoscale elements have been fabricated, integration solutions do not exist. Researchers have developed single molecules that emulate conventional circuit elements,[2–6] functional devices from conducting and semiconducting carbon nanotubes[7–9] and nanowires,[10,11] fluorescent nanocrystals,[12,13] supermolecular assemblies,[14,15] and molecular switches and transistors,[16,17] *inter alia*. However, the ability to position and join molecular components at molecularly accurate addresses, interface them with the outside world, and modify and manipulate molecules and molecular assemblies at the nanometer scale, is still in its infancy. It is generally accepted that molecular recognition and self-assembly techniques (e.g., those associated with the hybridization of two complementary single strands of DNA, deoxyribonucleic acid) can, in principle, be exploited to assemble complex nanoscale molecular objects from their building blocks, and could also provide a reliable approach for the attachment of molecular structures to solid surfaces such as metal electrodes.[18] This procedure may ultimately allow the integration of complex self-assembled structures into conventional electronic devices. Without the means to assemble and integrate nanoscale components, the promise of nanotechnology may never be fulfilled.

Many biological molecules are known to self-assemble complex structures, using very efficient and selective molecular recognition processes. In particular, the selective self-assembly and molecular recognition properties inherent to DNA might be exploited to engineer complex molecular networks of interlinked molecular circuit elements on a nanometer scale. The possible exploitation of biological processes for self-organization and self-assembly has therefore generated considerable interest. However, the assembly of large molecular networks and their integration into conventional electronics, in particular, is extremely challenging. In this chapter, we will concentrate solely on DNA as a means of facilitating self-assembly, although many other systems with highly sophisticated lock-and-key recognition properties are available.

DNA molecules can either be single- or double-stranded, and a schematic representation of the two conformations is given in Fig. 2. Each single strand of DNA consists of a backbone (five-carbon sugar molecules linked together via phosphodiester bonds) onto which a sequence of four different bases

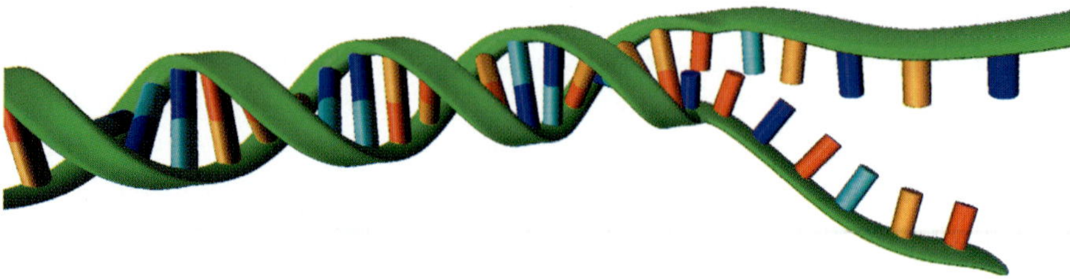

Fig. 2. Schematic illustration of double- and single-stranded DNA. The backbone is shown as a green ribbon to which the individual bases are attached. The four different DNA bases are indicated by different colors.

(adenine (A), thymine (T), cytosine (C), and guanine (G)) are attached. The backbone has a direction defined through the attachment points of the phosphodiester linker on the sugar molecules. One end of the backbone, the 5′ end, is terminated by a phosphate group and the other end, the 3′ end, by a hydroxy group. Double-stranded DNA is formed from two single strands of DNA paired together in opposite direction (i.e., one 5′ to 3′ and the other 3′ to 5′) held together by hydrogen bonds between the bases on opposite strands. However, the two strands only pair together if they are complementary, that is, if the As on one strand are opposite the Ts on the other strand, and the Gs opposite the Cs. This specific pairing, generally referred to as Watson–Crick basepairing, makes DNA a very powerful tool in self-assembly applications. The structure of double-stranded DNA is a right-handed double-helix[19] with a pitch of 10 basepairs (bp) per turn, a distance of 0.34 nm between bases, and a diameter of about 2 nm. The persistence length of double-stranded DNA, that is, the length over which the DNA molecules are locally straight, is of the order of 50 nm.[20] Double-stranded DNA can be as short as a few nanometers, but can also extend to several meters in length or more. Furthermore, a myriad of physical and chemical techniques are available to manipulate and modify DNA, for example more than 500 restriction enzymes are commercially available that cut DNA at specific basepair sequences.[21] These dimensions, and the modifications and manipulation tools, together with the readily available synthesis which allows short single-stranded DNA to be chemically synthesized, make DNA ideally suited for nanoscale applications.

Over the last decade or so, DNA has been used as a means to self-assemble various molecular-scale complexes. For example, gold nanocrystals have been assembled into networks and macroscopic materials,[22–24] and long DNA molecules have been attached to electrodes.[25] However, the controlled integration of organic and inorganic complexes with high spatial precision into existing semiconductor devices has only just begun.

One possible way of integrating molecular complexes into semiconductor environments is depicted in Fig. 3. The process can be separated into four individual and mostly independent tasks:

(1) An existing semiconductor electronic device, which can be fabricated using standard semiconductor processing technologies, is equipped with suitable surface metal electrodes where the molecular complex should interface with the semiconductor device.
(2) These metal interface points are functionalized with suitable anchor molecules. These anchor molecules are designed to interact only and very specifically with their respective counterpart on the molecular complex.
(3) The molecular object is functionalized with appropriate anchor groups at its interface points.

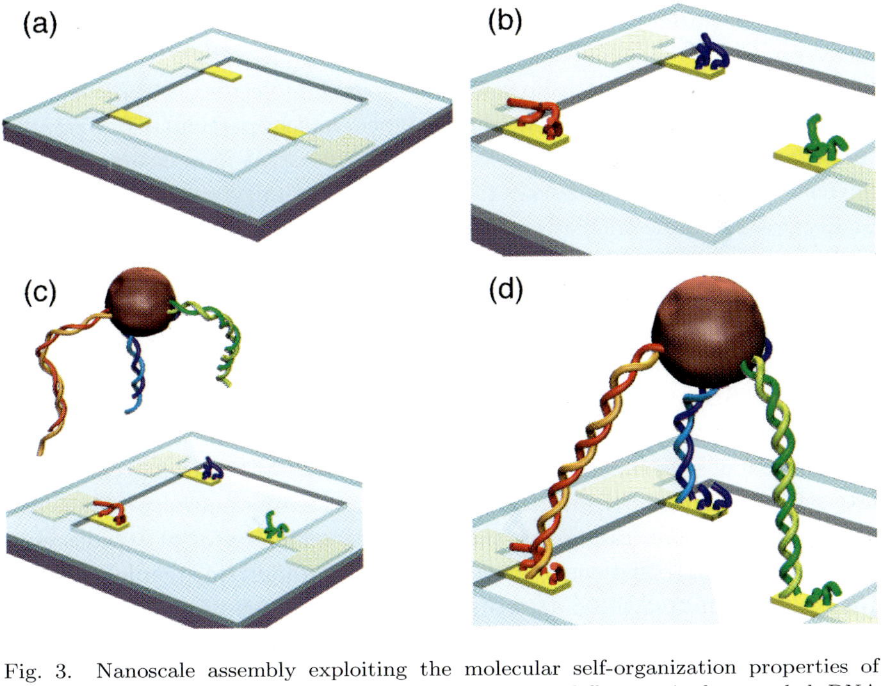

Fig. 3. Nanoscale assembly exploiting the molecular self-organization properties of DNA. The three electrodes are functionalized with different single-stranded DNA molecules (b) and the three arms of the molecular complex with the three corresponding complementary DNA oligonucleotides (c). The molecular complex is then assembled onto the device using the self-organizing properties of DNA (d).

(4) The molecular object is brought into close proximity with the semiconductor device and assembled onto the device in the correct orientation by exploiting the self-organizing functionality built into the individual interface points.

In this chapter, possible implementations of these steps to fabricate and integrate molecular complexes into existing, nanometer-scale environments, are discussed. Our efforts have been concentrated around DNA as the self-assembly promoter, and in the first part of this chapter, methods for functionalizing closely spaced metal electrodes with DNA oligonucleotides as anchor molecules are discussed. In the context of this chapter, DNA is not only a very powerful anchor molecule, but can also be used as a scaffold for the fabrication of complex molecular structures. In the second part of this chapter, techniques to fabricate branched structures consisting of native DNA are discussed. Finally, tools, and in particular electrokinetic tools, to position and orientate DNA molecules are discussed.

2. Functionalized Surfaces

Coating solid surfaces with functional biological molecules is of paramount importance in many areas of modern science, and in particular in bionanotechnology. Over the last few decades, self-assembled monolayers (SAMs) have played an important role in the functionalization of surfaces. SAMs are highly ordered molecular assemblies that form spontaneously when surfactants with a specific affinity of their headgroups to a substrate are adsorbed onto the substrate. Usually, SAMs are prepared from solution and allow coating of arbitrarily shaped, and not only planar, geometries. The majority of SAMs investigated belong to one of the two families: thiols (sulphur) on gold (R–SH + Au(111), where R denotes the rest of the molecule)[26] or silane-based systems on glass or silicon oxide (e.g., R–$SiCl_3$ + SiO_2).[27] In the context of this chapter, where the integration of molecular complexes into semiconductor environments is discussed, the interface between the two systems is generally present in the form of a set of metal electrodes connected to both the molecular complex and the underlying semiconductor. Hence, the preferred SAM system is from the thiol–gold family.

The formation of thiol SAMs on gold surfaces is well documented for macroscopic substrates.[28,29] However, most of the previous work is concerned with the formation of the SAM itself and not with methods to form two SAMs of different chemical, biological, or physical properties on the same substrate in close proximity to each other. If one wants to exploit this well-studied thiol–gold system for engineering-by-self-assembly applications, tools have to be employed to place different anchor molecules exhibiting different recognition properties selectively at specific sites within a particular nanometer-scale environment. Indeed, the ability to pattern a surface locally with different monolayers of anchor molecules in a well-controlled fashion and with a high spatial resolution has importance for applications ranging from the fabrication of high-density information storage devices, next-generation integrated circuits, high-resolution displays, biosensors, photonic bandgap devices, lab-on-a-chip devices, to micro-electromechanical systems (MEMS), *inter alia.* As discussed above, DNA, with its inherent self-assembly characteristics and very versatile recognition properties, is a good candidate to serve as molecular anchors that could be exploited to address some of the challenges of nanoscale engineering. DNA molecules can be attached to a surface via noncovalent bonds. DNA is highly negatively charged and is thus attracted to positively charged surfaces. However, a much more suitable way is attachment through thiol–gold chemistry. DNA oligonucleotides (short single-stranded DNA molecules) can be synthesized with a thiol group attached to either end via various different linker groups. A number of techniques are available for introducing short DNA oligonucleotides or other anchor molecules locally onto a surface, and a brief overview of some of the techniques is given below.

Micropipetting or microdrop-dispensing systems were one of the first tools available to deposit small amounts of biological samples onto spatially defined places. The technique was, to a large extent, developed for the fabrication of oligonucletide microchips.[30] The typical resolution is of the order of a few tens of microns, and volumes of about 1 nl can be dispensed. More recent developments have integrated electric fields into this pipetting technique to achieve a much better control of the liquid delivery.[31] Figure 4 shows the schematic setup of a nanopipet. An electrode is placed inside the pipet and a second one in the ionic solution outside the pipet. A small voltage can be applied between the two electrodes leading to an ionic current which in turn leads to the formation of small droplets, about 800 nm in diameter, at the pipet tip.

Micromachining allows the modification by mechanical means of a molecular layer on a solid substrate. Usually, a surgical scalpel blade or a sharp carbon fiber is used to scrape physically the molecular layer. Abbott *et al.*[32] used a combination of micromachining and molecular self-assembly to generate submicron-scale (0.1–1 μm) patterns of contrasting

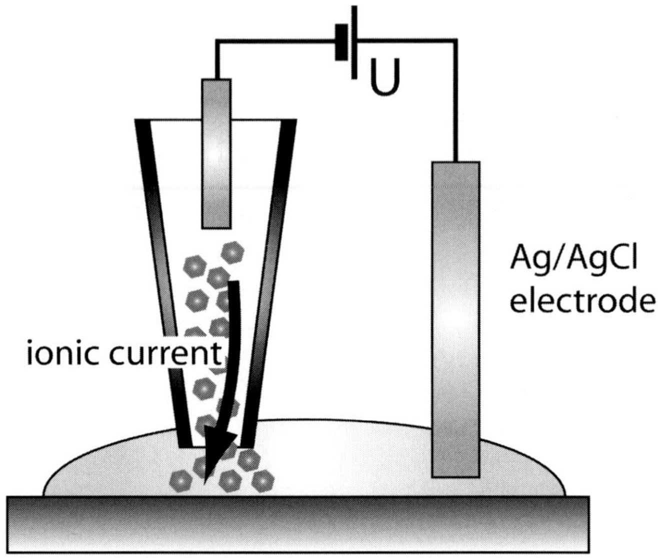

Fig. 4. Schematic of the nanopipet. The bore diameter is of the order of 100 nm, and by applying a negative bias between the internal electrode and the electrode outside the pipet, an ionic current is induced which forces the molecules out of the pipet (adapted from Ref. 31).

surface properties. They formed a hydrophilic SAM of acid-terminated alkanethiols on a bare gold substrate and removed thin lines of the SAM by micromachining. The substrate was then exposed to a second, methyl-terminated alkanethiol, leading to the formation of thin hydrophobic lines embedded into a monolayer of hydrophilic molecules.

Microcontact printing can be seen as a sophisticated and downscaled version of traditional rubber stamping.[33–35] The key element of this technique is the fabrication of a suitable elastomeric block containing the patterned relief to be transferred onto the surface. The most common material used to fabricate elastomeric stamps is poly(dimethylsiloxane) (PDMS). This technique has been used to form patterns of SAMs with submicron feature sizes on solid substrates. The elastomeric block is inked with suitable molecules (e.g., alkanethiols) and then pressed onto the solid substrate (e.g., gold), whereby the molecules are transferred from the stamp onto the surface to form an SAM. However, the extension of contact printing to produce sub-100-nm structures has been hindered by the low elastic modulus of PDMS and the diffusion of the low-molecular-weight inks during the printing. Recently, Whitesides and co-workers demonstrated the fabrication of

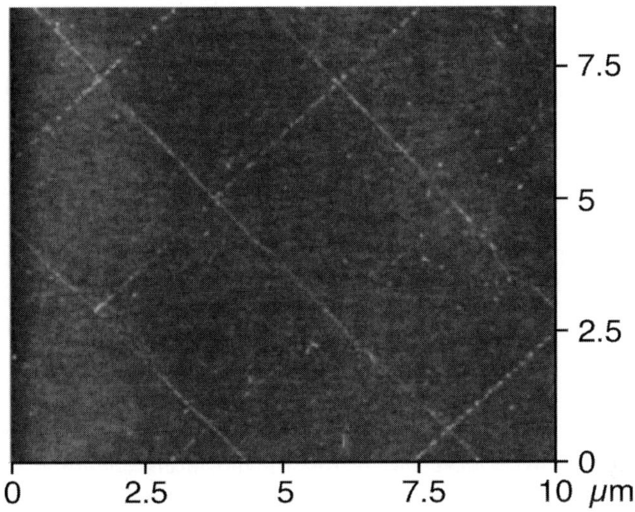

Fig. 5. Atomic force micrograph of a surface with lines of dendrimers printed using a double-layer PDMS stamp. Reprinted in part with permission from Ref. 38. Copyright (2003) American Chemical Society.

sub-100-nm features using a double-layer stamp composed of hard and soft PDMS.[36,37] Huck and co-workers overcame the diffusion problem by using 4th generation polyamidoamine dendrimers and demonstrated the printing of sub-50-nm features on silicon (see Fig. 5, also Ref. 38).

Patterning of molecular SAMs by optical lithography has been studied intensively over the past two decades.[39–42] Huang and Hemminger[40] demonstrated that alkanethiolates on gold can be photo-oxidized using radiation of 254 nm wavelength into alkanesulfonates, which can then be displaced easily by immersion in a dilute thiol solution.[43] This has been exploited for the fabrication of large, patterned SAMs with submicron features. However, the resolution of this technique is inherently limited owing to the diffraction limit of the UV light used in the photo-oxidation step. To reduce further the minimum feature size, Leggett and co-workers employed near-field scanning techniques in combination with a UV laser and demonstrated optically patterned SAM features of less than 50 nm.[44,45] Most photolithography of SAMs require UV light of 254 nm or shorter wavelength for efficient patterning, but these harsh conditions often lead to ozone-induced damage and nonspecific photodegradation of the SAMs.[46–48] Evans, Bushby and co-workers developed photocleavable SAMs based on nitro-benzyl photochemistry[49] which allows photopatterning at around

365 nm, that is, under much less harsh conditions, and hence much better suited for biological environments.

Nanographting is based on the same principles as micromachining (i.e., mechanically scraping off a molecular layer), but uses an AFM tip in contact mode instead of a sharp carbon fiber as in micromachining. The AFM tip removes the molecules from the surface which can result in sub-50-nm features of exposed substrate patterned into an SAM. Generally, this procedure is carried out under a liquid containing a suitable concentration of another molecular species which can adsorb onto the bare gold features as soon as they are graphted.[50,51] Dip-pen nanolithography[52–54] also employs an AFM tip and can be described as the down-scaled version of a fountain pen. The AFM tip is immersed in a solution containing molecules with a chemical affinity to the substrate to be patterned. The AFM tip is then brought into contact with and moved along the substrate surface. The molecules are transported from the tip to the substrate by capillary forces, which allows one to write directly patterns of molecules onto the substrate (see Fig. 6). The minimum achieved feature size (i.e., the width of the chemisorbed lines), is around 30 nm and thus similar to nanographting, but dip-pen nanolithography is a positive, direct-write, technique. This technique has been applied for patterning small features of alkanethiols,[52] thiolated DNA,[55] and proteins.[56]

An entirely different route is taken by using electrochemical methods. The thiol–gold bond is electrochemically active and Hsueh *et al.*[57] and

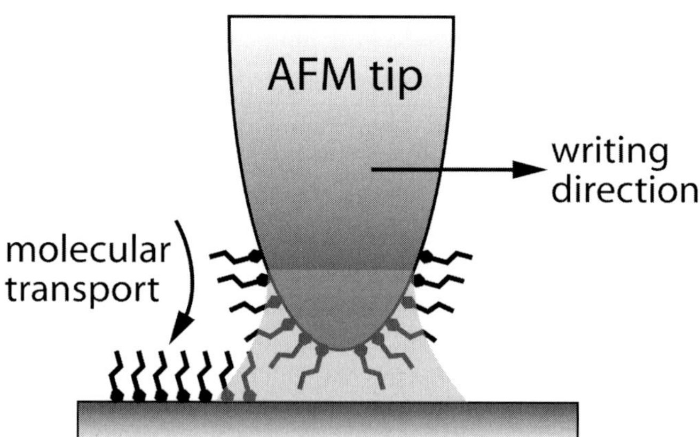

Fig. 6. Schematic illustration of dip-pen nanolithography. A substrate is patterned with molecules by transferring the molecules from the tip onto the substrate through capillary forces (adapted from Ref. 53).

Riepl *et al.*[58] used positive electrochemical potentials to facilitate the formation of alkanethiol SAMs on gold electrodes. In addition, they used negative electrochemical potentials to inhibit SAM formation on gold. On the other hand, Wang *et al.*[59] employed negative electrochemical potentials to force the growth of alkanethiol SAMs on their gold electrodes. However, the resolution of these approaches was $70\,\mu m$[57] and $6\,\mu m$,[59] respectively, far from the nanoscale, and thus not really suited for nanoscale patterning.

Local oxidation of the molecular layer via the application of an electric potential by a conductive AFM tip has been used to transform locally the hydrophobic CH_3 end-groups of an *n*-octadecyltricholosilane (OTS) SAM into hydrophilic hydroxyl groups.[60] The OH groups can be used to attach site-specifically another molecular layer onto the first one using selective silane chemistry. This technique has been employed to fabricate patterns of silver islands,[61] gold nanoparticles,[62] and gold clusters[63] on glass-like substrates. Photochemical methods other than the photolithography methods described above (e.g., light-directed solid-phase synthesis) have been demonstrated to be a highly efficient way for fabricating DNA oligonucleotide arrays on solid substrates (see Fig. 7, also Ref. 64).

Although all of the above methods have their strengths and advantages, none of these simultaneously meet the requirements of high spatial

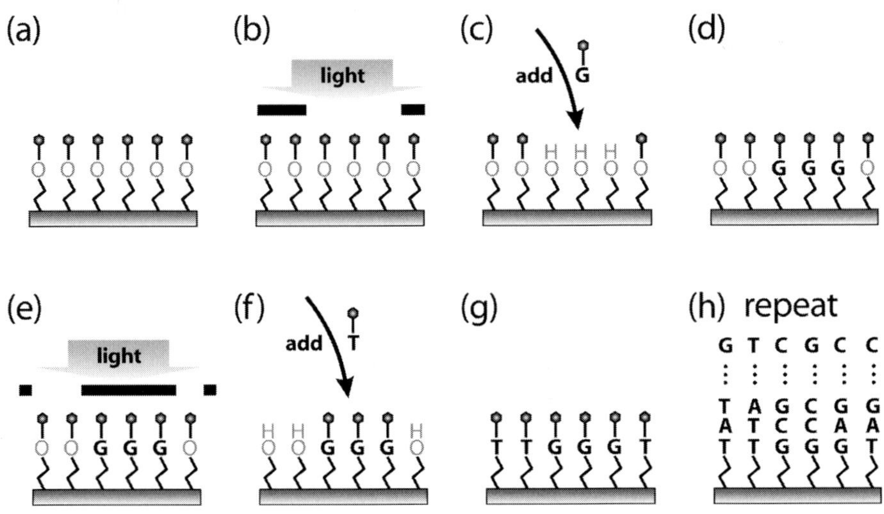

Fig. 7. Schematic illustration of a photochemical process to pattern a surface with different DNA oligonucleotides. Illumination of specific regions of the substrate causes a photochemical reaction and allows further chemistry to take place, while all other regions are protected (adapted from Ref. 64).

resolution, speed, integratability for large-scale processing, and the ability to coat different electrodes uniquely, which is required in the context of the integration of molecular functional elements into semiconductor devices. To address this challenge, we developed a method for selectively coating sub-50-nm-separated gold electrodes with different thiolated oligonucleotides in a controlled and simple manner.[65]

In conventional UV photolithography (see Fig. 8), a light-sensitive resist that initially covers the whole substrate surface is selectively removed by exposure to UV radiation and subsequent development to create specific, spatially well-defined regions of bare substrate, while keeping all other substrate areas covered. The minimum feature size that can be achieved depends on the wavelength of the light that is used (i.e., typically a few hundred nanometers). This is not sufficiently small to be used in the context of this chapter, nevertheless, the concept of patterning a resist to expose areas of the substrate that can then be coated with other molecules, is appealing.

Inspired by this well-developed process, we coat our electrode array with a molecular monolayer which acts as a molecular resist. Monolayers of thiol compounds, including thiolated oligonucleotides and alkanethiols, can be formed on gold surfaces by simply immersing the surface in an aqueous solution containing the thiolated molecules of interest.[28,66] Electrochemical studies show that the gold–sulphur bond formed during this spontaneous chemisorption process can undergo reductive cleavage at about

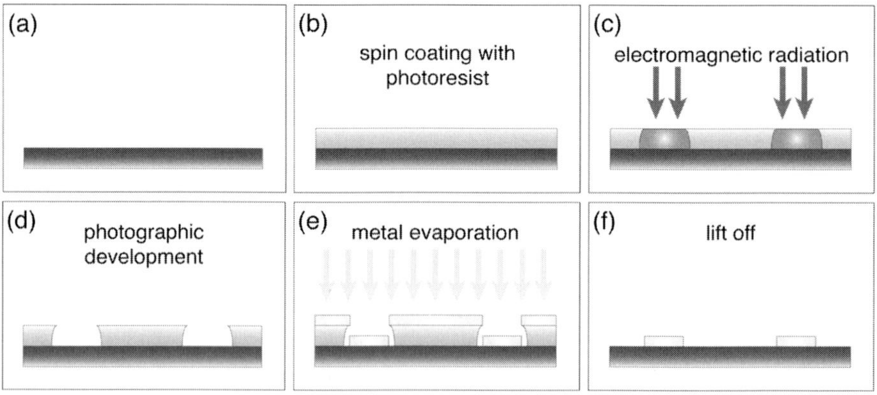

Fig. 8. Optical UV lithography used to pattern semiconductors. The minimum typical features that can be patterned using this technique are limited by diffraction of the UV light to about the wavelength of the light.

– 1 V versus an Ag/AgCl reference electrode, depending on the pH of the electrolyte.[17,67–70] These spontaneous chemisorption and electrochemical desorption mechanisms are exploited to control first the formation and then the selective removal of the molecular resist in the form of alkane monolayers on the electrode array. A schematic illustration of this process is shown in Fig. 9 (see also Ref. 65). In our case, a SAM of 6-mercapto-1-hexanol (MCH) molecules is formed on all electrodes to serve as a molecular resist, simply by immersing the whole electrode array in a solution containing 1 mM MCH. The molecular resist is "patterned" electrochemically, i.e., the SAM is removed from one or more individual electrodes while keeping it intact on all other electrodes by applying an electrochemical potential of – 1.4 V versus an Ag/AgCl reference electrode on the selected electrodes (Fig. 9(c)). Once the selected electrodes are exposed, they are coated with the thiolated anchor oligonucleotides of interest. This process is repeated for each distinct electrode or set of electrodes (Fig. 9(e)).

The repetitive process illustrated in Figs. 9(c)–9(f) suggests that the DNA oligonucleotide monolayers serve as molecular masks for subsequent coating steps. However, when thiolated DNA oligonucleotides are adsorbed onto a gold surface, the resulting monolayer is not perfect, since the

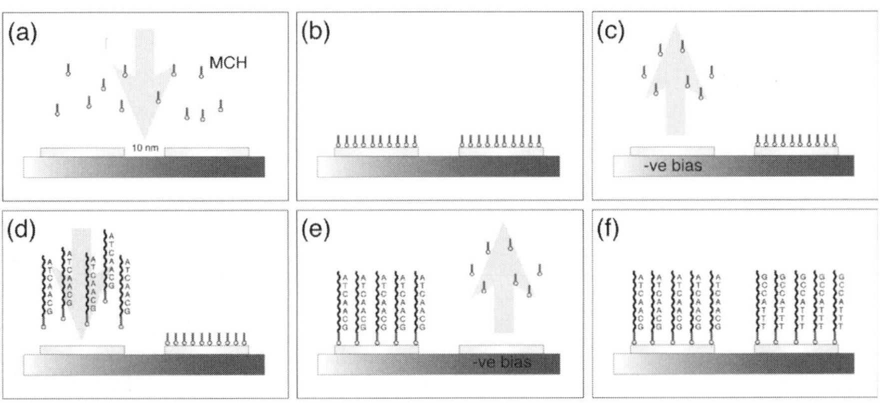

Fig. 9. Schematic illustration of the process employed to functionalize an array of electrodes with DNA anchor molecules. (a) The array is exposed to an aqueous solution of MCH to form a molecular resist (b). (c) The molecular resist is removed from selected areas by electrochemical means, and to obtain complete desorption of the molecular resist from a particular electrode, an electrochemical potential of – 1.4 V versus Ag/AgCl is applied to the electrode for 2 min while keeping all other electrodes at open circuit. (d) Thiolated DNA anchor molecules are attached to the exposed electrodes through spontaneous adsorption. The process in ((c), (d)) is repeated for other electrodes ((e), (f)).

oligonucleotides can not only bind via the terminal sulphur–gold bonds but also weakly via noncovalent amine bonds[71] and lie flat on the surface. This leads to disorder and defects in the monolayer that prevent it from serving as a perfect molecular resist in subsequent coating steps. Herne and Tarlov[66] showed that by immersing this imperfect monolayer in a solution of short thiolated molecules, the nonspecifically bound oligonucleotides are replaced with the short thiolated molecules bound to the surface by gold–sulphur bonds to produce a mixed monolayer with good resist characteristics.

Figures 10(a) and 10(b) show an electrode array comprising three pairs of opposing finger electrodes separated by gaps of less than 50 nm (see Fig. 10(d)), functionalized with two different thiolated DNA oligonucleotides serving as anchor molecules obtained using the above coating technique. Subsequent to the coating, DNA oligonucleotides with biotin, a small chemical group, attached to their ends, which are complementary to one of the two thiolated oligonucleotides on the array, are applied to the electrodes. Alkaline-phosphatase (AP), an enzyme that can cleave the colorimetric substrate BCIP/NBT, is then attached to the biotinylated oligonucleotides via an anti-biotin antibody. When the colorimetric substrate BCIP/NBT is applied to the electrode array, AP cleaves the BCIP/NBT which results in a dark precipitate on the electrodes where biotin and thus the complementary oligonucleotides are present.[72] In Figs. 10(a) and 10(b), the electrodes 1, 3, and 5, and 2, 4, and 6, respectively, appear darker than the other electrodes owing to the dark precipitate on the electrodes. We note that this colorimetric analysis does not only serve to assess the functionalization but

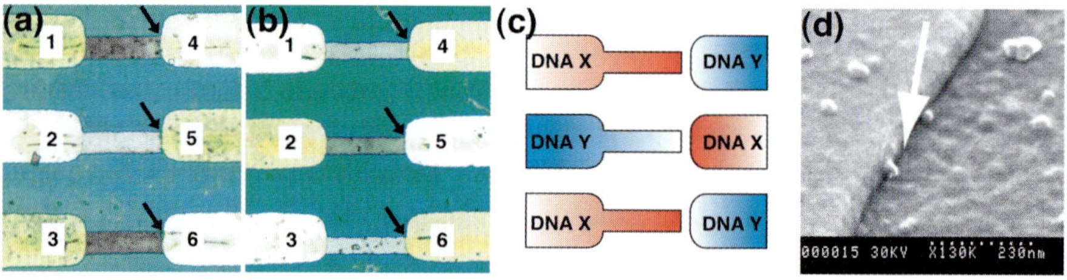

Fig. 10. ((a), (b)) Electrode arrays with nanoscale gaps separating opposing electrodes (indicated by black arrows). Electrodes 1, 3, and 5 are coated with oligonucleotides **X** and electrodes 2, 4, and 6 with oligonucleotides **Y**. The darker color of the electrodes (a) 1, 3, and 5 indicates the presence of surface-bound oligonucleotides **X**, and (b) of the electrodes 2, 4, and 6 the presence of oligonucleotides **Y**. (c) Schematic of the electrode array. (d) SEM image of the area between two opposing electrodes. The white arrow indicates the gap which appears as a dark line. Reprinted with permission from Ref. 65. Copyright (2003) American Chemical Society.

also demonstrates that the DNA anchor molecules retain their self-assembly properties. The dark precipitate is only present in the local environment of the biotin, which is only on the electrode if Watson–Crick basepairing between the DNA anchor molecule and the complementary oligonucleotide occurred.

In the above process, the DNA oligonucleotides used as anchor molecules were thiolated by attaching an SH group to the end of the DNA via a six-carbon-linker chain, SH–$(CH_2)_6$–R (C6). This is the most commonly used linker geometry, and although the linker chain is very flexible and keeps the anchor molecule far away from the surface, it also acts as an electrical insulator between the conducting surface and the DNA molecule. This is not necessarily a disadvantage, but in applications such as molecular electronics, where information in the form of electrical signals needs to be exchanged between the molecular complex and the solid substrate, an additional insulating barrier such as the $(CH_2)_6$ linker will certainly impede the use of DNA in this field. Conductivity experiments on conjugated molecular wires, for example, showed that increasing the linker length from an SH–CH_2 to an SH–$(CH_2)_3$ molecule increased the resistance by more than five orders of magnitude.[73]

The SH–$(CH_2)_6$ geometry is not the only linker possible. In an attempt to bring the anchor molecule closer to the surface while still maintaining the self-assembly properties of the DNA oligonucleotides, we investigated three other thiol configurations as well.[74] Figure 11 shows the chemical structures of the four different thiol arrangements. In addition to the six-carbon-linker chain, a similar arrangement but with only three carbons (C3) was used. This reduces the chain length to about half of the C6 configuration, while still maintaining some flexibility. In the third configuration, the thiophosphate arrangement, the thiol is incorporated into the 3′-phosphate group (TP). This brings the end of the DNA very close to the surface, but reduces the flexibility substantially and introduces steric restrictions. The fourth geometry employs phosphorothioates, where the sulfur atoms are directly incorporated into the backbone of the DNA between two bases (PT). We investigated the binding efficiency of single-stranded DNA, thiolated with the different thiol arrangements, onto gold. However, it is not only the binding of the thiolated DNA oligonucleotide to the gold that is important, but it is equally important that the DNA molecules retain their self-assembly properties once attached to the surface. Efficient and quantitatively similar thiol–gold binding and Watson–Crick basepairing was observed for the six- and the three-carbon-linker chains (C6 and C3). In

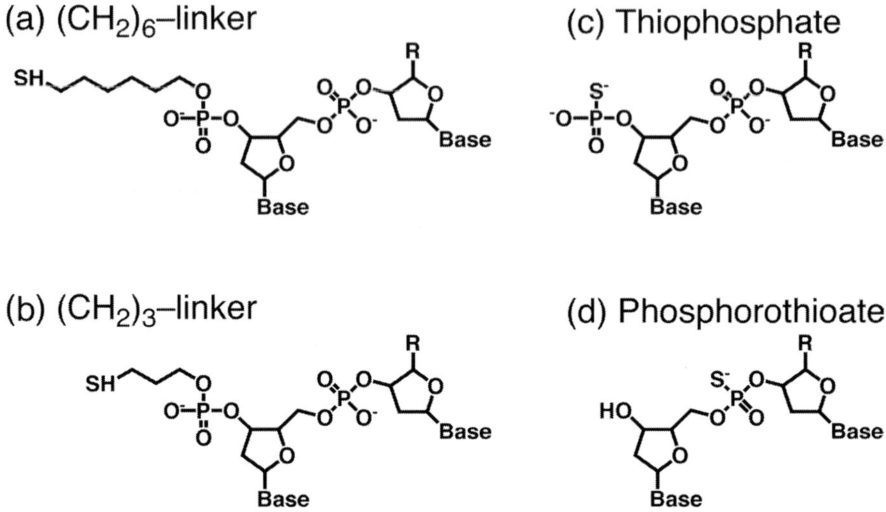

Fig. 11. The chemical structures of the four thiol arrangements investigated. (a) The thiol is attached to the DNA as a hexanethiol. (b) Similar to the hexanethiol configuration, except that the thiol is attached via a three-carbon-linker only (propanethiol). (c) The thiol is directly incorporated into the 3′-phosphate group (thiophosphate). (d) The thiol is incorporated into a nonterminal phosphate group (phosphorothioate). This latter configuration is not restricted to one thiol group per anchor molecule; an arbitrary number of phosphorothioates can be integrated along the backbone of a DNA molecule (adapted from Ref. 74).

contrast, neither the thiophosphate (Fig. 11(c)) nor the phosphorothioate (Fig. 11(d)) configuration allowed sufficient and specific surface binding while simultaneously retaining the self-assembly properties, although previous investigations reported at least partial binding.[75,76]

In conclusion, the method presented in this section to coat nanometer separated electrodes with different thiolated DNA oligonucleotides offers a convenient and reliable way for functionalizing the interface electrodes on semiconductor devices and to allow for the integration of molecular elements.

3. DNA-Based Branched Complexes

Many biological molecules are known to self-assemble into complex structures and networks, using their very efficient and selective molecular recognition processes. The use of biological molecules as scaffolds in the construction of functional devices has attracted significant attention over the past few decades and many different applications of molecular scaffolds have emerged. In particular, the self-organization and molecular recognition properties inherent to DNA discussed above have led to its use as a scaffold for various nanotechnologic self-assembly applications.

Liu *et al.*[77] demonstrated the use of short DNA molecules with a high affinity to certain proteins (DNA aptamers) to generate highly ordered protein arrays. In their work, proteins were linked via DNA aptamers to specific sites of the DNA scaffold, which consisted of linear DNA molecules. DNA has also been considered as a building block or wire for molecular electronics applications, but owing to the poor electrical conductivity of the DNA molecules, DNA has been used mainly as a scaffold for forming nanoscale wires, with metallic nanowires generated by depositing silver,[25] gold,[78,79] platinum,[80] palladium,[81] *inter alia*, onto the DNA scaffolds. Tanaka *et al.*[82] formed magnetic chains by assembling Cu^{2+} ions into a periodic arrangement by attaching them onto DNA scaffolds which were generated using artificial DNA bases to facilitate the Cu^{2+} attachment. Warner and co-worker utilized electrostatic binding of ligand-stabilized metal nanoparticles to the DNA backbone to generate extended linear chain-like structures, and even ribbon-like structures composed of nanoparticle chains assembled in a parallel fashion.[83]

These applications utilize DNA in its naturally occurring form, that is, as a linear molecule without any branch-points. However, a fundamental requirement for building more complex structures based on DNA scaffolds is the presence of branch-points in the scaffold. Otherwise, only point-to-point links are possible and although circles or knots may be formed, linking multiple linear molecules together will only produce (longer) linear molecules. Branched DNA structures do exist in nature in the form of Holliday junctions (see Fig. 12) which are four-way DNA junctions that form an important intermediate in genetic recombination.[86] Ever since this four-way DNA junction was proposed by Holliday in 1964, many studies have been carried out to investigate their structure and function.[87,88] The influence of base mismatches at the branch-point,[89,90] the effect of extra unpaired bases at the branch-point,[91–93] and the sequence-dependence of the conformation of the branch-point,[94] among others, have been studied. However, Seeman and co-workers were the first to exploit the self-assembly properties of DNA to generate branched DNA structures for applications involving engineering of structures at the nanometer scale.[95,96] A number of groups have fabricated three-arm[89,92] and four-arm[97] DNA structures, rigid DNA triangles,[98] and

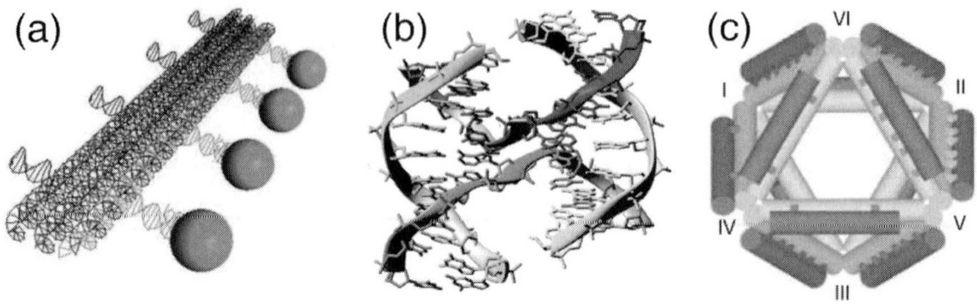

Fig. 12. Examples of DNA complexes. (a) DNA-aptamer-directed self-assembly of proteins onto a DNA scaffold. Reprinted in part with permission from Ref. 77. Copyright (2005) Wiley-VCH. (b) Center part of a Holliday-junction. Reprinted by permission from Macmillan Publishers Ltd.: *Nature Structural and Molecular Biology*,[84] copyright (1999). (c) Octahedron self-assembled from individual DNA molecules. Reprinted by permission from Macmillan Publishers Ltd.: *Nature*,[85] copyright (2004).

structures comprising three interlinked rings (Borromean rings).[99] Some of these three- and four-arm DNA structures were used, for example, to assemble two-dimensional (2D) DNA lattices[67,100] and DNA cubes.[101] Yan *et al.* assembled 2D lattices from branched DNA structures for use as a scaffold to generate ordered protein arrays.[102]

These examples demonstrate the versatility of DNA and the many different structures and geometries that can be assembled. However, these approaches all use synthetic DNA, that is short, chemically synthesized single-stranded DNA molecules, which inevitably limits the size of the constituent molecules. Although DNA synthesis has been developed into a robust, reliable technology, the maximum length of the molecules that can be produced is still of the order of 30–40 nm, which corresponds to about 100 bp. This length restriction limits the range of applications for which DNA scaffolds could be advantageous. For example, any program in which it is desired to integrate the DNA assembly into structures patterned by state-of-the-art nanofabrication tools, such as electron-beam or focused ion-beam lithography, will require DNA scaffolds several tens of nanometers in dimension. This is difficult to achieve using only synthetic DNA.

However, several groups have addressed this problem using different approaches. Cooper and Hagerman,[103] for example, constructed small four-arm junctions from synthetic DNA which were then extended by attaching longer double-stranded DNA molecules to the individual arms employing an enzymatic process (ligation) to study the structure and function of Holliday junctions. Although this approach generates branched, pure DNA complexes of variable lengths over a large length scale, a high yield in the ligation process during the extension step is difficult to achieve, generally resulting in a low overall yield for the complete assembly process. An entirely different approach was taken by Keren *et al.*[104] They used proteins called RecA to promote stable three-arm junction formation in DNA fragments with very similar, often identical, sequences (homologous DNA). RecA is the major protein responsible for the homologous recombination process in *Escherichia coli* (*E.coli*, a common bacteria), a protein-mediated process by which two DNA fragments of the same sequence form a triple-helix structure.[105–107] Two long DNA fragments with different end-sequences but homologous middle-sequences, can therefore be assembled into a branched complex using RecA. Shih *et al.*[85] combined a plasmid-derived 1.7 kb single-stranded DNA fragment with five short synthetic DNA oligonucleotides, which facilitate the correct folding of the long single-stranded DNA, to form DNA octahedron complexes. Other groups have investigated the use of large tree- or dendrimer-like structures.[108,109] These constructs were generated from Y-shaped, three-arm DNA junctions, which were themselves assembled from synthetic DNA oligonucleotides designed such that each arm has a single-stranded overhang. This Y-shaped DNA junction can then bind three other, different, Y-shaped DNA junctions via the single-stranded overhangs on the arms to form the second generation of the dendrimer. This process can be repeated many times.[109] A slightly different approach is used for the generation of chemical dendrimers.[110] Here, the branching of the structure is obtained through the use of small multi-branched molecules. Synthetic single-stranded DNA oligonucleotides are attached to the junction arms, and similar to the DNA dendrimers, the complementarity of these oligonucleotides is exploited to assemble the individual building blocks into large dendrimeric structures.

However, for applications where large branched complexes made exclusively from DNA are required, for example for DNA scaffolds, none of these methods can be used. They rely on synthetic DNA to introduce branching, on subsequent extension of the junction arms by ligation, on junction-formation-promoting proteins, or on small branched molecules to introduce branching.

To address this problem, we have chosen a different approach to construct large branched molecular complexes, based entirely on natural DNA building blocks.[111] Inspired by the assembly methods generally used to generate small DNA junctions, that is, self-assemble appropriate synthetic single-stranded DNA fragments into the desired geometry, we generate large fragments of single-stranded DNA and assemble them into the desired geometry by exploiting their self-organization capabilities. However, large fragments of single-stranded DNA cannot be generated synthetically as these methods are limited to around 100 bp, and therefore another method is required. In principle, enzymatic methods can be exploited to tackle this challenge. It is worth noting that when using this approach, the size of the branched DNA complexes is limited only by the ability of the enzymes used to process the constituent linear DNA molecules. A further advantage of using DNA chemistry in this way is the large enzymatic toolkit available to manipulate the molecules.[21]

Figure 13 shows a schematic illustration of the simplest branched structure, a three-armed construct comprising one branch-point. Each double-stranded arm is built from two different, reverse complement (RC) individual DNA components. For example, arm C's left strand (labeled c in Fig. 13) is part of component γ, whereas the right strand (labeled c-RC) is part of component α. Therefore, each component (α, β, γ) contains two pieces (indicated by different colors) which will each form one-half of an arm of the DNA complex. Each component has a downstream region in reverse complement (RC) orientation to the upstream region of another component, and in the case of the three-arm complex α binds β and γ; β binds γ and α; and γ binds α and β. Upon self-assembly, the only possible geometry that can be formed is the one shown in Fig. 13.

Figure 14 shows typical AFM images of the single branch-point complex, together with a double branch-point complex constructed using a similar

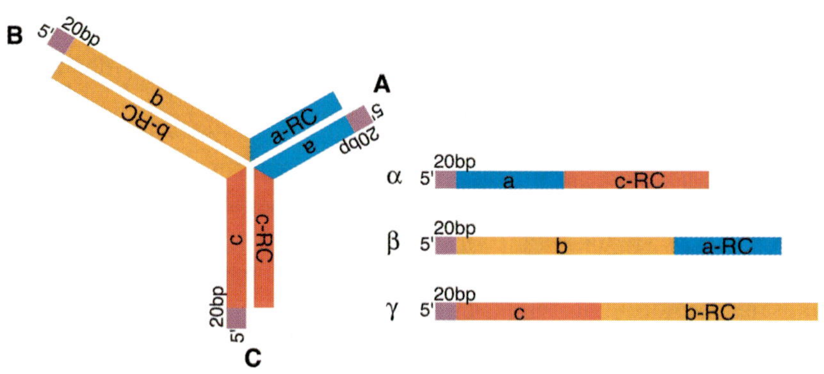

Fig. 13. Design schematic of a simple, large three-way DNA complex. The individual components (α, β and γ are generated from double-stranded DNA fragments using a selective enzymatic digestion process that only digests one of the two strands while leaving the other one intact. The components are then assembled into the final product. Note that the individual components are designed such that each arm possesses a single-stranded overhang of 20 DNA bases which can be used to integrate this DNA complex into existing environments (adapted from Ref. 111).

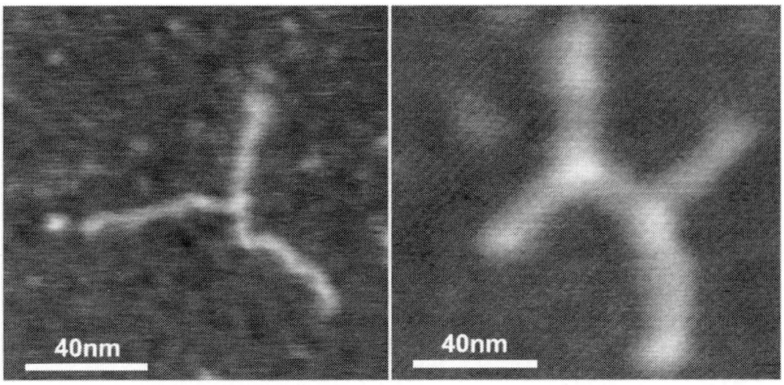

Fig. 14. Atomic force micrographs of branched DNA structures. (a) A three-arm, single branch-point structure assembled according to the design schematic in Fig. 13. (b) A four-arm, double branch-point structure. The design is very similar to the single branch-point structure, but is assembled from four individual single-stranded DNA fragments. Reprinted in part with permission from Ref. 111. Copyright (2006) American Chemical Society.

procedure. This method is not limited to one or two branch-points, or the number of arms. The design and assembly of more complicated structures is very similar to the three-arm single branch-point structure, and the number of individual single-stranded components required is given by the number of arms of the complex.

The individual components (α, β, and γ) shown in Fig. 13 are designed such that each arm possesses a single-stranded overhang of 20 DNA bases at the 5′-end. In the previous section, a method for functionalizing small, closely spaced electrodes with short single-stranded DNA anchor molecules was discussed. The single-stranded DNA overhangs on the branched DNA complex can now be designed to be RC to the anchor molecules on the electrode, and the complexes can be applied to the surface and integrated into existing environments by exploiting the self-assembly properties inherent to DNA.

These branched DNA complexes, integrated into semiconductor devices, can then serve as scaffolds for functionalizing the devices with other nanoscale elements, such as proteins or nanoparticles, by providing a means for positioning with high spatial resolution.

4. Manipulation of DNA by Electric Fields

The controlled manipulation of particles and molecules is a fundamental prerequisite for the field of nanotechnology. For many applications, the ability to manipulate molecules or particles on the molecular scale is essential to control their position and orientation, and hence their efficacy. Furthermore, manipulation techniques provide an important tool to study fundamental properties of molecular systems, for example, the direct and controlled manipulation of single molecules of DNA has contributed enormously to the understanding of the mechanical properties of DNA.[112]

The most prominent manipulation tool is certainly scanning probe microscopy.[113,114] Among the many different scanning probe techniques, atomic force microscopy (AFM) is arguably the most promising in terms of controlled manipulation of molecules. Several groups have used this technique to manipulate molecules, for example DNA, and to study their force-extension behavior.[115–118] AFM offers an unprecedented spatial resolution among the different manipulation techniques, but requires expensive infrastructure and cannot be integrated easily into nanoscale applications. An alternative to scanning probe techniques are optical[112,119,120] and magnetic[121–124] tweezers, where a dielectric or magnetic bead, respectively, is attached to the molecule. A laser beam is used to trap the dielectric particle and by moving the laser, the particle, and thus the molecule can be manipulated. A very similar approach is used in the case of magnetic tweezers, where the magnetic particle is manipulated using a spatially confined, highly inhomogenous magnetic field. A different approach was taken to study the mechanical properties of polymers by manipulation through hydrodynamic force techniques.[125–127] While the scanning probe and the tweezer-based techniques apply a point-like force to the molecules, usually at its end, hydrodynamic drag forces act along the whole length of the molecules.

Recently, AC electrokinetic manipulation techniques have received considerable attention as an alternative to the techniques mentioned above. When a polarizable particle or molecule, such as DNA, in an electrolyte is exposed to an AC electric field, a variety of forces act on the molecule. As a result of the redistribution of the charges at the molecule–electrolyte interface, a dipole is induced in the molecule itself,[128] which, in a non-uniform field, leads to a dielectrophoretic force on the molecule. Depending on whether the permittivity of the particle ε_p or the permittivity of the surrounding medium ε_m is larger, the dielectrophoresis is positive ($\varepsilon_p > \varepsilon_m$) or negative ($\varepsilon_p < \varepsilon_m$). The dielectrophoretic force is proportional to $\mathbf{E}\cdot\mathrm{grad}(\mathbf{E})$, and in the case of DNA in aqueous solution (positive dielectrophoresis), the net time-averaged dielectrophoretic force causes a movement of the DNA molecules in the direction of the highest value of the product **E**·grad(**E**), that is usually toward the electrode edge, where the field and the field gradient are highest.[128,129] In addition, a torque is exerted on the induced dipole which aligns the dipole with the electric field.[130] This phenomenon is known as electro-orientation. It is worth noting that in contrast to the dielectric force, the torque is proportional to the electric field rather than its gradient and can be finite even in a uniform electric field.

Dielectrophoresis has been applied to various systems. The controlled manipulation of submicron particles was demonstrated by separating herpes simplex and tobacco mosaic viruses from a mixture into two distinct populations,[132] by separating 93-nm-diameter latex beads of different dielectric properties,[133,134] and by the collection of 14-nm-diameter carboxylate-modified latex beads.[135] Further, dielectrophoretic forces have been used to separate and manipulate cells [136,137] and bacteria,[138,139] and to concentrate DNA molecules.[140,141] Other applications include site-specific attachment of DNA,[142] measurement of nuclease activity by measuring the decrease in elongated DNA length over reaction time.[142] Further, molecular surgery, where the elongated and positioned DNA was cut using a UV laser beam[142] and immobilized restriction enzymes, DNaseI and HindIII.[143] This may have applications in the direct sequencing of DNA.[144,145]

The variety of the different applications of AC electrokinetic manipulation demonstrates the large potential of this technique. However, a detailed understanding of the behavior of DNA, and in particular of surface-immobilized DNA, when exposed to high-frequency AC electric fields, and the orientation and elongation of DNA as a result of dielectrophoretic force and torque, needs to be well understood to realize the full potential of the technique. Over the past few years, we have investigated the elongation of surface-tethered DNA on exposure to an AC electric field to gain a deeper understanding of the underlying phenomena.[146–148] A possible mechanism was proposed in Ref. 149 and discussed in detail in Ref. 147, and is illustrated in Fig. 15.

Charged polymeric molecules under the influence of external forces, for example DNA exposed to an AC electric field, can be described by the worm-like chain (WLC) model.[20] The WLC model is the continuous limit of the Kratky–Porod (KP) model,[150] which describes the polymeric molecules as a succession of N segments of length l, that is, in the WLC model, $l \to 0$. DNA is highly polarizable, and when exposed to an electric field, a dipole is induced along the backbone of the DNA. The electric field exerts a torque τ on this induced dipole which causes the individual DNA segments to align with the electric field. The alignment of the dipole and thus the DNA

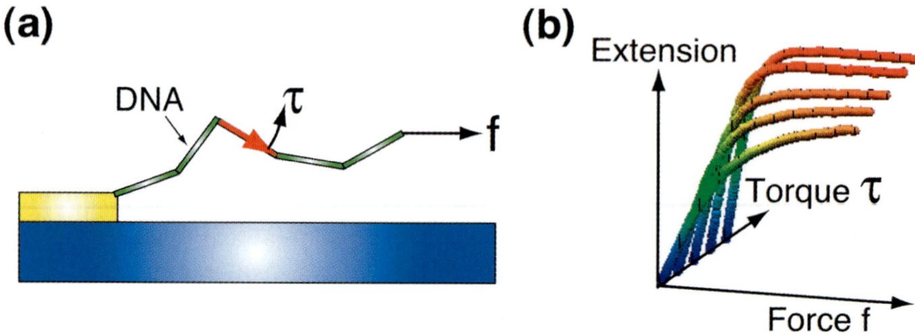

Fig. 15. (a) The DNA molecules are modeled as long strings of connected short segments. When exposed to an electric field, a dipole is induced in each segment (indicated by the red arrow), and the resulting electrokinetic torque τ aligns the segments with the electric field. However, the probability of parallel and anti-parallel alignment of the segments with the electric field is equal and therefore no elongation is expected in this case. To achieve elongation, an additional bias force f, for example a point force pulling on the free end of the DNA is required so that a forward, parallel alignment of the segments is favored over an anti-parallel alignment. (b) The resulting extension as a function of the torque and the bias force (data taken from Ref. 131).

segment is either parallel or anti-parallel with the electric field, as both the orientations are energetically equal. Therefore, this would not result in elongation, and an additional bias force f, for example a point force pulling on the free end of the DNA, is required such that a forward, parallel alignment of the DNA segments with the electric field is favored over an anti-parallel alignment.

The elongation of a polymeric molecule as a result of an external applied electric field has been calculated[131] and is shown in Fig. 15(b). It can be seen that elongation only occurs if the torque is supplemented by a bias force, which could be provided by the dielectrophoretic force. As noted above, though, the dielectrophoretic force points toward the ends of the molecules that are tethered to the surface and therefore would not result in elongation. However, the DNA molecules are not in vacuum, they are surrounded by a solvent. The applied AC electric field can also lead to fluid flow in the solvent, which in turn can result in movement of the molecule.[125,130,133,147,151,152] We have investigated the fluid flow that occurs in our system and a schematic illustration is given in Fig. 16. We found that the fluid flow decreases with increasing AC electric field frequency but was qualitatively similar for all conditions used in our studies.[147]

We tethered fluorescently labeled DNA of various lengths (48 kilobasepairs (kb) (contour length 21 μm), 35 kb (15 μm), 25 kb (11 μm), and 15 kb

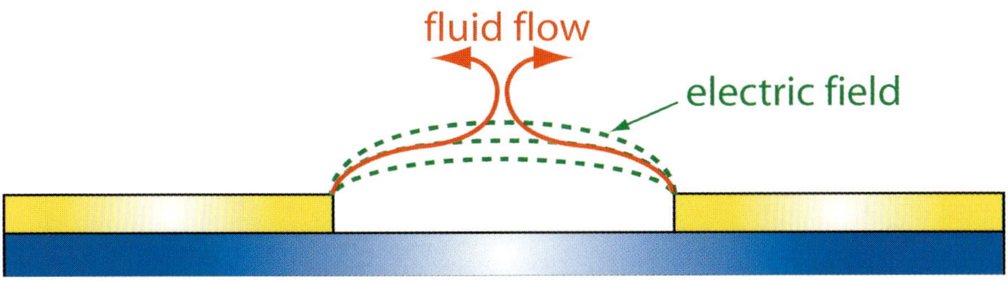

Fig. 16. Side view of the electrode setup. The electric field is indicated by the green dashed line and the induced fluid flow by the red arrows. The fluid flow and the electric field lines are nearly perpendicular at the midpoint of the gap between the electrodes.[147] Reused with permission from Ref. 148. Copyright 2006, American Institute of Physics.

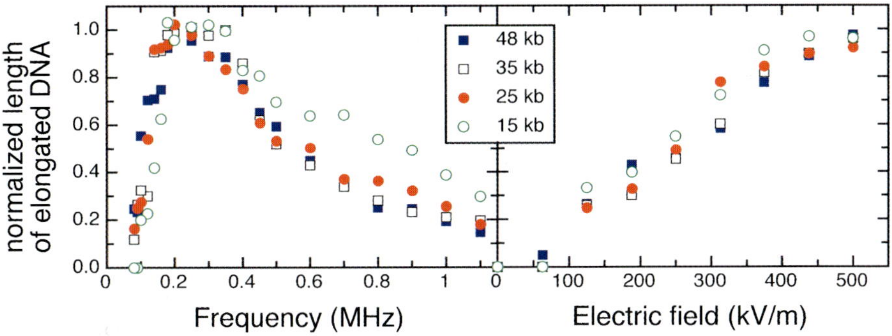

Fig. 17. Normalized length of elongated DNA versus frequency and electric field. Regardless of the length of the DNA, the normalized length follows a universal curve. Reused with permission from Ref. 147. Copyright 2005, American Institute of Physics.

(6.5 μm)[a]) on one of two opposing finger-like electrodes separated by gaps of 40, 30, and 20 μm.[147] The lengths of the elongated DNA molecules were measured as a function of applied electric field and frequency and the results are shown in Fig. 17. The data are displayed as normalized length, that is, the measured length of the molecule divided by the contour length. To achieve measurable elongation, electric fields of several hundred kV/m are required. This is not surprising as the forces acting on the molecules must overcome Brownian motion, which counteracts the process of aligning the individual DNA segments. The elongation reaches a maximum at around 200–300 kHz, regardless of the length of the DNA. Above that, the phase

[a]The length of DNA increases on fluorescent labeling. The length of unlabeled 48 kb is 16.5 μm,[154] but the intercalation of the fluorophore YOYO-I at a dye:basepair ratio of 1:8 leads to a 20% increase in length.[119]

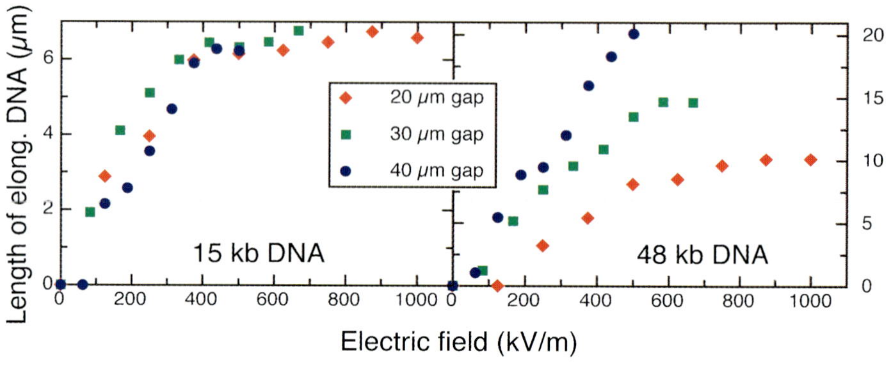

Fig. 18. DNA elongated across different-sized gaps. The 15-kb, 6.5-μm DNA shows the same behavior for all gaps, whereas the elongation of the 48-kb, 21-μm DNA strongly depends on the gap size. Reused with permission from Ref. 147. Copyright 2005, American Institute of Physics.

of the induced dipole starts lagging behind the AC electric field and the effective torque decreases rapidly.[130,146] Below around 100 kHz, the effective field is reduced owing to screening effects of the electrodes by the electrolyte.[151,153] An interesting fact is that, regardless of the length of the DNA, the normalized elongation follows a universal behavior, that is, the elongation only depends on the electric field and its frequency but not on the length of the DNA, suggesting that the total force acting on each segment is uniform over most of the molecule.

Another interesting result is shown in Fig. 18. It may be expected that reducing both the gap size and the applied electric potential in such a way that the electric field stays constant, does not change the elongation of the DNA molecules.[b] Figure 18 shows the elongation of DNA at a frequency of 300 kHz across electrode gaps of 40, 30, and 20 μm, and it would be expected that this should lead to full elongation at sufficiently strong electric fields for all gap sizes. Indeed, this is the case for 15 kb DNA (contour length 6.5 μm), but not at all for 48 kb DNA (21 μm). For the long DNA molecules, the maximum elongation that can be achieved is only half the gap size, that is, for a 20-μm gap, the maximum elongation is 10 μm. Regardless of the strength of the electric field, it is not possible to stretch the molecules beyond the mid-point of the gap.

As discussed above, to achieve elongation, a (small) bias force is required that pulls in the direction of elongation. The dielectric force works in the

[b]Modeling of the electric field showed that the field is only strongly inhomogenous at the electrode edges, while a nearly constant field was found across most of the gap.[147]

wrong direction but the force acting on the molecules through viscous drag of the induced fluid flow points in the right direction (see Fig. 16). Furthermore, the force owing to the fluid flow changes direction at the mid-point of the gap. It is this that leads to the restricted elongation seen for the 48-kb-DNA in Fig. 18(b).

It has been shown that fluid flow alone can lead to elongation of DNA (Perkins *et al.*, 1995). In our setup, if the fluid flow were the major effect responsible for the elongation, the DNA would follow the fluid flow. In this case, for long DNA elongated across a small gap, the DNA is expected to show an upturn at the mid-point of the gap following the flow of the solvent. On the other hand, if the fluid flow was providing only a bias force and the major contribution was owing to the alignment of the individual segments by the electrokinetic torque, no upturn of the DNA at the mid-point of the gap is expected. The two different cases cannot be distinguished by top-view 2D imaging.[147] To address this, we performed three-dimensional (3D) imaging of elongation-restricted systems, using confocal microscopy techniques.[148]

Figure 19 shows fluorescently labeled 48-kb-DNA (contour length 21 μm) elongated across a 20-μm-wide gap. Although the elongation is restricted to about one half of the contour length of the DNA, the DNA only extents to the mid-point of the gap. No indication of an upturn is observed, in agreement with the model of elongation, where the DNA molecules are elongated as a result of the electrokinetic torque aligning the individual segments, supplemented with a bias force that favors forward alignment. Furthermore, the contour of the area containing the elongated DNA closely resembles the shape of the electric field lines.

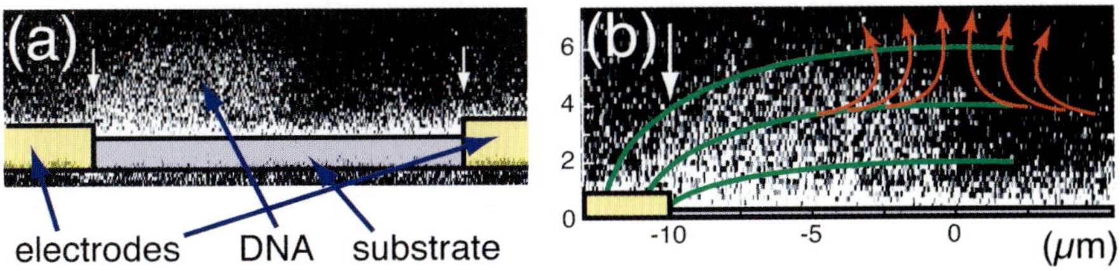

Fig. 19. (a) Three-dimensional imaging of 21-μm-long DNA elongated across a 20-μm-wide gap. The image was taken with a laser-scanning confocal microscope. The white arrows indicate the gold electrode edge. (b) The same situation as in (a); the green lines indicate the electric field lines and the red arrows the induced fluid flow. There is no evidence that the DNA follows the red arrows. Reused with permission from Ref. 148. Copyright 2006, American Institute of Physics.

5. Concluding Remarks and Future Directions

Nanotechnology and engineering at the nanometer scale has considerably advanced over the past few years, but despite many successes, the field is still in its infancy. A myriad of techniques to fabricate fascinating functional nanometer-scale complexes has been developed, including several DNA-based nanostructures. However, many of their properties, including electrical, magnetic, catalytic, and optical properties, *inter alia*, have not yet been studied systematically or in great detail. In addition, although many of these complexes can generally be fabricated in large numbers, generic and reliable ways of interconnecting them into larger, multi-functional networks, and integrating them in a controlled manner into macroscopic environments, have not yet been developed. Very advanced tools exist to manipulate these complexes one by one. They are invaluable in connection with the study of fundamental properties of these complexes, but less well suited for their assembly and interconnection. A modern-day semiconductor-integrated circuit may contain 100 million individual transistors; the ability to assemble and connect nanoscale elements rapidly and reliably is going to be crucial for this technology to approach the sophistication of the semiconductor industry. Molecular self-assembly, a mechanism that exploits sophisticated lock-and-key recognition, is found in many biological systems, and is a very promising candidate to fulfill this task. It can be expected that engineering by molecular self-assembly will have a prominent place in this field over the next decade and beyond.

Future devices and materials will most likely consist of a mixture of "hard" and "soft" materials, i.e., a combination of solid substrates such as semiconductors and metals, and biological complexes such as proteins, viruses, and DNA. Although molecular complexes have been attached to solid substrates, the properties of such interfaces is not well investigated and understood to date. Considering that such interfaces are among the most fundamental and new aspects in this field, this area will represent a new frontier in this field and it can be anticipated that many new phenomena will emerge.

Acknowledgments

C. Wälti is grateful for the support by the EPSRC and the Swiss National Science Foundation and would like to acknowledge many of his colleagues who were involved at various stages in the work presented, in particular P. Tosch, W. A. Germishuizen, R. Wirtz, A. G. Davies, C. F. Kaminski, D. M. D. Bailey, A. E. Cohen, M. A. Buswell, D. A. Williams, A. P. J. Middelberg, and M. Pepper.

References

1. A. G. Davies, Quantum electronics: The physics and technology of low-dimensional electronic systems into the new millennium, *Philos. Trans. R. Soc. A* **358**, 151–172 (2000).
2. J. Chen, W. Wang, M. A. Reed, A. M. Rawlett, D. W. Price and J. M. Tour, Room-temperature negative differential resistance in nanoscale molecular junctions, *Appl. Phys. Lett.* **77**, 1224–1226 (2000).
3. H. Park, J. Park, A. K. L. Lim, E. H. Anderson, A. P. Alivisatos and P. L. McEuen, Nanomechanical oscillations in a single-C-60 transistor, *Nature* **407**, 57–61 (2000).
4. J. Park, A. N. Pasupathy, J. L. Goldsmith, C. Chang, Y. Yaish, J. R. Petta, M. Rinkoski, J. P. Sethna, H. D. Abruna, P. L. McEuen and D. C. Ralph, Coulomb blockade and the Kondo effect in single-atom transistors, *Nature* **417**, 722–726 (2002).
5. W. J. Yoon, S. Y. Chung, P. R. Berger and S. M. Asar, Room-temperature negative differential resistance in polymer tunnel diodes using a thin oxide layer and demonstration of threshold logic, *Appl. Phys. Lett.* **87**, 203506/1–3 (2005).
6. D. K. James and J. M. Tour, Molecular wires, *Top. Curr. Chem.* **257**, 33–62 (2005).
7. S. J. Tans, A. R. M. Verschueren and C. Dekker, Room-temperature transistor based on a single carbon nanotube, *Nature* **393**, 49–52 (1998).
8. A. Bachtold, P. Hadley, T. Nakanishi and C. Dekker, Logic circuits with carbon nanotube transistors, *Science* **294**, 1317–1320 (2001).
9. T. Rueckes, K. Kim, E. Joselevich, G. Y. Tseng, C. L. Cheung and C. M. Lieber, Carbon nanotube-based nonvolatile random access memory for molecular computing, *Science* **289**, 94–97 (2000).
10. Y. Huang, X. F. Duan, Y. Cui, L. J. Lauhon, K. H. Kim and C. M. Lieber, Logic gates and computation from assembled nanowire building blocks, *Science* **294**, 1313–1317 (2001).
11. X. F. Duan, Y. Huang, Y. Cui, J. F. Wang and C. M. Lieber, Indium phosphide nanowires as building blocks for nanoscale electronic and optoelectronic devices, *Nature* **409**, 66–70 (2001).
12. M. Bruchez, M. Moronne, P. Gin, S. Weiss and A. P. Alivisatos, Semiconductor nanocrystals as fluorescent biological labels, *Science* **281**, 2013–2016 (1998).
13. A. Sukhanova, L. Venteo, J. Devy, M. Artemyev, V. Oleinikov, M. Pluot and I. Nabiev, Highly stable fluorescent nanocrystals as a novel class of

labels for immunohistochemical analysis of paraffin-embedded tissue sections, *Lab. Invest.* **82**, 1259–1261 (2002).
14. E. Winfree, F. Liu, L. A. Wenzler and N. C. Seeman, Design and self-assembly of two-dimensional DNA crystals, *Nature* **394**, 539–544 (1998).
15. F. M. Menger, Supramolecular chemistry and self-assembly, *Proc. Natl. Acad. Sci. USA* **99**, 4818–4822 (2002).
16. S. J. Tans and C. Dekker, Molecular transistors — Potential modulations along carbon nanotubes, *Nature* **404**, 834–835 (2000).
17. D. I. Gittins, D. Bethell, D. J. Schiffrin and R. J. Nichols, A nanometre-scale electronic switch consisting of a metal cluster and redox-addressable groups, *Nature* **408**, 67–71 (2000).
18. J. M. Lehn, *Supramolecular Chemistry: Concepts and Perspectives* (VCH, Weinheim, 1995).
19. J. D. Watson and F. H. C. Crick, A Structure for Deoxyribose Nucleic Acid, *Nature* **171**, 737–738 (1953).
20. J. F. Marko and E. D. Siggia, Stretching DNA, *Macromolecules* **28**, 8759–8770 (1995).
21. R. J. Roberts and D. Macelis, REBASE — Restriction enzymes and methylases, *Nucleic Acids Res.* **29**, 268–269 (2001).
22. C. A. Mirkin, R. L. Letsinger, R. C. Mucic and J. J. Storhoff, A DNA-based method for rationally assembling nanoparticles into macroscopic materials, *Nature* **382**, 607–609 (1996).
23. A. P. Alivisatos, K. P. Johnsson, X. Peng, T. E. Wilson, C. J. Loweth, M. P. Bruchez, Jr. and P. G. Schultz, Organization of nanocrystal groups using DNA, *Nature* **382**, 609–611 (1996).
24. C. J. Loweth, W. B. Caldwell, X. Peng, A. P. Alivisatos and G. Schultz, DNA-based assembly of gold nanocrystals, *Angew. Chem. Int. Ed.* **38**, 1808–1812 (1999).
25. E. Braun, Y. Eichen, U. Sivan and G. Ben-Yoseph, DNA-templated assembly and electrode attachment of a conducting silver wire, *Nature* **391**, 775–778 (1998).
26. R. G. Nuzzo and D. L. Allara, Adsorption of bifunctional organic disulfides on gold surfaces, *J. Am. Chem. Soc.* **105**, 4481–4483 (1983).
27. J. Sagiv, Organized monolayers by adsorption. 1. Formation and structure of oleophobic mixed monolayers on solid surfaces, *J. Am. Chem. Soc.* **102**, 92–98 (1980).
28. A. Ulman, Formation and structure of self-assembled monolayers, *Chem. Rev.* **96**, 1533–1554 (1996).
29. F. Schreiber, Structure and growth of self-assembling monolayers, *Prog. Surf. Sci.* **65**, 151–256 (2000).
30. G. Yershov, V. Barsky, A. Belgovskiy, E. Kirillov, E. Kreindlin, I. Ivanov, S. Parinov, D. Guschin, A. Drobishev, S. Dubiley and A. Mirzabekov, DNA analysis and diagnostics on oligonucleotide microchips, *Proc. Natl. Acad. Sci. USA* **93**, 4913–4918 (1996).
31. A. Bruckbauer, L. M. Ying, A. M. Rothery, D. J. Zhou, A. I. Shevchuk, C. Abell, Y. E. Korchev and D. Klenerman, Writing with DNA and protein

using a nanopipet for controlled delivery, *J. Am. Chem. Soc.* **124**, 8810–8811 (2002).
32. N. L. Abbott, J. P. Folkers and G. M. Whitesides, Manipulation of the wettability of surfaces on the 0.1-micrometer to 1-micrometer scale through micromachining and molecular self-assembly, *Science* **257**, 1380–1382 (1992).
33. A. Kumar and G. M. Whitesides, Features of gold having micrometer to centimeter dimensions can be formed through a combination of stamping with an elastomeric stamp and an alkanethiol ink followed by chemical etching, *Appl. Phys. Lett.* **63**, 2002–2004 (1993).
34. A. Kumar and G. M. Whitesides, Patterned condensation figures as optical diffraction gratings, *Science* **263**, 60–62 (1994).
35. M. Geissler, A. Bernard, A. Bietsch, H. Schmid, B. Michel and E. Delamarche, Microcontact-printing chemical patterns with flat stamps, *J. Am. Chem. Soc.* **122**, 6303–6304 (2000).
36. T. W. Odom, V. R. Thalladi, J. C. Love and G. M. Whitesides, Generation of 30–50 nm structures using easily fabricated, composite PDMS masks, *J. Am. Chem. Soc.* **124**, 12112–12113 (2002).
37. T. W. Odom, J. C. Love, D. B. Wolfe, K. E. Paul and G. M. Whitesides, Improved pattern transfer in soft lithography using composite stamps, *Langmuir* **18**, 5314–5320 (2002).
38. H. W. Li, B. V. O. Muir, G. Fichet and W. T. S. Huck, Nanocontact printing: A route to sub-50-nm-scale chemical and biological patterning, *Langmuir* **19**, 1963–1965 (2003).
39. M. J. Tarlov, D. R. F. Burgess and G. Gillen, UV photopatterning of alkanethiolate monolayers self-assembled on gold and silver, *J. Am. Chem. Soc.* **115**, 5305–5306 (1993).
40. J. Y. Huang and J. C. Hemminger, Photooxidation of thiols in self-assembled monolayers on gold, *J. Am. Chem. Soc.* **115**, 3342–3343 (1993).
41. J. M. Calvert, M. S. Chen, C. S. Dulcey, J. H. Georger, M. C. Peckerar, J. M. Schnur and P. E. Schoen, Deep ultraviolet patterning of monolayer films for high-resolution lithography, *J. Vac. Sci. Technol. B* **9**, 3447–3450 (1991).
42. H. Sugimura, T. Hanji, O. Takai, T. Masuda and H. Misawa, Photolithography based on organosilane self-assembled monolayer resist, *Electrochim. Acta* **47**, 103–107 (2001).
43. M. J. Tarlov and J. G. Newman, Static secondary ion mass-spectrometry of self-assembled alkanethiol monolayers on gold, *Langmuir* **8**, 1398–1405 (1992).
44. S. Q. Sun, K. S. L. Chong and G. J. Leggett, Nanoscale molecular patterns fabricated by using scanning near-field optical lithography, *J. Am. Chem. Soc.* **124**, 2414–2415 (2002).
45. S. Sun, M. Montague, K. Critchley, M. S. Chen, W. J. Dressick, S. D. Evans and G. J. Leggett, Fabrication of biological nanostructures by scanning near-field photolithography of chloromethylphenylsiloxane monolayers, *Nano Lett.* **6**, 29–33 (2006).

46. H. Sugimura, K. Ushiyama, A. Hozumi and O. Takai, Micropatterning of alkyl- and fluoroalkylsilane self-assembled monolayers using vacuum ultraviolet light, *Langmuir* **16**, 885–888 (2000).
47. M. M. Ferris and K. L. Rowlen, Direct evidence of ozone as the active oxidant in photooxidation of alkanethiols on SERS-active silver, *Appl. Spectrosc.* **54**, 664–668 (2000).
48. W. J. Dressick, C. S. Dulcey, M. S. Chen and J. M. Calvert, Photochemical studies of (aminoethylaminomethyl)phenethyltrimethoxysilane self-assembled monolayer films, *Thin Solid Film* **285**, 568–572 (1996).
49. K. Critchley, J. P. Jeyadevan, H. Fukushima, M. Ishida, T. Shimoda, R. J. Bushby and S. D. Evans, A mild photoactivated hydrophilic/hydrophobic switch, *Langmuir* **21**, 4554–4561 (2005).
50. S. Xu and G. Y. Liu, Nanometer-scale fabrication by simultaneous nanoshaving and molecular self-assembly, *Langmuir* **13**, 127–129 (1997).
51. P. V. Schwartz, Meniscus force nanografting: Nanoscopic patterning of DNA, *Langmuir* **17**, 5971–5977 (2001).
52. R. D. Piner, J. Zhu, F. Xu, S. H. Hong and C. A. Mirkin, "Dip-pen" nanolithography, *Science* **283**, 661–663 (1999).
53. C. A. Mirkin, S. H. Hong and L. Demers, Dip-pen nanolithography: Controlling surface architecture on the sub-100 nanometer length scale, *ChemPhysChem.* **2**, 37–39 (2001).
54. M. Su, X. G. Liu, S. Y. Li, V. P. Dravid and C. A. Mirkin, Moving beyond molecules: Patterning solid-state features via dip-pen nanolithography with sol-based inks, *J. Am. Chem. Soc.* **124**, 1560–1561 (2002).
55. L. M. Demers, D. S. Ginger, S. J. Park, Z. Li, S. W. Chung and C. A. Mirkin, Direct patterning of modified oligonucleotides on metals and insulators by dip-pen nanolithography, *Science* **296**, 1836–1838 (2002).
56. K. B. Lee, S. J. Park, C. A. Mirkin, J. C. Smith and M. Mrksich, Protein nanoarrays generated by dip-pen nanolithography, *Science* **295**, 1702–1705 (2002).
57. C. C. Hsueh, M. T. Lee, M. S. Freund and G. S. Ferguson, Electrochemically directed self-assembly on gold, *Angew. Chem. Int. Ed.* **39**, 1228–1231 (2000).
58. M. Riepl, V. M. Mirsky and O. S. Wolfbeis, Electrical control of alkanethiols self-assembly on a gold surface as an approach for preparation of microelectrode arrays, *Mikrochim. Acta* **131**, 29–34 (1999).
59. J. Wang, M. Jiang, A. M. Kawde and R. Polsky, Electrochemically induced deposition of thiol-based monolayers onto closely spaced microelectrodes, *Langmuir* **16**, 9687–9689 (2000).
60. R. Maoz, E. Frydman, S. R. Cohen and J. Sagiv, Constructive nanolithography: Inert monolayers as patternable templates for *in situ* nanofabrication of metal–semiconductor–organic surface structures — A generic approach, *Adv. Mater.* **12**, 725–732 (2000).
61. R. Maoz, E. Frydman, S. R. Cohen and J. Sagiv, Constructive nanolithography: Site-defined silver self-assembly on nanoelectrochemically patterned monolayer templates, *Adv. Mater.* **12**, 424–430 (2000).

62. S. Hoeppener, R. Maoz, S. R. Cohen, L. F. Chi, H. Fuchs and J. Sagiv, Metal nanoparticles, nanowires and contact electrodes self-assembled on patterned monolayer templates — A bottom-up chemical approach, *Adv. Mater.* **14**, 1036–1042 (2002).
63. S. T. Liu, R. Maoz, G. Schmid and J. Sagiv, Template guided self-assembly of [Au_{55}] clusters on nanolithographically defined monolayer patterns, *Nano Lett.* **2**, 1055–1060 (2002).
64. S. P. A. Fodor, J. L. Read, M. C. Pirrung, L. Stryer, A. T. Lu and D. Solas, Light-directed, spatially addressable parallel chemical synthesis, *Science* **251**, 767–773 (1991).
65. C. Wälti, R. Wirtz, W. A. Germishuizen, D. M. D. Bailey, M. Pepper, A. P. J. Middelberg and A. G. Davies, Direct selective functionalisation of nanometre separated gold electrodes with DNA oligonucleotides, *Langmuir* **19**, 981–984 (2003).
66. T. M. Herne and M. J. Tarlov, Characterization of DNA probes immobilized on gold surfaces, *J. Am. Chem. Soc.* **119**, 8916 (1997).
67. C. A. Widrig, C. Chung and M. D. Porter, The electrochemical desorption of n-alkanethiol monolayers from polycrystalline Au and Ag electrodes, *J. Electroanal. Chem.* **310**, 335–359 (1991).
68. D. E. Weisshaar, B. D. Lamp and M. D. Porter, Thermodynamically controlled electrochemical formation of thiolate monolayers at gold — Characterization and comparison to self-assembled analogs, *J. Am. Chem. Soc.* **114**, 5860–5862 (1992).
69. S. Imabayashi, D. Hobara, T. Kakiuchi and W. Knoll, Selective replacement of adsorbed alkanethiols in phase-separated binary self-assembled monolayers by electrochemical partial desorption, *Langmuir* **13**, 4502–4504 (1997).
70. M. J. Esplandiu, H. Hagenstrom and D. M. Kolb, Functionalized self-assembled alkanethiol monolayers on Au(111) electrodes: 1. Surface structure and electrochemistry, *Langmuir* **17**, 828–838 (2001).
71. D. V. Leff, L. Brandt and J. R. Heath, Synthesis and characterization of hydrophobic, organically-soluble gold nanocrystals functionalized with primary amines, *Langmuir* **12**, 4723–4730 (1996).
72. R. Wirtz, C. Wälti, W. A. Germishuizen, M. Pepper, A. P. J. Middelberg and A. G. Davies, High sensitivity colorimetric detection of DNA hybridisation on a gold surface with high spatial resolution, *Nanotechnology* **14**, 7–10 (2003).
73. T. Ishida, W. Mizutani, Y. Aya, H. Ogiso, S. Sasaki and H. Tokumoto, Electrical conduction of conjugated molecular SAMs studied by conductive atomic force microscopy, *J. Phys. Chem. B* **106**, 5886–5892 (2002).
74. R. Wirtz, C. Wälti, P. Tosch, M. Pepper, A. G. Davies, W. A. Germishuizen and A. P. J. Middelberg, Influence of the thiol position on the attachment and subsequent hybridization of thiolated DNA on gold surfaces, *Langmuir* **20**, 1527–1530 (2004).
75. T. Hianik, V. Gajdos, R. Krivanek, T. Oretskaya, V. Metelev, E. Volkov and P. Vadgama, Amperometric detection of DNA hybridization on a gold

surface depends on the orientation of oligonucleotide chains, *Bioelectrochemistry* **53**, 199–204 (2001).

76. M. Nakayama, T. Ihara, K. Nakano and M. Maeda, DNA sensors using a ferrocene-oligonucleotide conjugate, *Talanta* **56**, 857–866 (2002).
77. Y. Liu, C. Lin, H. Li and H. Yan, Aptamer-directed self-assembly of protein arrays on a DNA nanostructure, *Angew. Chem. Int. Ed.* **44**, 4333–4338 (2005).
78. J. K. N. Mbindyo, B. D. Reiss, B. R. Martin, C. D. Keating, M. J. Natan and T. E. Mallouk, DNA-directed assembly of gold nanowires on complementary surfaces, *Adv. Mater.* **13**, 249–255 (2001).
79. O. Harnack, W. E. Ford, A. Yasuda and J. M. Wessels, Tris(hydroxymethyl)phosphine-capped gold particles templated by DNA as nanowire precursors, *Nano Lett.* **2**, 919–923 (2002).
80. M. Mertig, L. C. Ciacchi, R. Seidel, W. Pompe and A. Vita, DNA as a selective metallization template, *Nano Lett.* **2**, 841–844 (2002).
81. J. Richter, R. Seidel, R. Kirsch, M. Mertig, W. Pompe, J. Plaschke and H. K. Schackert, Nanoscale palladium metallization of DNA, *Adv. Mater.* **12**, 507–511 (2000).
82. K. Tanaka, A. Tengeiji, T. Kato, N. Toyama and M. Shionoya, A discrete self-assembled metal array in artificial DNA, *Science* **299**, 1212–1213 (2003).
83. M. G. Warner and J. E. Hutchison, Linear assemblies of nanoparticles electrostatically organized on DNA scaffolds, *Nat. Mater.* **2**, 272–277 (2003).
84. M. Lombardia, A. Gonzalez, R. Eritja, J. Aymami, F. Azorin and M. Coll, Crystal structure of a DNA Holliday junction, *Nat. Struct. Biol.* **6**, 913–917 (1999).
85. W. M. Shih, J. D. Quispe and G. F. Joyce, A 1.7-kilobase single-stranded DNA that folds into a nanoscale octahedron, *Nature* **427**, 618–621 (2004).
86. R. Holliday, Mechanism for gene conversion in fungi, *Genet. Res.* **5**, 282–304 (1964).
87. N. C. Seeman and N. R. Kallenbach, DNA branched junctions, *Annu. Rev. Biophys. Biomol. Struct.* **23**, 53–86 (1994).
88. D. M. J. Lilley, Structures and interactions of helical junctions in nucleic acids, *Oxford Handbook of Nucleic Acid Structure*, ed. S. Neidle (Oxford University Press, Oxford, 1999), pp. 471–498.
89. M. Zhong, M. S. Rashes and N. R. Kallenbach, Effect of T–T base mismatches on 3-arm DNA junctions, *Biochemistry* **32**, 6898–6907 (1993).
90. M. Zhong, M. S. Rashes, N. B. Leontis and N. R. Kallenbach, Effects of unpaired bases on the conformation and stability of 3-arm DNA junctions, *Biochemistry* **33**, 3660–3667 (1994).
91. N. B. Leontis, W. Kwok and J. S. Newman, Stability and structure of 3-way DNA junctions containing unpaired nucleotides, *Nucleic Acid. Res.* **19**, 759–766 (1991).
92. J. B. Welch, D. R. Duckett and D. M. J. Lilley, Structures of bulged 3-way DNA junctions, *Nucleic Acid. Res.* **21**, 4548–4555 (1993).
93. F. Stuhmeier, J. B. Welch, A. I. H. Murchie, D. M. J. Lilley and R. M. Clegg, Global structure of three-way DNA junctions with and without additional

unpaired bases: A fluorescence resonance energy transfer analysis, *Biochemistry* **36**, 13530–13538 (1997).
94. R. Assenberg, A. Weston, D. L. N. Cardy and K. R. Fox, Sequence-dependent folding of DNA three-way junctions, *Nucleic Acid. Res.* **30**, 5142–5150 (2002).
95. N. C. Seeman, Nucleic-acid junctions and lattices, *J. Theor. Biol.* **99**, 237–247 (1982).
96. N. C. Seeman, At the crossroads of chemistry, biology, and materials: Structural DNA nanotechnology, *Chem. Biol.* **10**, 1151–1159 (2003).
97. M. Zhong, M. S. Rashes, L. A. Marky and N. R. Kallenbach, T–T base mismatches enhance drug-binding at the branch site in a 4-arm DNA junction, *Biochemistry* **31**, 8064–8071 (1992).
98. D. Liu, M. Wang, Z. Deng, R. Walulu and C. Mao, Tensegrity: Construction of rigid DNA triangles with flexible four-arm DNA junctions, *J. Am. Chem. Soc.* **126**, 2324–2325 (2004).
99. C. Mao, W. Sun and N. C. Seeman, Assembly of Borromean rings from DNA, *Nature* **386**, 137–138 (1997).
100. N. C. Seeman, The use of branched DNA for nanoscale fabrication, *Nanotechnology* **2**, 149–159 (1991).
101. J. Chen and N. C. Seeman, Synthesis from DNA of a group with the connectivity of a cube, *Nature* **350**, 631–633 (1991).
102. H. Yan, S. H. Park, G. Finkelstein, J. H. Reif and T. H. LaBean, DNA-templated self-assembly of protein arrays and highly conductive nanowires, *Science* **301**, 1882–1884 (2003).
103. J. P. Cooper and P. J. Hagerman, Gel electrophoretic analysis of the geometry of a DNA four-way junction, *J. Mol. Biol.* **198**, 711–719 (1987).
104. K. Keren, M. Krueger, R. Gilad, G. Yoseph, U. Sivan and E. Braun, Sequence-specific molecular lithography on single DNA molecules, *Science* **297**, 72–76 (2002).
105. S. C. West, Enzymes and molecular mechanisms of genetic-recombination, *Annu. Rev. Biochem.* **61**, 603–640 (1992).
106. T. Koller, E. Di Capua and A. Stasiak, Complexes of RecA protein with single stranded DNA, *Mechanisms of DNA Replication and Recombination* (Alan R. Liss, Inc., New York, 1983).
107. M. M. Cox, Recombinational DNA repair in bacteria and the RecA protein, *Prog. Nucl. Acid. Res. Mol. Biol.* **63**, 311–366 (2000).
108. T. H. LaBean, H. Yan, J. Kopatsch, F. R. Liu, E. Winfree, J. H. Reif and N. C. Seeman, Construction, analysis, ligation, and self-assembly of DNA triple crossover complexes, *J. Am. Chem. Soc.* **122**, 1848–1860 (2000).
109. Y. Li, Y. D. Tseng, S. Y. Kwon, L. D'Espaux, J. S. Bunch, P. L. McEuen and D. Luo, Controlled assembly of dendrimer-like DNA, *Nat. Mater.* **3**, 38–42 (2004).
110. M. S. Shchepinov, K. U. Mir, J. K. Elder, M. D. Frank-Kamenetskii and E. M. Southern, Oligonucleotide dendrimers: Stable nano-structures, *Nucleic Acid. Res.* **27**, 3035–3041 (1999).

111. P. Tosch, C. Wälti, A. P. J. Middelberg and A. G. Davies, Generic technique to generate large branched DNA complexes, *Biomacromolecules* **7**, 677–681 (2006).
112. C. Bustamante, Z. Bryant and S. B. Smith, Ten years of tension: Single-molecule DNA mechanics, *Nature* **421**, 423–427 (2003).
113. G. Binnig and H. Rohrer, Scanning tunneling microscopy, *Helv. Phys. Acta* **55**, 726–735 (1982).
114. G. Binnig, C. F. Quate and C. Gerber, Atomic force microscope, *Phys. Rev. Lett.* **56**, 930–933 (1986).
115. W. Bowen, R. W. Lovitt and C. Wright, Application of atomic force microscopy to the study of micromechanical properties of biological materials, *Biotechnol. Lett.* **22**, 893–903 (2000).
116. E. Florin, V. Moy and H. Gaub, Adhesion force between individual ligand-receptor pairs, *Science* **264**, 415–417 (1994).
117. H. G. Hansma, Surface biology of DNA by atomic force microscopy, *Annu. Rev. Phys. Chem.* **52**, 71–96 (2001).
118. M. Rief, F. Oesterhelt, B. Heymann and H. E. Gaub, Single molecule force spectroscopy on polysaccharides by atomic force microscopy, *Science* **275**, 1295–1297 (1997).
119. M. L. Bennink, O. D. Scharer, R. Kanaar, K. Sakata-Sogawa, J. M. Schins, J. S. Kanger, B. G. de Grooth and J. Greve, Single-molecule manipulation of double-stranded DNA using optical tweezers: Interaction studies of DNA with RecA and YOYO-1, *Cytometry* **36**, 200–208 (1999).
120. M. D. Wang, H. Yin, R. Landick, J. Gelles and S. M. Block, Stretching DNA with optical tweezers, *Biophys. J.* **72**, 1335 (1997).
121. F. Amblard, B. Yurke, A. Pargellis and S. Leibler, A magnetic manipulator for studying local rheology and micromechanical properties of biological systems, *Rev. Sci. Instrum.* **67**, 1–10 (1996).
122. C. Gosse and V. Croquette, Magnetic tweezers: Micromanipulation and force measurement at the molecular level, *Biophys. J.* **82**, 3314–3329 (2002).
123. C. Haber and D. Wirtz, Magnetic tweezers for DNA micromanipulation, *Rev. Sci. Instrum.* **71**, 4561–4570 (2000).
124. T. R. Strick, J. F. Allemand, V. Croquette and D. Bensimon, Physical approaches to the study of DNA, *J. Stat. Phys.* **93**, 647–672 (1998).
125. T. T. Perkins, D. E. Smith, R. G. Larson and S. Chu, Stretching of a single tethered polymer in a uniform-flow, *Science* **268**, 83–87 (1995).
126. S. Smith, L. Finzi and C. Bustamante, Direct mechanical measurements of the elasticity of single DNA molecules by using magnetic beads, *Science* **258**, 1122–1126 (1992).
127. R. M. Zimmermann and E. C. Cox, DNA stretching on functionalized gold surfaces, *Nucleic Acid. Res.* **22**, 492–497 (1994).
128. H. A. Pohl, *Dielectrophoresis* (Cambridge University Press, Cambridge, 1978).
129. T. B. Jones, *Electromechanics of Particles* (Cambridge University Press, Cambridge, 1995).

130. M. P. Hughes, AC electrokinetics: Applications for nanotechnology, *Nanotechnology* **11**, 124 (2000).
131. A. E. Cohen, Nanoscale Mechanics, PhD thesis, University of Cambridge (2003).
132. H. Morgan and N. G. Green, Dielectrophoretic manipulation of rod-shaped viral particles, *J. Electrostat.* **42**, 279–293 (1997).
133. N. G. Green, A. Ramos and H. Morgan, AC electrokinetics: A survey of sub-micrometre particle dynamics, *J. Phys. D* **33**, 632–641 (2000).
134. N. G. Green and H. Morgan, Dielectrophoretic separation of nano-particles, *J. Phys. D* **30**, 41–44 (1997).
135. T. Muller, A. Gerardino, T. Schnelle, S. G. Shirley, F. Bordoni, G. de Gasperis, R. Leoni and G. Fuhr, Trapping of micrometre and sub-micrometre particles by high-frequency electric fields and hydrodynamic forces, *J. Phys. D* **29**, 340–349 (1996).
136. J. Gimsa, A comprehensive approach to electro-orientation, electrodeformation, dielectrophoresis, and electrorotation of ellipsoidal particles and biological cells, *Bioelectrochemistry* **54**, 23–31 (2001).
137. J. Suehiro and R. Pethig, The dielectrophoretic movement and positioning of a biological cell using a three-dimensional grid electrode system, *J. Phys. D* **31**, 3298–3305 (1998).
138. Y. Huang, K. L. Ewalt, M. Tirado, R. Haigis, A. Forster, D. Ackley, M. J. Heller, J. P. O'Connell and M. Krihak, Electric manipulation of bioparticles and macromolecules on microfabricated electrodes, *Anal. Chem.* **73**, 1549–1559 (2001).
139. G. H. Markx, P. A. Dyda and R. Pethig, Dielectrophoretic separation of bacteria using a conductivity gradient, *J. Biotechnol.* **51**, 175–180 (1996).
140. C. L. Asbury, A. H. Diercks and G. Van der Engh, Trapping of DNA by dielectrophoresis, *Electrophoresis* **23**, 2658–2666 (2002).
141. M. Washizu, S. Suzuki, O. Kurosawa and T. Nishizaka, Molecular dielectrophoresis of biopolymers, *IEEE Trans. Ind. Appl.* **30**, 835 (1994).
142. M. Washizu, O. Kurosawa, I. Arai, S. Suzuki and N. Shimamoto, Applications of electrostatic stretch-and-positioning of DNA, *IEEE Trans. Ind. Appl.* **31**, 447–456 (1995).
143. T. Yamamoto, O. Kurosawa, H. Kabata, N. Shimamoto and M. Washizu, Molecular surgery of DNA based on electrostatic micromanipulation, *IEEE Trans. Ind. Appl.* **36**, 1010–1017 (2000).
144. H. Oana, M. Ueda and M. Washizu, Visualization of a specific sequence on a single large DNA molecule using fluorescence microscopy based on a new DNA-stretching method, *Biochem. Biophys. Res. Co.* **265**, 140–143 (1999).
145. H. Kabata, W. Okada and M. Washizu, Single-molecule dynamics of the EcoR1 enzyme using stretched DNA: Its application to *in situ* sliding assay and optical DNA mapping, *Jpn. J. Appl. Phys.* **39**, 7164–7171 (2000).

146. W. A. Germishuizen, C. Wälti, R. Wirtz, M. B. Johnston, M. Pepper, A. G. Davies and A. P. J. Middelberg, Selective dielectrophoretic manipulation of surface-immobilized DNA molecules, *Nanotechnology* **14**, 896–902 (2003).
147. W. A. Germishuizen, P. Tosch, A. P. J. Middelberg, C. Wälti, A. G. Davies, R. Wirtz and M. Pepper, Influence of alternating current electrokinetic forces and torque on the elongation of immobilized DNA, *J. Appl. Phys.* **97**, 014702/1–7 (2005).
148. C. Wälti, P. Tosch, A. G. Davies, W. A. Germishuizen and C. F. Kaminski, Establishment of the ac electrokinetic elongation mechanism of DNA by three-dimensional fluorescent imaging, *Appl. Phys. Lett.* **88**, 153901/1–3 (2006).
149. A. E. Cohen, Force-extension curve of a polymer in a high-frequency electric field, *Phys. Rev. Lett.* **91**, 235506/1–4 (2003).
150. T. R. Strick, M. Dessinges, G. Charvin, N. H. Dekker, J. F. Allemand, D. Bensimon and V. Croquette, Stretching of macromolecules and proteins, *Rep. Prog. Phys.* **66**, 1–45 (2003).
151. A. Ramos, H. Morgan, N. G. Green and A. Castellanos, AC electrokinetics: A review of forces in microelectrode structures, *J. Phys. D* **31**, 2338–2353 (1998).
152. H. Morgan and N. G. Green, *AC Electrokinetics: Colloids and Nanoparticles* (Research Studies Press, Baldock, Hertfordshire, UK, 2003).
153. R. Paul and K. Kaler, Theory of electrode polarization in dielectrophoresis and electrorotation, *J. Colloid Interf. Sci.* **194**, 225–238 (1997).
154. B. Ladoux and P. S. Doyle, Stretching tethered DNA chains in shear flow, *Europhys. Lett.* **52**, 511–517 (2000).

Christoph Wälti studied at ETH Zürich, Switzerland, where he graduated with a degree in Physics in 1995. He obtained his PhD from the Laboratory for Solid State Physics, also at ETH Zürich, in 2000. After a short period as a postdoctoral research assistant at the same institution, he was awarded a Swiss-Marie-Curie Fellowship which allowed him to explore new challenges in the field of bionanotechnology. He moved to the University of Cambridge in 2001 where he spent a few years as a PostDoctoral Research Fellow at the Cavendish Laboratory and the Department of Chemical Engineering. In 2003, Christoph was awarded an EPSRC Advanced Research Fellowship and in 2004 he moved to the University of Leeds, where he is based in the School of Electronic and Electrical Engineering. Christoph's scientific interests are in the area of molecular nanotechnology and bioelectronics with a particular focus on the development of techniques that interface the biological with the electronics world. Christoph is married and enjoys photography and skiing.

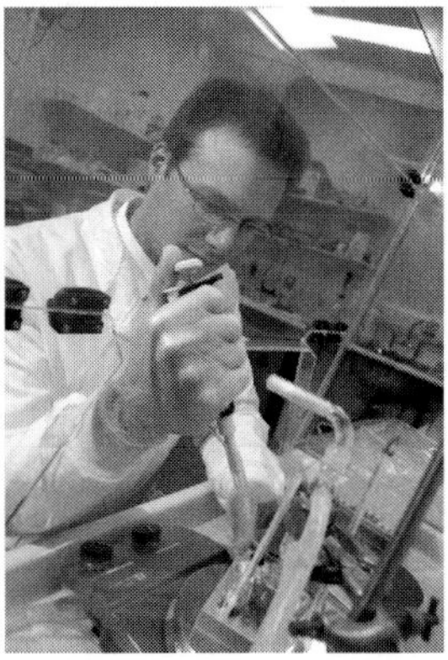

CHAPTER 7

EXPLORING TUNNEL TRANSPORT THROUGH PROTEIN AT THE MOLECULAR LEVEL

Jason J. Davis*, Nan Wang, Wang Xi, and Jianwei Zhao

Chemistry Research Laboratory, Department of Chemistry
University of Oxford
Mansfield Road, Oxford, OX1 3TA, UK

Significant advances in molecular electronics require an ability to reliably analyze the relationship between molecular and electronic characteristics. The spatial resolution of conductive proximal probes has much to offer in this regard. Previous work has shown that protein molecules can be controllably assembled on planar electrode surfaces and subsequently trapped between the conductive surfaces presented by a metal-coated AFM probe and an underlying planar substrate. This chapter reviews our recent efforts in analyzing the conductance of these metalloprotein junctions and includes discussions of relevant work in the area.

Keywords: Metalloprotein, tunneling, molecular electronics, conductive probe, AFM.

1. Introduction

The field of molecular electronics can be broadly defined as an attempt to utilize molecular components in the construction of electronic circuitry. It not only represents a defining technological stage in the miniaturization of computing components, but also provides a promising new method for high-speed signal processing and communication, novel associative and neural architecture, and miniaturized, high-sensitivity sensory devices.

*Corresponding author.

In recent years much progress, largely driven by advances in self-assembly and microfabrication, has been made in analyzing the characteristics of nanoscale electronic junctions into which molecules of interest have been deposited.

From both a fundamental and potentially applied sensory perspective, the integration of native or engineered biomolecules into nanometer-scale circuitry is of considerable interest. The field of bioelectronics has, until very recently, been largely dominated by the prospects of high molecular conductance in deoxyribonucleic acid (DNA) and the subsequent intense debate this has generated.[1,2] The difficulties associated with carrying out reliable transport measurements on structurally complex species have been spectacularly highlighted by the variant results obtained with this, comparatively simple, biomolecular structure.[1,3,4]

The primary role played by transition metal containing proteins, metalloproteins, in facilitating facile current flow through biological media makes a direct analysis of their electronic properties intriguing. In many, the molecular structure is specifically designed so as to facilitate both efficient electron tunneling and highly specific recognition of partner species; the latter may feasibly be utilized in programmed "bottom-up" assembly. In many specific cases, the metal prosthetic group exists in two stable oxidation states (i.e., redox switchable), and can be substituted with negligible change in geometric structure. This facilitates studies in which the role of this center can be defined and may provide a mechanism by which molecular conductance is controllable (switchable). The surface immobilization of these molecules may be assayed by spectroscopic (UV/visible, ellipsometry, internal reflection absorption and fluorescence, IR and Raman, cyclic voltammetry) or mass-sensing technologies such as those based on quartz crystal microbalance (QCM) or microcantilever fabrication. These surface and electron transfer assays rely on bulk measurements and are, thus, inherently averaging. In considering the generation of "molecular devices" it is important that we understand not only the stochastic characteristics inherent in a molecular population, but also the influence of environment on these. Such information can only be acquired through experiments at the molecular scale. Here we review our efforts to analyze the conductance characteristics of metalloproteins at a molecularly resolved scale by, principally, conductive probe atomic force microscopy (CP-AFM) and electrochemical scanning tunneling microscopy (EC-STM).

2. Molecular Electronics

Solid-state metal–insulator–metal (MIM) junctions, where electrons flow between (usually) metallic surfaces separated by insulating films, form the basis of light-emitting diodes[5] and field-effect transistors,[6,7] and are the most controlled means of scrutinizing conductance on this scale. There are, in essence, two different approaches to the construction of MIM junctions. One is lithographic in nature and results in planar junctions set on an insulating surface (and will be discussed no further here) and one is vertical, constructed within the confines of an appropriately configured scanning probe microscope.

The scanning tunneling microscope (STM) consists, in essence, of a metallic wire electrode (ultimately terminating in a single atom) electronically coupled to a planar conductive substrate. By mounting either the tip or the underlying substrate on a voltage-responsive piezoelectric ceramic crystal (piezo), it is possible to laterally scan one with respect to the other with exceedingly good accuracy. In simultaneously measuring the current generated by electrons tunneling between the two surfaces when they approach each other sufficiently closely ($< 3\,\text{nm}$), it is possible to both generate atomic/molecular level surface images and analyze the effect of surface-mounted molecules on this passage of electrons. The latter offers a potentially powerful means of scrutinizing the electronic properties of single molecules under a variety of controllable conditions. Contrast in adsorbate STM images has been found to be dominated by topography unless electronic effects are strong. This arises largely because of the extreme (exponential) sensitivity of the current to minute changes within the confines of the tunnel gap; when the tip scans over "high" regions of a molecule, the current transiently increases since the tunneling decay coefficient through the structure is likely to be less than that of the "empty" (vacuum, air, or fluid-filled) gap — this leads to "positive contrast" in that the adsorbate appears (correctly) to be higher than the substrate surface on which it sits. Though biomolecules are classically insulators (in the sense that they contain no energetically accessible delocalized electron density), it is possible to generate currents "through" them during the process of data acquisition and thereby to obtain high (molecular-level) resolution images (see Figs. 1 and 2).[8] If one wishes to characterize samples under electrolytic solution (while observing dynamic or electrochemical processes, or simply maintaining the immobilized molecules in a more fully hydrated state), then it is

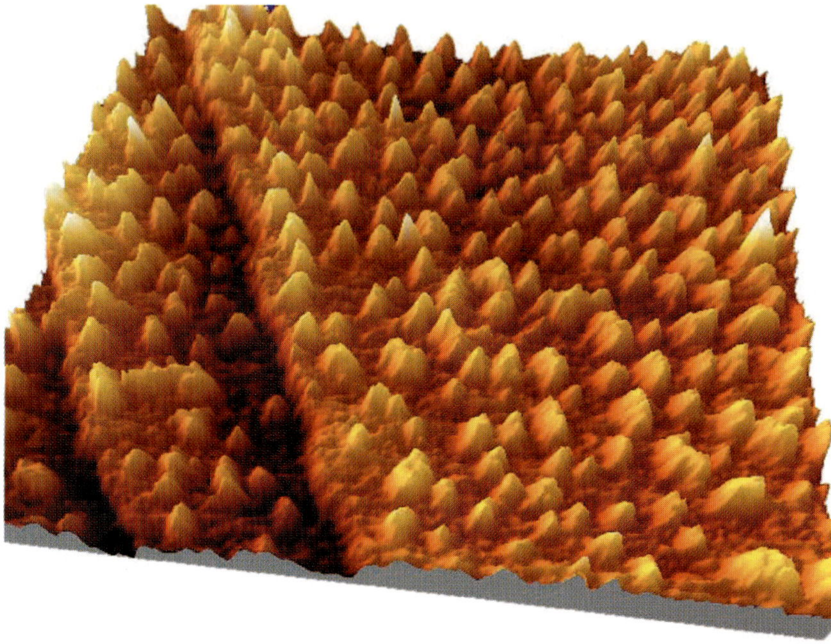

Fig. 1. Tunneling image of an array of self-assembled blue copper proteins on an Au[III] electrode surface. The image (approximately 140 nm × 140 nm) was acquired under a water-glycerol mix at room temperature at a tunneling set point of 75 pA (200 mV). These robust molecular layers are generated by using the strong thiol–gold-bonding interactions attainable by adding, genetically, cysteine amino acids to the surface of the metalloprotein. The surface density of these layers can be measured by voltammetric methods to be 1–2 × 10^{13} molecules/cm^2.

necessary to insulate the imaging tip (tunneling currents are typically in the nanoampere range and this would be swamped by capacitative current generated should a bare wire be exposed to electrolyte) with polymeric or waxen material.[9]

Though capable of extremely high spatial resolution and electronic measurement under both fluid and electrolyte, the analyses available from these STM configurations are somewhat complicated by the largely unknown relative positions (and so interactions) between probe and molecule of interest. Atomic force microscopes (AFMs) operate through a calibrated force interaction between probe and surface and, in doing so, facilitate studies where potential molecular perturbations may be controlled and, indeed, quantified. If the AFM probe is coated with a metallic film (usually evaporatively), it may be utilized as an electrode and reliably be brought into

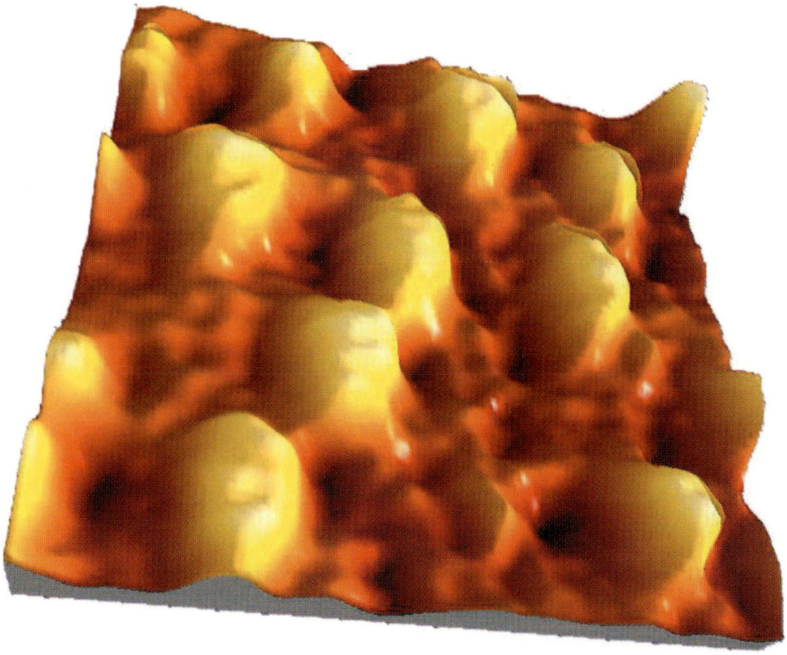

Fig. 2. *In situ* electrochemical STM (10-mM phosphate-buffered saline, pH 7.5; − 0.3 V, 300 pA) of yeast iso-cytochrome *c* molecules adsorbed onto a bare gold electrode surface. Each molecule is approximately 3 nm in diameter and contains one redox-addressable heem group. The role played by the latter in mediating tunneling, under appropriate experimental conditions, can be utilized in "gating" conductance electrochemically (Fig. 3).

mechanical contact with the top surface of orientated and immobilized molecules. As is inherent in AFM, the compressional forces imparted to molecules can be monitored with a high level of precision. This is likely to be particularly important in regard to the interpretation of transport measurements; there is no reason to expect the electronic structure of molecules which are both surface confined and under many atmospheres of imposed pressure (the pressure under an AFM probe or an evaporated metallic film *can* run into several thousands of atmospheres) to be unaltered from those of the "native" structure. In such devices a single or low number of molecules is/are sandwiched between the tip electrode and an underlying planar electrode. The electrical characteristics, namely current–voltage profiles, of such junctions can then be analyzed under conditions of controlled molecular compressibility (imposed tip pressure). At higher force the probe physically penetrates surface-confined layers, with a resulting sharp drop in measured resistance. Both to minimize electrolytic/ionic contributions to current and to eliminate the possibility of unwanted increases in applied force, it is necessary to carry out these measurements under either nonionic fluid (organic solvent) or low-humidity ambient or inert atmosphere.

Such CP-AFM configurations have enabled very low-current measurements to be performed under a variety of conditions on surface-confined adsorbates of interest.[10,11] These have included carbon nanotubes (CNTs),[12] organic crystals/molecules,[11,13] and gold nanowires.[14,15] To reliably assay *molecular* properties (as opposed to the electronic characteristics associated with the experimental configuration), it is highly beneficial to employ more than one investigative approach. We will focus herein on CP-AFM and EC-STM junctions.

3. Assembling Proteins at Electroactive Surfaces

To acquire topological or spectroscopic information on single (bio) molecules in real time, these molecules must be stationary, at least on the "timescale of interaction" with the measuring probe. To achieve this, one can tap into the great deal of effort which has been invested in developing methods whereby biomolecular structure can be robustly but nondestructively immobilized. Specific control is achievable by covalently tethering biomolecules through either chemical modification of the underlying substrate surface or of the biomolecule itself. The unchecked physical adsorption of biomolecules at metallic interfaces, a process governed by a complex interplay of (primarily) electrostatic, hydrophobic, and dispersion forces, typically occurs with a loss of native structure and so biological function.

The blue copper protein, azurin, plays an important role in respiratory and photosynthetic electron transport chains,[16] and has been the subject of numerous bioelectroanalytical studies.[17–20] The protein contains a surface-exposed disulfide moiety which may be utilized in robustly anchoring the molecule to gold electrode surfaces. A more directable immobilization may similarly be achieved by engineering cysteine residues into the surface, facilitating the formation of high-density redox-addressable monolayers on gold surfaces; these may be electrically or topographically addressed with high spatial resolution (Fig. 1).[18,19,21]

4. Protein Tunnel Transport Probed in an STM Junction

The efficiency with which metalloproteins, in particular, mediate electron tunneling lies central to sustaining life-dependent processes such as aerobic respiration and photosynthesis. Facile electron transfer reactions between donor and acceptor structures (typically both redox-active sites) separated by distances well beyond noncovalent atomic contact distances (van der Waals' contact) are ubiquitous in nature. In such circumstances, the weak electronic coupling is mediated (usually nonadiabatically; the energy barrier to electronic transfer greatly exceeds thermal energy) by the intervening protein medium and is dependent on two terms: (i) an electronic term exponentially variant on distance and characterized by a decay constant, β and (ii) a thermodynamic term largely representative of the free-energy-driving force associated with electron transfer. In recent years, two methods by which the electronic coupling parameter is calculated have been proposed and vigorously discussed. The first is known as the "tunneling pathway model" in which the electron movement is considered to be confined to specific through-bond and through-space routes (each with a specific tunneling decay, see below). In the second, "average packing density" model, specifically optimized routes are deemed to be less important than the total tunneling distance. The interested reader may refer to the work of Page *et al.* and Regan and Onuchic.[22,23]

In the classical picture of electron tunneling across a square barrier, transmission probability (and therefore current) scales exponentially according to the relation transmission directly proportionate to $e^{-\beta L}$ where β is known as the tunneling decay constant and L is the barrier width (e.g., the fully extended molecular height in a self-assembled monolayer, SAM, a two-dimensional (2D) film, one molecule thick, assembled covalently or noncovalently at an interface). The greater the value of β for any particular medium, the greater the fall-off in current with distance, that is, the less "electronically conductive" the medium. Values of β have been reliably determined for a number of systems: through space (no medium) $\beta \sim 3.4\,\text{Å}$ and alkyl hydrocarbon $\beta \sim 1.0\,\text{Å}$.[24] If one treats a protein molecule as single homogenous entity composed of α-helical and β-sheet components, the decay coefficients concerned have been determined to be $\beta \sim 1.4\,\text{Å}$ and $\beta \sim 1.1\,\text{Å}$, respectively.[22,25,26] For unsaturated or delocalized systems tunneling decays are, predictably, as low as $0.1\,\text{Å}$.[27] The protein matrix, thus, exhibits tunneling characteristics comparable to those of a saturated hydrocarbon, though the perturbation of this model when metallic centers are added, be those redox-active or not, is unclear. In summary, then, though protein-based ion channels are able to mediate large ionic current flow, protein molecules themselves are (classically) large band-gap insulators with β values and resistivities in the range of 1.1–1.4 Å and 10^{15}–10^{18} Ωm, respectively.

In recent years, several groups have demonstrated the ability to attain high-contrast tunneling images of metalloproteins and enzymes under a variety of controllable conditions.[8,17,28,29] Though both natural and probe-induced conformational motion are resolution limiting, the tunneling profiles of individual molecules can be resolved under controllable electrolytic conditions (Fig. 2). The potential ability of metal centers to modulate conductance in these configurations has been of some considerable interest.[30] Under conditions of appropriate surface potential and robust electronic coupling, there exist good reasons to believe that the electronic conductance of these proteins be both significant and potentiostatically gateable.[17,30,31] Despite this, in most cases, the role of the metal in imaging across a broad range of potentials has been unclear. Figure 3(b) shows

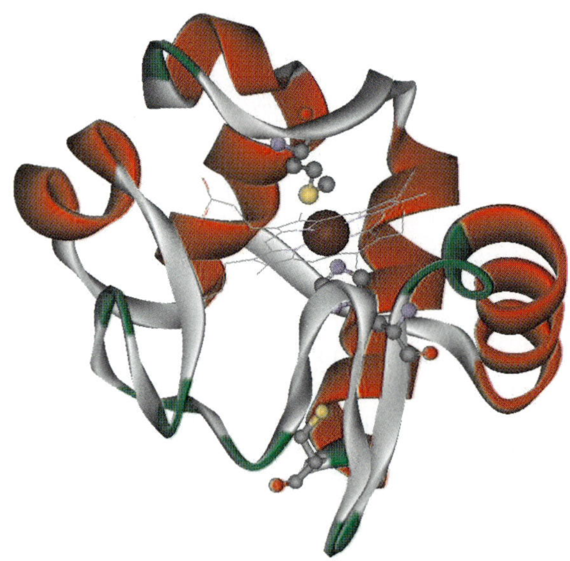

Fig. 3. (a) Structure of yeast iso-cytochrome *c* (pdb IC:1YCC, Ref. 32). The ~3.5-nm diameter, 13-kDa, protein has a heem group which is electrochemically switchable. This particular form of cytochrome *c* has a solvent-exposed cysteine residue which may usefully be utilized in anchoring the molecule to gold- or sulphur-presenting surfaces. The distance between the thiol and the buried edge of the heem is approximately 1.6 nm.

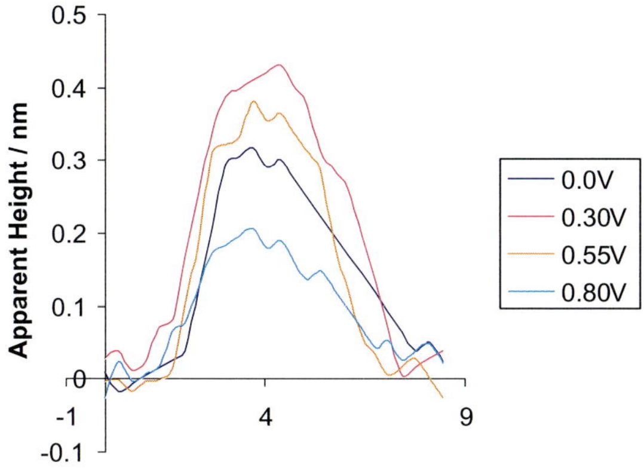

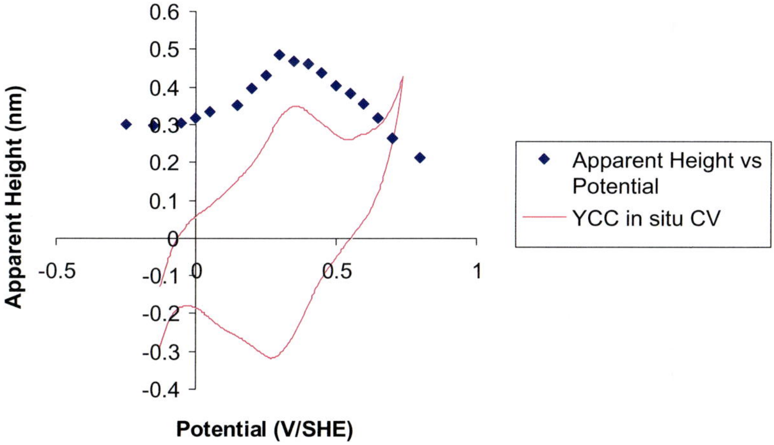

Fig. 3. (b) (Top) Graphic representation of the apparent molecular height/cross-section (averaged over 20 individual molecules) of iso-cytochrome *c* molecules immobilized on a gold, as a function of underlying electrode, surface potential. The latter is modulated with respect to a standard electrochemical reference through the use of a bipotentiostat. The relationship between the underlying surface potential at which molecular height (and so "conductance") is greatest and the electrochemically derived response is clear (bottom). YCC is yeast cytochrome *c*. Potential (V/SHE) indicates the voltage axes, which denote the electrode potential, in volts, reference to the standard hydrogen electrode.

the variance in STM-derived height of a molecule of iso-yeast cytochrome *c* (Fig. 3(a)) immobilized on a gold electrode surface under phosphate-buffered saline. As is evident in the figure, the current through the molecule, at fixed bias, *is* gateable by modulating the underlying gold electrode surface potential but the effect of surface potential is relatively small (the change in current is some way short of an order of magnitude). If the redox accessible iron-based orbitals play a dominant role in conductance, then one might expect the apparent conductance modulation to be significantly larger. These and other observations are consistent with current transport being predominantly nonresonant under most experimental conditions and appears to be consistent with observations made in CP-AFM configurations (see below) where direct tip–substrate current flow is dominant.[30,33] It is highly likely that the role of the metal center depends very sensitively on the degree to which the electrodes about it are able to couple to its density of states. It is noteworthy, then, that good voltammetric responses are obtained only with difficulty with these proteins on bare electrode surfaces (Fig. 3(b)). Extrapolation may lead to a suggestion that, under conditions where facile and robust voltammetry (electrode–redox site coupling) can be achieved, for example, by immobilizing the protein on a thiol-based adlayer, the role of the redox-switchable group may be considerably greater. A recent report by Ulstrup *et al.* is supportive of this concept.[34]

5. Assaying Protein Conductance in CP-AFM Configurations

A schematic representation of a CP-AFM configuration is shown in Fig. 4. Such setups offer both highly spatially resolved electronic characterization and independent force calibration. The latter facilitates studies into the effects of mechanical perturbation on the charge transport characteristics of confined molecules.[35]

5.1. *Tunnel transport under conditions of low to moderate load*

Current–voltage profiles recorded under conditions where the metallic probe contacts azurin molecules with forces $< 3\,\mathrm{nN}$ show negligible current flow at bias voltages $< 4\,\mathrm{V}$. At higher voltages, charge accumulation leading, ultimately, to dielectric breakdown, occurs. Assuming minimal protein compression at these forces, the breakdown voltage observations can lead to an approximation of the dielectric strength of the protein at between 1.0 and 1.4 GV/m. This range compares well to that shown by silicon dioxide, SiO_2 (0.8–1.3 GV/m),[36] being somewhat smaller than the 2.0 GV/m

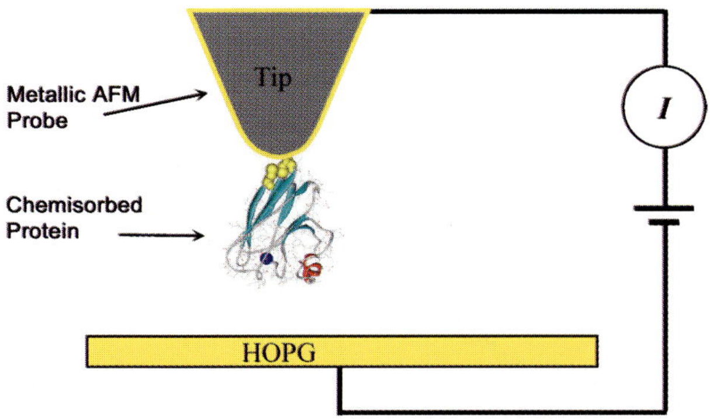

Fig. 4. Schematic representation of a conducting probe–protein–HOPG junction. In a typical CP-AFM configuration, a bias voltage is applied between a metal-coated AFM probe and the underlying planar substrate. By adsorbing a molecule of interest on either surface, and then bringing them together while monitoring lever deflection, it is possible to establish robust junctions from which the conductance properties of the "trapped" molecule can be established. In this case, the protein is chemisorbed onto the probe. HOPG (highly-orientated pyrolytic graphite) is an atomically flat graphite sheet at which the hydrophobic basal surfaces are exposed.

determination for alkyl monolayers on either metal[37] or semiconductor,[38] surfaces but greater than the values determined for alkyl bilayers sandwiched between a mercury drop and silver plate ($\sim 0.5\,\mathrm{GV/m}$).[39]

On slightly increasing the mechanical interaction between the electrode and protein (by increasing the calibrated vertical force applied) to $> 3\,\mathrm{nN}$, stable tunnel transport can be maintained.[30,35,40] In the low-bias regime ($\pm 0.5\,\mathrm{V}$) the current–voltage relationship is approximately linear and can be used in an estimation of molecular resistance at specific imposed force (Fig. 5). At the minimal forces required to establish tunneling without breakdown (3–5 nN) zero-bias resistances are observed to be 45–60 GΩ, a range bearing some similarity to tunnel resistance set points used in the nondestructive STM imaging of these (and comparable) molecular adsorbates.[8,9,41] The resistance across hexanethiol, octanethiol, decanethiol, and dodecanethiol SAMs obtained under equivalent conditions of vertical force and humidity are $140 \pm 50\,\mathrm{M\Omega}$, $1.5 \pm 0.7\,\mathrm{G\Omega}$, $10 \pm 5\,\mathrm{G\Omega}$, and $40 \pm 8\,\mathrm{G\Omega}$, respectively, values bearing close similarity to those previously determined (Fig. 6).[42,43] Fitting these resistance variations to an exponential tunneling equation, $R = R_0 \exp(\beta d)$ (where R_0 is an effective contact resistance and d is the junction length), allows the exponential distance decay factor β to be determined as 1.2 ± 0.1 per methylene

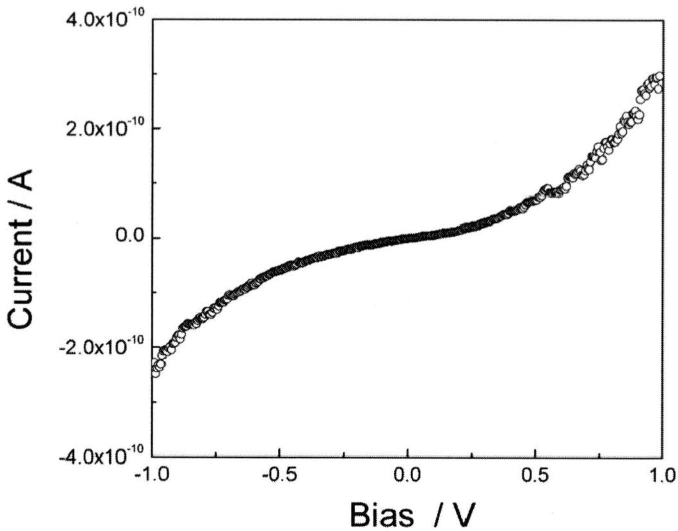

Fig. 5. A representative current–voltage curve obtained at a scan speed of 4 V/s and a junction force of 6 nN. Though the current shows some asymmetry about zero bias, this is minimal ($< 5\%$).

group, a value in reasonable agreement with previous reports under comparable experimental conditions.[41] To rule out the possibility that current could be dominated by conduction across the surface of a protein-associated water film in the case of the azurin assays, the effects of humidity on conductance were analyzed. Within a 20–60% humidity range junction resistance values varied by $< 10\%$. In view of the stability and reproducibility of these observations, and consideration of the relative magnitudes of tip radius and protein diameter, it is likely that these observations are those associated with single protein molecule tunnel transport (see analysis below).

Under voltage conditions where the traversing electrons will have insufficient energy to access molecular orbitals on the molecule, the tunneling process is nonresonant (i.e., the electrons are not energetically resonant with orbital states) and well described by the superexchange model in which the electron transfer rate decreases exponentially with distance.[44] Assuming that nonresonant superexchange tunneling is the dominant mechanism of charge transfer across these junctions (see below), the greater resistance of the protein junctions over the saturated alkyl SAMs may be assignable to differences in both molecular height (tunnel distance) and tunnel barrier.

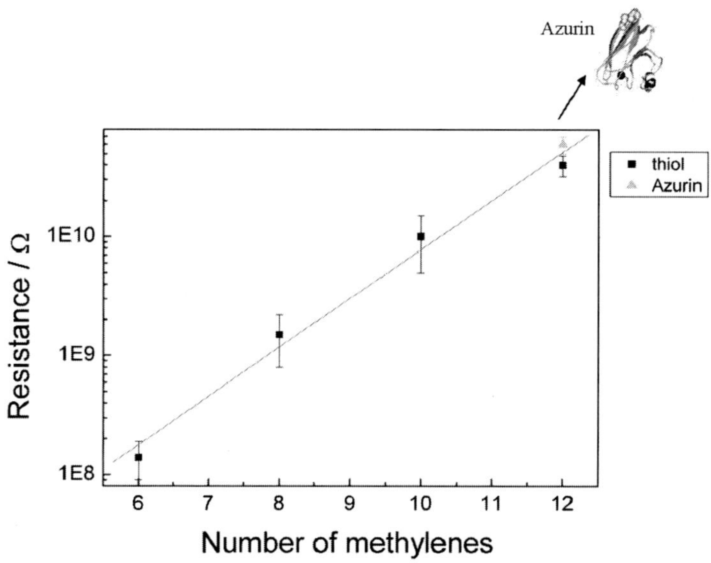

Fig. 6. Comparative force–resistance profiles of azurin and a range of alkyl thiol SAMs. Data was obtained at junction forces in the 5–7 nN range at 40–50% humidity. Extrapolation of the data to zero chain length gives a contact resistance of some 100 kΩ. The protein conductance, specified by the green triangle, is approximately equivalent to that of a C_{12}-alkyl thiol under these experimental conditions.

If one assumes a tunnel barrier comparable to that through the saturated alkyl systems[45] (see below), extrapolation of the data in Fig. 6 leads to an estimated protein tunneling distance of approximately 1.5 nm, suggesting that, even in this force regime, appreciable molecular compression takes place. The validity of this may be directly assessed through controlled force molecular imaging. Figure 7 shows the force-induced variation in azurin height as determined by fluid-phase (contact mode) AFM imaging under aqueous solution. Though the data shows some scatter, it is clear that the protein is compressed to something approaching half of its unperturbed height within force limits required in the establishment of reliable electrical contact. At the minimum force required for stable image acquisition (~ 300 pN), the measured protein height is ~ 2.8 nm, a value lying close to its crystallographic size.

A simple theoretical analysis of transport through these junctions has been carried out previously.[40,46] The electron transfer between two conducting electrodes separated by a thin dielectric layer can be well described by the Simmons model[47] in which current density, i, is

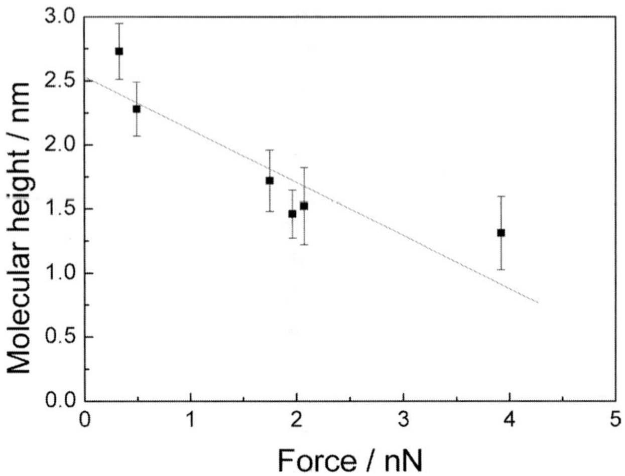

Fig. 7. The variation in molecular height with imaging force. Data taken from contact-mode imaging under 18.2 MΩ cm water, with silicon nitride tip ($k = 0.06$). A $\sim$5-μM solution of azurin in acetate buffer (pH 4.65) was spin-coated onto a freshly annealed gold substrate, then rinsed with buffer and dried under nitrogen. Protein height was averaged over 90 individual molecules, with standard deviations below 10%.

given by:

$$i = \frac{e^2}{2\pi h L^2}\left\{\left(\varphi_0 - \frac{V}{2}\right)\exp\left[-K\left(\varphi_0 - \frac{V}{2}\right)^{1/2}\right] - \left(\varphi_0 + \frac{V}{2}\right)\exp\left[-K\left(\varphi_0 + \frac{V}{2}\right)^{1/2}\right]\right\}, \quad (1)$$

where

$$K = \frac{4\pi L}{h}(2me)^{1/2} \quad (2)$$

and L is the barrier length, φ_0 is the mean barrier height, and e and m are the electron charge and mass, respectively. Due to the inherent asymmetry of conductive probe (CP-AFM) junctions, a more realistic treatment is to divide the voltage drop into two nonequivalent components, in which case the net current density is given by:

$$i = \frac{e^2}{2\pi h L^2}\{(\varphi_0 - \alpha V)\exp[-K(\varphi_0 - \alpha V)^{1/2}] - [\varphi_0 + (1-\alpha)V] \times \exp[-K(\varphi_0 + (1-\alpha)V)^{1/2}]\}\,. \quad (3)$$

Equation (3) provides a relationship between the variance of tunneling current with bias and three structure-related variables, namely, φ_0, L, and the asymmetry factor α.

A comparison of experimental and simulated current–voltage data is shown in Fig. 8. At low force the data are highly symmetrical about the voltage origin (asymmetry typically $< 4\%$) and well described by Eq. (1) (the simulated data tracks the experimentally observed current variations to within 0.5% across the associated voltage range, demonstrating the applicability of a three-variable Simmons model to such configurations).[46] Best fits across 20 data sets at $\sim 6\,\text{nN}$ are obtained with $\alpha = 0.5$, $L = 12\,\text{Å}$, and $\varphi_0 = 1.1\,\text{eV}$. This barrier height corresponds to a tunneling coefficient (β) of 1.1 Å, a value which compares reasonably well with those derived from bulk, solution-phase, studies.[20,22,26] As we have discussed previously, current–voltage profiles become noticeably more asymmetric on excursion to regions of higher compressional force where Eq. (3) becomes more appropriate.[46] In conclusion, CP-AFM configurations facilitate a direct analysis of protein conductance well describable by a nonresonant Simmons tunneling model in which the molecule is treated as a structureless dielectric. *In situ* imaging and simulation studies suggest that stable transport is only achievable under conditions of considerable molecular compression.

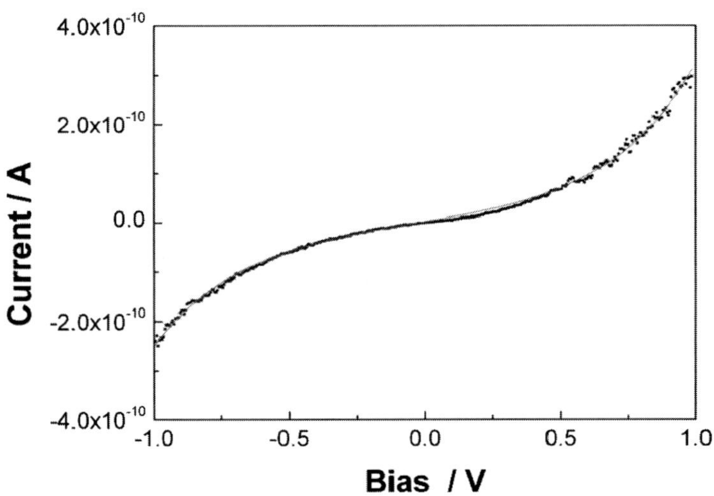

Fig. 8. A representative comparison of experimental (black dots) and Simmons simulated (line) current–voltage characteristics at 6 nN. The fit is smooth and good (relative fitting error is $\pm 0.5\%$) implying that the conductance properties of this "metal–insulator–metal" junction are well described by the Simmons model.

5.2. *Modulation of protein conductance under moderate load*

As noted earlier, the inherent mechanical flexibility of AFM-based molecular junctions can be utilized in an examination of compression-based changes in conductance. In the case of azurin CP-AFM configurations, stable and reproducible tunneling characteristics are attainable across a 3–80 nN range enabling the effects of variant applied load on the conductance of an azurin metalloprotein junction to be examined (Figs. 9 and 10). The highly reproducible decrease in resistance with load can, in the first instance, be ascribed to either a load-dependent decrease in tunneling barrier height, or tunneling distance (or a combination of both).

Though it is impossible to deconvolute the force-dependent contribution of probe–probe contact area to the charge transfer characteristics of these junctions, variation in this parameter is expected to bring with it only minimal change in current.[46] Though the predictable variations in tunnel distance, L, with increasing load, can account for some of the conductance modulation observed, we have been unable to simulate these changes through variance of L or φ_0 alone. In considering barrier height changes under compression (Table 1) it is useful to make reference to previous work in which metalloprotein distance decay parameters have been determined in bulk, solution-phase, experiments.

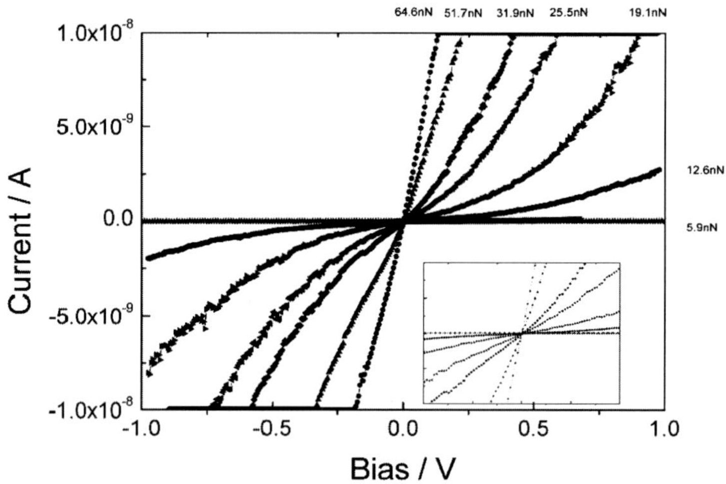

Fig. 9. Representative current–voltage behavior obtained in the -1 to 1 V bias window (scan speed 4 V/s) at various applied forces. Inset shows current–voltage traces in the low ($\pm$ 100 mV) bias regime.

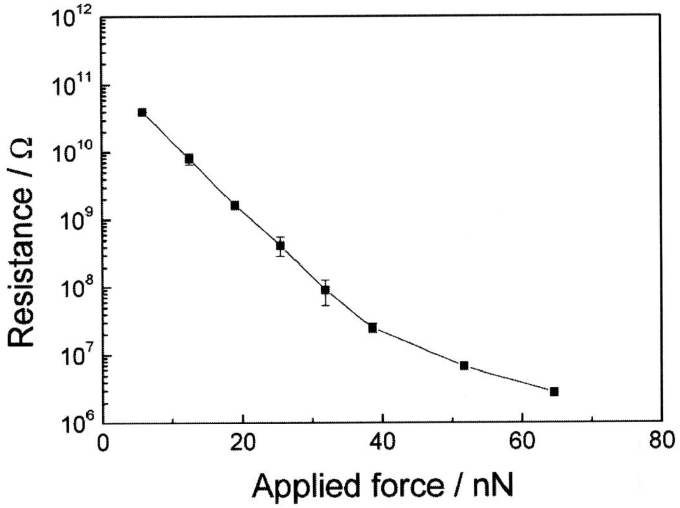

Fig. 10. Force-dependent junction resistance as evaluated from the slopes of 20 current–voltage curves at varying forces (5.9, 12.6, 19.1, 25.5, 31.9, 38.6, 51.7, and 64.6 nN) in the low-bias regime.

Table 1. Barrier height and length variations with imposed force as obtained from nonlinear least squares fitting of current–voltage profiles by a Simmons model.

Force (nN)	φ_0 (eV)	L (Å)
5.9	1.11	11.9
12.6	1.02	11.0
19.1	1.00	10.3
25.5	0.92	9.6
31.5	0.78	9.0

Dutton *et al.* have considered the effects of atomic packing, quantified by an atomic packing density coefficient, ρ, on the ability of a protein matrix to mediate electron transfer, and given an empirical expression to account for structure-dependent variations of the tunneling coefficient, β,[22]

$$\beta = \rho \times 0.9 + (1 - \rho) \times 2.8\,. \tag{4}$$

The association of electron transfer barrier height, φ_0 (in eV) with tunnel decay coefficient, β in Å, is as through Eq. (6)[47,48]:

$$\beta = 1.025\varphi_0^{1/2}\,. \tag{5}$$

In principle ρ can range from 1, corresponding to a fully packed medium ($\beta \sim 0.9$ Å), to a value of 0, corresponding to the interstitial space in the protein structure ($\beta = 2.8$ Å).[22,25,49,50] φ_0 is therefore correlated with the atomic packing density through the following equation:

$$\varphi_0 = (\rho \times 0.88 + (1 - \rho) \times 2.73)^2 , \tag{6}$$

that is, the higher the packing density, the lower the barrier height. Under large constant force, the protein is compressed and this leads to an associated increase in atomic packing density and decrease in barrier height. In the regime of large molecular compression counterbalancing repulsive forces may lead to structural deformation in the horizontal plane where there is no geometric restriction. The associated increase in protein volume will be associated with a regime of decreasing atomic packing density or increasing barrier height. The trends here are qualitatively predictable, supported by molecular-dynamics simulations and have been previously outlined.[46] In summary, a Simmons analysis of the force-dependent current–voltage characteristics has suggested that the tunnel barrier and length vary from 1.1 to 0.7 eV and 12 to 9 Å, respectively, across a force range of 5 to 40 nN. In the regime of greater molecular compression, current–voltage profiles are steep and cannot be adequately used in fitting.

5.3. *Accessing the metallic states: Negative differential resistance*

Much of the previously discussed data is supportive of electrons moving through protein tunnel junctions by a predominantly nonresonant mechanism. Though azurin contains a redox-accessible center in these immobilized configurations,[19] the current–voltage plots in the medium (± 1 V) bias regime discussed above are fully consistent with a nonresonant tunneling process between the metallic contacts in which protein-based molecular states are not accessed at the bias voltages applied (no "molecular features" are evident in either raw data or derivative current–voltage plots in an analysis of many hundreds of scans). These spectroscopic characteristics are broadly supportive of observations made in a tunneling imaging mode; though increased current flow in the vicinity of the metal center has occasionally been reported in metalloprotein tunneling imaging, in our studies image contrast has been largely independent of overpotential across a broad range.[29,51,52] The apparent dominance of nonresonant direct tip–substrate transport in CP-AFM or EC-STM junctions is likely to be related to either marked molecular compression or comparatively poor electronic coupling between the buried redox moieties and the electrodes. In CP-AFM configurations, in particular, it is clear that, under force conditions required to establish stable conductance, both tunnel distance and barrier may be reduced to comparatively low values. Direct tunneling is then likely to be facile. At reduced applied force (greater tunnel distance) it is reasonable to suggest that direct tunneling will become less efficient.

Despite these observations and comments, a considerable number of tunneling spectroscopy studies have now been published in which the ability of tunnel electrons to access molecular states in an STM configuration has, *apparently*, been evident. In tunneling spectroscopic analyses, these molecular resonances are most usefully depicted within (dI/dV) versus V plots in which the former parameter is directly related to the density of electronic states (DOS). By changing the energy range accessible to the tunneling electrons (through the applied bias) it should be possible to modulate the ability of these to access orbital states within the junction and, therefore, the resistance. Of the spectroscopic features which may result, negative differential resistance (NDR), has, on the basis of its possible application within switchable devices, attracted much attention.[53–56] Though NDR spectroscopic features have been assigned to numerous mechanisms,[54,57,58] those associated with Fermi-level frontier molecular orbital resonance are perhaps the most attractive from the perspective of relating tunnel transport to redox switchable characteristics.[30,59] In imaging experiments, the effects of aligning the Fermi level of the underlying electrode to molecular orbital states on mediating increased current flow through the molecule has been strikingly evident in a number of cases.[60,61] Figure 11 shows a current–voltage plot of Cu-azurin obtained at a force of $\sim 4\,\text{nN}$ which displays NDR characteristics (an initial rise in current followed by a sharp decrease as bias is progressively augmented). Though the "peak heights" associated with these apparent electronic resonances show some experimental variation, they are invariably associated with 300–700% increase in current above that extrapolated from the "nonresonant" regime. A statistical analysis of "peak maxima" demonstrates considerable fluctuation in the voltage at which current maxima are observed; specifically, the peak center at negative bias is observed at $-2.2 \pm 0.7\,\text{V}$ and $2.75 \pm 1.0\,\text{V/s}$ at positive bias. We, and others, have noted these fluctuations previously.[59,62]

The apparent NDR features are reproducible and only observable in the force window between breakdown (poor tip–molecule contact) and large molecular compression (the onset of direct tunneling). With azurin, metal

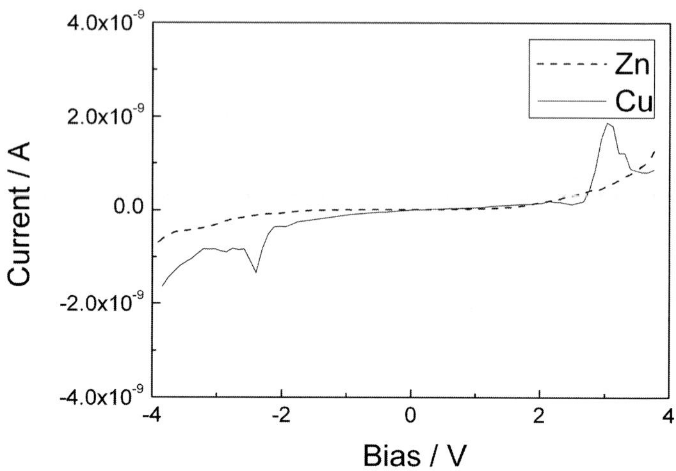

Fig. 11. Comparative current–voltage characteristics of the zinc and copper forms of azurin at junction forces in the 3–5 nN range. While zero bias resistances of the proteins are very comparable, apparent NDR is observed only with the copper form. Only the copper protein has redox-addressable metallic states.

substitution can be achieved with minimal structural perturbation and this, accordingly, facilitates studies in which the precise role of this site in mediating charge flow can be clarified. The assignment of these features to resonant accessing of the copper states is, then, strengthened by observations made in comparative experiments carried out with the zinc protein (Fig. 11). In the low bias regime, an analysis of several hundred comparative current–voltage profiles indicates, within experimental variation, no significant differences in Cu-azurin or Zn-azurin conductance. Significantly, the apparent resonances observed with the copper protein at high bias are not observed with the (redox-inactive) zinc protein. The electronic coupling of these proteins to the underlying planar electrode surface is usefully assessed electrochemically; in the case of azurin molecules immobilized directly onto gold surfaces well-defined (if nonideal) electrochemical responses — from which molecular coverage, electron transfer kinetics, and half-wave potentials may be obtained — are attainable. Though the potential at which transmission through the confined electrochemically active molecules is observed to increase has, in some cases, been usefully related to the electrochemically determined redox potential,[61,63–65] we, to date, have been unable to find any sensible correlation between these potentials and those associated with the spectroscopic features resolvable in tunneling conductance assessments.

6. Conclusions

The difficulties associated with attaining robust (ideally chemisorbed) mechanical contact between nanometer-scale electrodes and molecule(s) of interest, and thereby acquiring reliable transport data, remain. Within this context, the molecular scaling and mechanical flexibility provided by conductive AFM junctions is exceptionally useful. The inherent recognition and redox activity associated with a number of metalloproteins makes a study of their tunneling characteristics highly interesting. In chemisorbing such proteins to a metal-coated AFM probe, many of the effects of lateral drift on molecular analysis are removed (lateral drift of the tip with respect to the atomically flat substrate does not lead to loss of electronic coupling to the sandwiched protein; if the molecule of interest is chemisorbed onto the substrate, lateral drift effects will lead to loss of interaction with the probe). From an electronic perspective, it is clear that a protein fold, though classically highly resistive, is able to act as a conduit for an appreciable current flow within such junctions. The measured giga ohm resistance of a $\sim$3-nm protein examined by us lies in the range of a C_{12}-saturated alkyl thiol SAM, the uncompressed height of which rests in reasonable agreement with simulation-determined (and imaging supported) protein compression data in this force regime.

At the minimum force required for stable force feedback ($\sim 2\,\text{nN}$) in an ambient or inert atmosphere, environment electronic coupling between the confined protein and the electrodes is unstable and dielectric breakdown is observed at both bias extremes. Under conditions of very low force, it is reasonable to consider the junction as comprising two sequential tunnel barriers (one at the tip surface and the second at the substrate surface) across which most of the applied field will be dropped. If the associated interfacial resistances are high (in considerable excess of the quantum unit of resistance, $26\,\text{k}\Omega$) charge accumulation on the molecule, followed by dielectric breakdown when the bias exceeds a critical value, occurs.[66] At higher force (5–80 nN), observations are consistent with a fall in contact resistance and the onset of direct, nonresonant, tip–substrate tunneling with associated distance decay and tunnel lengths which are force-dependent. In this regime, current–voltage profiles are well described by a Simmons model in which the protein is treated as a featureless (uniform) tunnel barrier. The simulated variance of tunnel distance with force correlates well with image height variations across an equivalent force range.

Transport across metal–molecule–metal junctions more generally depends strongly on the position of the metallic Fermi energies with respect to the frontier molecular orbitals of the molecule. When the latter are not energetically accessible (the zinc protein), electron transport is expected to occur by nonresonant superexchange tunneling. Though, in the case of azurin, the copper-based orbital states are likely to be potentially accessible, it appears that this occurs only within a narrow force window where electrical contact is robust but (direct) superexchange tunneling is inefficient (the barrier to direct tunneling is high). Specifically, in the regime of stable contact at low force, current–voltage profiles show characteristics potentially assignable to an ability of the electrons to access copper-based molecular states when the applied bias is high. Electronic movement then occurs via transient occupancy of copper-based DOS. At higher compressional force, the barrier to direct tunneling falls and these spectroscopic features vanish. Under these conditions metal substitution in the protein leads to negligible change in current–voltage characteristics. Though careful EC-STM experiments do further support the role of metallic/heem prosthetic groups in enhancing conductance at specific bias voltages, this effect is both confined to the regime of weak tip–surface (direct) electronic coupling and appears to be comparatively small (molecular conductance changes by some way less than an order of magnitude).

Though the subject of ongoing experimentation, the adequate correlation between barrier heights determined in this study and those resolved in bulk (solution-phase) work is interesting. The work described herein further sheds light on the important consideration of mechanical/environmental perturbation of subsequently reported "molecular" properties.

Acknowledgments

The authors wish to acknowledge the financial support of the EPSRC and thank Prof. Gerard Canters, Leiden Institute of Chemistry, The Netherlands, for supplying azurin.

References

1. H. W. Fink and C. Schonenberger, Electrical conduction through DNA molecules, *Nature* **398**, 407–410 (1999).
2. D. Porath, A. Bezryadin, S. D. Vries and C. Dekker, Direct measurement of electrical transport through DNA molecules, *Nature* **403**, 635 (2000).

3. L. Cai, H. Tabata and T. Kawai, Self-assembled DNA networks and their electrical conductivity, *Appl. Phys. Lett.* **77**, 3105–3106 (2000).
4. A. J. Storm, J. V. Noort, S. D. Vries and C. Dekker, Insulating behavior for DNA molecules between nanoelectrodes at the 100 nm length scale, *Appl. Phys. Lett.* **79**, 3881–3883 (2001).
5. N. C. Greenham and R. H. Friend, *Solid State Physics* (Academic Press, San Diego, 1995).
6. F. Garnier, G. Horowitz, X. Peng and D. Fichou, An all-organic\"soft\" thin-film transistor with very high carrier mobility, *Adv. Mater.* **2**, 592–594 (1990).
7. A. Aviram and M. A. Ratner, Molecular rectifiers, *Chem. Phys. Lett.* **29**, 277–283 (1974).
8. J. J. Davis and H. A. O. Hill, The scanning probe microscopy of metalloproteins and metalloenzymes, *Chem. Commun.*, 393–401 (2002).
9. J. J. Davis, Molecular bioelectronics, *Phil. Trans. Roy. Soc. Lond. Ser. A — Math. Phys. Eng. Sci.* **361**, 2807–2825 (2003).
10. A. V. Tivanski and G. C. Walker, Ferrocenylundecanethiol self-assembled monolayer charging correlates with negative differential resistance measured by conducting probe atomic force microscopy, *J. Am. Chem. Soc.* **127**, 7647 (2005).
11. T. Kelley, E. Granstrom and C. Frisbie, Conducting probe atomic force microscopy: A characterization tool for molecular electronics, *Adv. Mater.* **11**, 261 (1999).
12. P. D. Pablo, M. T. Martinez, J. Colchero, J. Gomez-Herrero, W. K. Mager, A. M. D. Benito, E. Munoz and A. M. Baro, Electrical characterization of single-walled carbon nanotubes with scanning force microscopy, *Mater. Sci. Eng. C* **15**, 149–151 (2001).
13. Z. X. Cui, X. Zarate, J. Tomfohr, O. F. Sankey, A. Primak, A. L. Moore, T. A. Moore, D. Gust, G. Harris and S. M. Lindsay, Making electrical contacts to molecular monolayers, *Nanotechnology* **13**, 5 (2002).
14. A. Bietsch, M. A. Schneider, M. E. Welland and B. Michel, Electrical testing of gold nanostructures by conducting atomic force microscopy, *J. Vacuum Sci. Technol. B* **18**, 1160–1170 (2000).
15. A. Bietsch and B. Michel, Size and grain-boundary effects of a gold nanowire measured by conducting atomic force microscopy, *Appl. Phys. Lett.* **80**, 3346–3348 (2002).
16. E. T. Adman, *Adv. Protein Chem.* **42**, 145–197 (1991).
17. Q. Chi, J. Zhang, J. E. T. Andersen and J. Ulstrup, Ordered assembly and controlled electron transfer of the blue copper protein azurin at gold (111) single-crystal substrates, *J. Phys. Chem. B* **105**, 4669 (2001).
18. J. J. Davis, D. Bruce, G. W. Canters, J. Crozier and H. A. O. Hill, Genetic modulation of metalloprotein electron transfer at bare gold, *Chem. Commun.*, 576–577 (2003).
19. L. Andolfi, D. Bruce, S. Cannistraro, G. W. Canters, J. J. Davis, H. A. O. Hill, J. Crozier, M. P. Verbeet, C. L. Wrathmell and Y. Astier, The electrochemical characteristics of blue copper protein monolayers on gold, *J. Electroanal. Chem.* **565**, 21–28 (2004).

20. K. Fujita, N. Nakamura, H. Ohno, B. S. Leigh, K. Niki, H. B. Gray and J. H. Richards, Mimicking protein–protein electron transfer: Voltammetry of pseudomonas aeruginosa azurin and the thermus thermophilus cua domain at ö-derivatized self-assembled-monolayer gold electrodes, *J. Am. Chem. Soc.* **126**, 13954 (2004).
21. J. J. Davis, C. M. Halliwell, H. A. O. Hill, G. W. Canters, M. C. Van Amsterdam and M. P. Verbeet, Protein adsorption at a gold electrode studied by *in situ* scanning tunnelling microscopy, *New. J. Chem.* **22**, 1119–1123 (1998).
22. C. C. Page, C. C. Moser, X. Chen and P. L. Dutton, Natural engineering principles of electron tunnelling in biological oxidation–reduction, *Nature* **402**, 47 (1999).
23. J. J. Regan and J. N. Onuchic, Electron transfer: From isolated molecules to biomolecules, *Adv. Chem. Phys.* **107**, 497–553 (1999).
24. J. F. Smalley, S. W. Feldberg, C. E. D. Chidsey, M. R. Linford, M. D. Newton and Y.-P. Liu, The kinetics of electron transfer through ferrocene terminated alkanethiol monolayers on gold, *J. Phys. Chem.* **99**, 13141–13149 (1995).
25. D. N. Beratan, J. N. Onuchic, J. R. Winkler and H. B. Gray, Electron tunnelling pathway in protein, *Science* **258**, 1740 (1992).
26. H. B. Gray and J. R. Winkler, Electron transfer in proteins, *Ann. Rev. Biochem.* **65**, 537–561 (1996).
27. V. Grosshenny, A. Harriman and R. Ziessel, Electronic energy transfer across ethynyl-bridged RuIIOsII terpyridyl complexes, *Angew. Chem. Int. Ed. Eng.* **34**, 1100–1102 (1995).
28. J. J. Davis, H. A. O. Hill and A. M. Bond, The application of electrochemical scanning probe microscopy to the interpretation of metalloprotein voltammetry, *Coord. Chem. Rev.* **200**, 411–442 (2000).
29. E. P. Friis, J. E. T. Andersen, Y. I. Kharkats, A. M. Kuznetsov, R. J. Nichols, J. D. Zhang and J. Ulstrup, An approach to long-range electron transfer mechanisms in metalloproteins: *in situ* scanning tunneling microscopy with submolecular resolution, *Proc. Natl. Acad. Sci. USA* **96**, 1379 (1999).
30. J. J. Davis, D. A. Morgan, C. L. Wrathmell, D. N. Axford, J. Zhao and N. Wang, Molecular bioelectronics, *J. Mater. Chem.* **15**, 2160–2174 (2005).
31. Q. Chi, O. Farver and J. Ulstrup, Long-range protein electron transfer observed at the single-molecule level: *in situ* mapping of redox-gated tunneling resonance, *Proc. Natl. Acad. Sci. USA* **102**, 16203–16208 (2005).
32. G. V. Louie and G. D. Brayer, High-resolution refinement of yeast iso-1-cytochrome *c* and comparisons with other eukaryotic cytochromes *c*, *J. Mol. Biol.* **214**, 527–555 (1990).
33. J. J. Davis, N. Wang, D. Morgan, T. Zhang and J. Zhao, Metalloprotein tunnel junctions: Compressional modulation of barrier height and transport mechanism, *Faraday Discuss Chem. Soc.* **131**, 167–179 (2005).
34. T. Albrecht, A. Guckian, J. Ulstrup and J. G. Vos, Transistor-like behavior of transition metal complexes, *Nano Lett.* **5**, 1451–1455 (2005).
35. J. W. Zhao and J. J. Davis, Force dependent metalloprotein conductance by conducting atomic force microscopy, *Nanotechnology* **14**, 1023–1028 (2003).

36. C. Raynaud, Silica films on silicon carbide: A review of electrical properties and device applications, *J. Non-Cryst. Solid.* **280**, 1–31 (2001).
37. D. Wold and C. Frisbie, Fabrication and characterization of metal–molecule–metal junctions by conducting probe atomic force microscopy, *J. Am. Chem. Soc.* **123**, 5549 (2001).
38. J. W. Zhao and K. Uosaki, Dielectric properties of organic monolayers directly bonded on silicon probed by current sensing atomic force microscope, *Appl. Phys. Lett.* **83**, 2034–2036 (2003).
39. R. Holmlin, R. Haag, M. Chabinye, R. Ismagilov, R. Terfort, M. Rampi and G. Whitesides, Electron transport through thin organic films in metal–insulator–metal junctions based on self-assembled monolayers, *J. Am. Chem. Soc.* **123**, 5075 (2001).
40. J. Zhao and J. J. Davis, Molecular electron transfer of protein junctions characterised by conducting atomic force microscopy, *Coll. Surf. B: Biointerf.* **40**, 189–194 (2005).
41. J. J. Davis, C. L. Wrathmell, J. Zhao and J. Fletcher, The tunnelling conductance of molecularly ordered metalloprotein arrays, *J. Mol. Recog.* **17**, 167–173 (2004).
42. D. J. Wold and C. D. Frisbie, Formation of metal–molecule–metal tunnel junctions: Microcontacts to alkanethiol monolayers with a conducting AFM Tip, *J. Am. Chem. Soc.* **122**, 2970 (2000).
43. J. Zhao and K. Uosaki, Formation of nanopatterns of a self-assembled monolayer (SAM) within a SAM of different molecules using a current sensing atomic force microscope, *Nano Lett.* **2**, 137–140 (2002).
44. H. M. McConnell, Intramolecular charge transfer in aromatic free radicals, *J. Chem. Phys.* **35**, 508 (1961).
45. D. J. Wold, R. Haag, M. A. Rampi and C. D. Frisbie, Distance dependence of electron tunneling through self-assembled monolayers measured by conducting probe atomic force microscopy: Unsaturated versus saturated molecular junctions, *J. Phys. Chem. B* **106**, 2813 (2002).
46. J. Zhao, J. Davis, M. Sansom and A. Hung, Exploring the electronic and mechanical properties of protein by conducting atomic force microscopy, *J. Am. Chem. Soc.* **126**, 5601–5609 (2004).
47. J. G. Simmons, Generalized formula for the electric tunnel effect between similar electrodes separated by a thin insulating film, *J. Appl. Phys.* **34**, 1793–1803 (1963).
48. V. Mujica and M. Ratner, Current–voltage characteristics of tunneling molecular junctions for off-resonance injection, *Chem. Phys.* **264**, 365–370 (2001).
49. C. C. Moser, J. M. Keske, K. Warncke, R. S. Farid and P. L. Dutton, Nature of biological electron transfer, *Nature* **355**, 796 (1992).
50. L. H. Guo, J. S. Facci and G. McLendon, Distance dependence of electron transfer rates in bilayers of a ferrocene Langmuir–Blodgett monolayer and a self-assembled monolayer on gold, *J. Phys. Chem.* **99**, 8458 (1995).
51. J. Davis, H. Hill, A. Kurz, C. Jacob, W. Maret and B. Vallee, A scanning tunnelling microscopy study of rabbit metallothionein, *J. Phys. Chem. Commun.* **1**, 12–20 (1998).

52. P. Facci, D. Alliata and S. Cannistraro, Potential-induced resonant tunneling through a redox metalloprotein investigated by electrochemical scanning probe microscopy, *Ultramicroscopy* **89**, 291–298 (2001).
53. J. Chen, W. Wang, M. A. Reed, A. M. Rawlett, D. W. Price and J. M. Tour, Room-temperature negative differential resistance in nanoscale molecular junctions, *Appl. Phys. Lett.* **77**, 1224–1226 (2000).
54. F.-R. F. Fan, R. Y. Lai, J. Cornil, Y. Karzazi, J.-L. Brédas, L. Cai, L. Cheng, Y. Yao, D. W. Price, Jr., S. M. Dirk, J. M. Tour and A. J. Bard, Electrons are transported through phenylene-ethynylene oligomer monolayers via localized molecular orbitals, *J. Am. Chem. Soc.* **126**, 2568–2573 (2004).
55. A. M. Rawlett, T. J. Hopson, L. A. Nagahara, R. K. Tsui, G. K. Ramachandran and S. M. Lindsay, Electrical measurements of a dithiolated electronic molecule via conducting atomic force microscopy, *Appl. Phys. Lett.* **81**, 3043–3045 (2002).
56. R. A. Wassel, G. M. Credo, R. R. Fuierer, D. L. Feldheim and C. B. Gorman, Attenuating negative differential resistance in an electroactive self-assembled monolayer-based junction, *J. Am. Chem. Soc.* **126**, 295–300 (2004).
57. J. Gaudioso, L. Lauhon and W. Ho, Vibrationally mediated negative differential resistance in a single molecule, *Phys. Rev. Lett.* **85**, 1918–1921 (2000).
58. N. Sutin, B. S. Brunschwig and C. Creutz, *J. Phys. Chem. B* **107**, 10687 (2003).
59. C. Gorman, R. Carroll and R. Fuierer, Negative differential resistance in patterned electroactive self-assembled monolayers, *Langmuir* **17**, 6923–6930 (2001).
60. N. J. Tao, Probing potential-tuned resonant tunneling through redox molecules with scanning tunneling microscopy, *Phys. Rev. Lett.* **76**, 4066 (1996).
61. W. Han, E. Durantini, T. Moore, A. Moore, D. Gust, P. Rez, G. Leatherman, G. Seely, N. Tao and S. Lindsay, STM contrast, electron-transfer chemistry, and conduction in molecules, *J. Phys. Chem. B* **101**, 10719–10725 (1997).
62. C. B. Gorman, Y. He and R. L. Carroll, The influence of headgroup on the structure of self-assembled monolayers as viewed by scanning tunneling microscopy, *Langmuir* **17**, 5324–5328 (2001).
63. R. L. McCreery, Molecular electronic junctions, *Chem. Mater.* **16**, 4477 (2004).
64. W. Haiss, H. V. Zalinge, S. J. Higgins, D. Bethell, H. Hoebenreich, D. J. Schiffrin and R. J. Nichols, Redox state dependence of single molecule conductivity, *J. Am. Chem. Soc.* **125**, 15294 (2003).
65. U. Mazur and K. W. Hipps, Resonant tunneling bands and electrochemical reduction potentials, *J. Phys. Chem.* **99**, 6684 (1995).
66. K. K. Likharev, *Proc. IEEE* **87**, 606 (1999).

Jason Davis was born in London and graduated from Kings College with numerous prizes and first class honors in chemistry in 1993. He then moved to the Inorganic Chemistry Laboratory at Oxford where he obtained a

PhD in 1998 (with Prof. Allen Hill). He was subsequently awarded an Extraordinary Junior Research Fellowship at the Queens College in 1998 and became a Tutor in Inorganic Chemistry and PostDoctoral Researcher in Chemistry. Later that year he was awarded the Royal Society University Research Fellowship and subsequently became a Lecturer in Inorganic Chemistry at Jesus College (2001). He was appointed to a University Lectureship at Oxford and a Tutorial Fellowship at Christ Church in 2003. He has published more than 60 papers in leading journals and current scientific interests are focused on biomolecular electronics, single-molecule and sensor-linked optics, molecular recognition, and interfacial/nanoparticle-based sensing.

CHAPTER 8

TWO FRONTIERS OF ELECTRONIC ENGINEERING: SIZE AND FREQUENCY

John Cunningham

School of Electronic and Electrical Engineering
University of Leeds, Woodhouse Lane
Leeds, LS2 9JT, UK

This chapter discusses two limitations which restrict the design of modern electronic circuits. The first is how small a circuit can be made, which is ultimately governed by the size of the individual electrons inside a circuit. Physicists are probing this limit, and in doing so are finding ways to control with great precision the movement of single electrons. The second limitation is the operational frequency of electronic systems, which is currently restricted to frequencies below approximately 1 THz. At frequencies in the terahertz range, the concepts governing electronic circuits design start to merge with concepts from the field of photonics. Experiments which are probing these two limitations are discussed in the context of their implications for future microelectronic circuit design.

Keywords: Nanoscale, semiconductor, terahertz, picosecond, electronics.

1. Introduction: Size and Frequency Limits for Modern Electronic Systems

The field of microelectronics has existed for just less than 60 years, and in that short time it has seen the complexity of circuits increase from a single transistor, to today's thousand million transistors on a single chip. Gordon Moore, one of the founders of Intel, made a famous observation in the mid-1960s that the complexity of integrated circuits, as represented by the number of transistors which they contain, approximately doubles every two years. To keep increasing the complexity of circuits, the microelectronics industry has since its inception been engaged on a program of miniaturization, with the goal to produce the maximum number of transistors in the minimum area. It is a testament to the amazing progress of this approach that, as this text is being written, manufacturing lines are tooling up for mass production of silicon-based circuits in which the smallest features have a size of just 45 nm, which corresponds to the diameter of $\sim$150 atoms. Despite advances in miniaturization, circuits remain much larger than the atomic systems from which they are ultimately composed (see Fig. 1(a)).

Two limits constrain the future development of electronic circuits however. The first is set by the size of electrons themselves. Numerous experiments have confirmed the quantum mechanical description of electrons, which requires them to have both wave- and particle-like components to their behavior. Their wave-like properties can be directly shown in diffraction and focusing experiments (see Ref. 1). To consider how small one can make an electronic circuit, a fundamental length scale is set by this wavelength. In a metal, the electron wavelength is usually of the order of several nanometers, while in some semiconductor systems it can be hundreds of

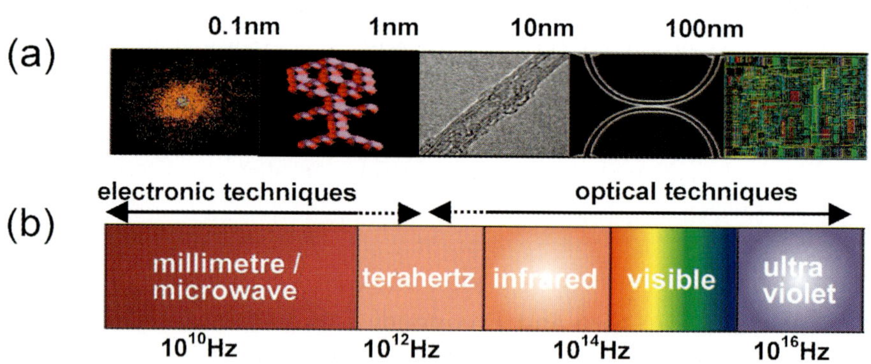

Fig. 1. Two limits for modern electronic systems. (a) The size of the system which contains electrons; from left to right in order of increasing size: simulation of electron density around a single atom, carbon monoxide atoms arranged on a platinum surface, a multiwalled carbon nanotube, a semiconductor structure fabricated using electron beam lithography, and a conventional electronic circuit formed using optical lithography. (b) The frequency of operation; the upper frequency range for electronic circuits lies in the low terahertz range, where electronic concepts merge with photonics.

times larger. The microelectronics industry has developed a set of tools to make electronic circuits with constituent components which are smaller than 1 μm; these tools have allowed physicists to explore the regime where the size circuit of components is similar to the wavelength of the electrons flowing through them. In some of these circuits, exact control of single electrons has been achieved and a host of new physical effects are being discovered. These effects will need to be well understood if further miniaturization of transistor components is to bc pursued. The first section of this chapter will discuss these single-electron effects, putting them into their physical context.

A second limit that is constraining the development of microelectronics concerns the operational frequency of circuits. The maximum operational frequency governs the maximum possible rate for transfer of digital information in the circuit. High-frequency electronic components form the backbone of the wireless telecommunications industry, and systems built from them are responsible for transferring an increasingly large portion of data traffic on the Internet. There is great need for increase in the bandwidth of these systems to allow faster connection speeds between computer systems, and there is therefore a desire to increase the transmission frequency available. Above the microwave frequency range, which extends from ~ 1 to 300 gigahertz (GHz), techniques for generating and transmitting electronic signals in circuits become increasingly difficult, but the demand for higher data transfer rates in the information technology industries is a major incentive to overcome these problems. The current upper limit for an all-electronic approach to signal generation is the terahertz (THz) frequency range; which lies between 300 GHz and 10 THz (see Fig. 1(b)). However, very few electronic devices are capable of emitting a useful signal above ~ 300 GHz. At higher frequencies, optical concepts such as refractive index become more relevant than the electronic concepts of resistance and conductivity. The properties of comparatively few materials have been completely characterized in the terahertz range, compounding the problems and adding complexity to circuit designs which would be considered trivial at lower frequencies. The excitation and propagation of terahertz frequency electrical waves therefore represents a second limit for conventional microelectronics. Despite the generation and transmission problems, terahertz circuits are starting to be investigated in earnest, and the second half of this chapter is devoted to a discussion of some of these systems and their possible practical uses.

2. Single Electronics

2.1. *Confining electrons*

To measure single electrons, a starting point is to restrict the movement of electrons within a solid to a two-dimensional (2D) plane. This can be achieved by using a semiconductor growth technique called molecular beam epitaxy (MBE). Using MBE, single atomic layers of crystalline semiconductors can be deposited, one by one. Figure 2(a) shows an MBE machine, in which a number of heated cells, each of which contains a different element, is pointed toward a crystal substrate. A flux of atoms is emitted by each cell, and the individual atoms in the flux arrange themselves on the substrate into atomic layers. As the flux is continued, a second and third layer of atoms will grow, and so on, until an almost perfect three-dimensional (3D) crystal is formed. An extremely high-vacuum environment must be maintained during growth so that contaminants do not affect the formation of the atomic layers into a crystal on the substrate. If several cells containing different elements are used simultaneously, such as gallium and

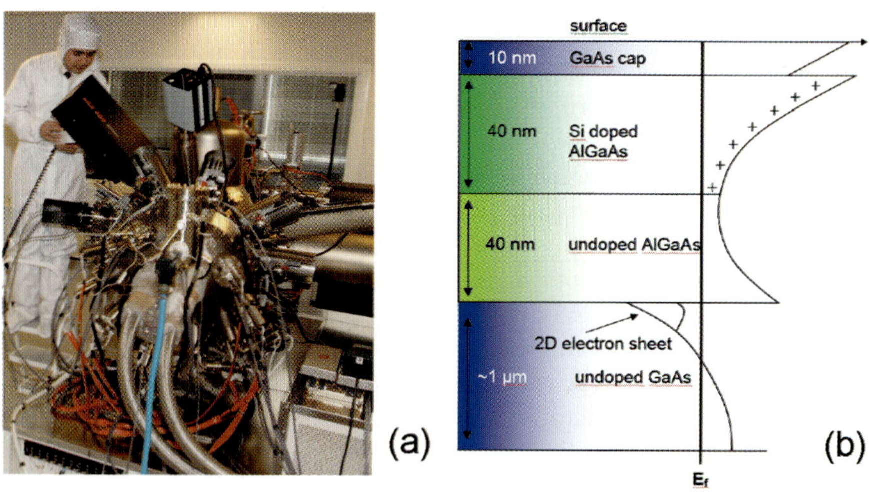

Fig. 2. (a) Photograph of a typical MBE machine, used to grow atomic layers of semiconductor material. (b) Schematic of MBE-grown GaAs/AlGaAs heterostructure wafer structure, and its associated energy profile as a function of depth into the wafer. Electrons from the n-type donar layer migrate to the interface between the undoped AlGaAs and GaAs substrate interface, where confinement causes them to form a 2D electron sheet.

arsenic for example, the growth of compound semiconductors, such as gallium arsenide, is made possible.

If the constituent elements in the atomic flux are abruptly changed during the growth of the crystal, an atomically smooth interface between the two layers can be formed; this is termed a *heterojunction.* Heterojunctions formed at the interface between gallium arsenide (GaAs) and its alloys with aluminium (AlGaAs) have been investigated extensively over the last quarter of a century. The similar intcr-atomic spacings in GaAs and AlGaAs produce a crystal interface during growth which is stress-free and therefore low in crystal defects such as gaps in the lattice, extra atoms, or other imperfections. An example of a layer structure which has been grown by MBE to produce a heterojunction using GaAs and AlGaAs is shown in Fig. 2(b). To explain this structure, we must first mention the concept of doping, which is the incorporation of impurity atoms during MBE growth to alter the electrical properties of the crystal in a controlled manner. In this case, the incorporation of silicon during growth of an AlGaAs layer provides electrons available for conduction. After growth of the layer structure shown in Fig. 2(b), these free electrons in the silicon-doped AlGaAs layer move down through the undoped AlGaAs layer, to become confined at the interface between this layer and the adjacent GaAs substrate. This confinement dictates, as quantum mechanics correctly predicts, that their energy must take on some discrete value in the confinement direction; the continuum of possible energy which is available to free electrons becomes condensed to just a few possible energy levels. The nature of these energy levels depend on how compressed the electrons are by the confining walls. The more confined, the smaller their associated wavelength and the higher their energy will be. The confined electrons form a 2D electron sheet, an important property of which is its electrical mobility. This mobility is high when the electrons undergo few large-angle scattering events as they move through the system. Large-angle scattering events occur when electrons hit charged atoms in the lattice. The high mobility found in a heterojunction results from the spatial separation of the free electrons at the interface of the crystal layers from the charged dopant atoms. This can increase the distance an electron travels before being scattered to several microns. The confined electrons in a heterojunction have an exceptionally high mobility (in excess of $1 \times 10^6\,\mathrm{cm^2/V/s}$), provided the temperature of the lattice is low enough to allow scattering from atomic vibrations to be neglected. To ensure that this is the case, heterojunctions are usually measured inside low-temperature ($< 4\,\mathrm{K}$) apparatus.

When the plane of the electrons is further confined, a system can be created which is smaller than the distance between the large-angle scattering events. This is referred to as the ballistic regime, in which electrons can move smoothly through the system. This further confinement, beyond that achieved in the heterojunction, can be imposed by a number of techniques. One of the most frequently used is the manipulation of the electron density in the heterojunction by applying a voltage to metallic contacts, known as gates, on the crystal surface. This is the split-gate technique, which is shown in Fig. 3(a). A split-gate device is formed by two metal contacts, deposited on the semiconductor surface above the heterojunction, which are separated by a submicron gap. When a negative voltage is applied to the gates with respect to the electron sheet, it repels electrons below the gate, leaving electrons only in the submicron gap between the gates and

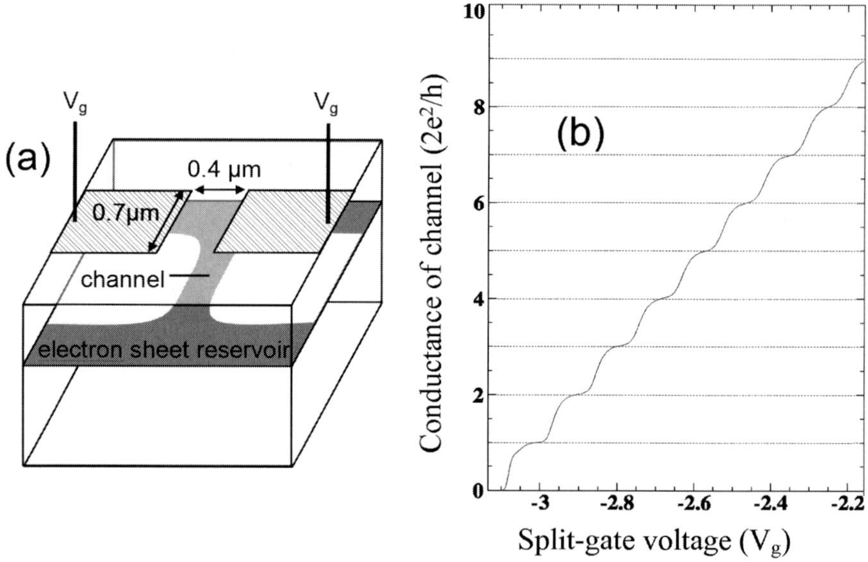

Fig. 3. (a) Schematic diagram of a split-gate which forms a one-dimensional channel from an underlying 2D electron sheet in a GaAs/AlGaAs heterostructure. Hatched areas are the split-gate on the surface of the wafer, gray areas are the 2D electron sheet forming the reservoir. White areas underneath the split-gate indicate depletion of electrons from these regions to form a narrow confined channel for the electrons. (b) Quantized conductance plateaux measured in one such split-gate defined 1D channel, shown as a function of voltage applied to the split-gate, with data taken at $T = 300\,\text{mK}$. Data are corrected for a small series resistance ($600\,\Omega$) chosen to align the plateaux with integer multiples of $2\,e^2/h$, caused by the ohmic contacts and 2D reservoirs.

in the reservoirs on either side. The ballistic and quantum nature of the electron transport in the channel formed between the reservoirs, which is sometimes known as a quantum-wire, is readily shown by measuring its conductance as the applied split-gate voltage is altered. The conductance moves through a series of quantized steps, as the increasing applied voltage narrows the channel. The clear steps, shown in Fig. 3(b), are a result of the sequential removal of confined electron energy levels within the channel as it is made increasingly narrow. It is a result of the width of the channel being of a similar size to the electron wavelength (see Ref. 2 for a full explanation of this effect, which involves a cancellation between two quantities in the equation which defines the conductance; the group velocity and the density of states in the channel).

In a split-gate, the electrons are confined in two directions, but full 3D constriction of electrons is also possible. This can be achieved in several ways, including etching the electron sheet into a pillar shape,[3] or by completely confining a region of the electron sheet using negative voltages applied to surface gates, in a modification of the split-gate technique.[4] It is necessary to use this degree of confinement to observe single-electron effects. A schematic of one such system is shown in Fig. 4(a).

Surface gates are arranged to form, after application of a negative voltage, electrical barriers to the transmission of electrons from one side of the device to the other. Between these barriers, a puddle of single electrons is trapped. Such an arrangement is commonly called a quantum-dot, and its electrical conduction properties are very different to the quantum-wires obtained using the split-gate technique as described above. The energy associated with adding a single electron to the puddle in-between the barriers is called the charging energy of the dot (see Fig. 4(b)), and is around 1 meV for typical geometries (see Ref. 7). To move either into or out of the electron puddle from outside, the energy of electrons in the 2D reservoirs forming the leads to the dot must be close to a vacant energy level within the dot. When this occurs, electrons can flow into or out of the structure. When the energy of electrons in the reservoirs is not aligned with an energy level in the dot, current cannot flow through the device, and this high-resistance state is known as the Coulomb-blockade regime. The system of gates which define the dot can include one gate which can be independently held at negative voltages with respect to the others — a *plunger gate*. As the voltage on the plunger gate is adjusted, it moves the system consecutively between Coulomb-blockaded and conducting states as single electrons are expelled from the dot. This gives rise to oscillations in

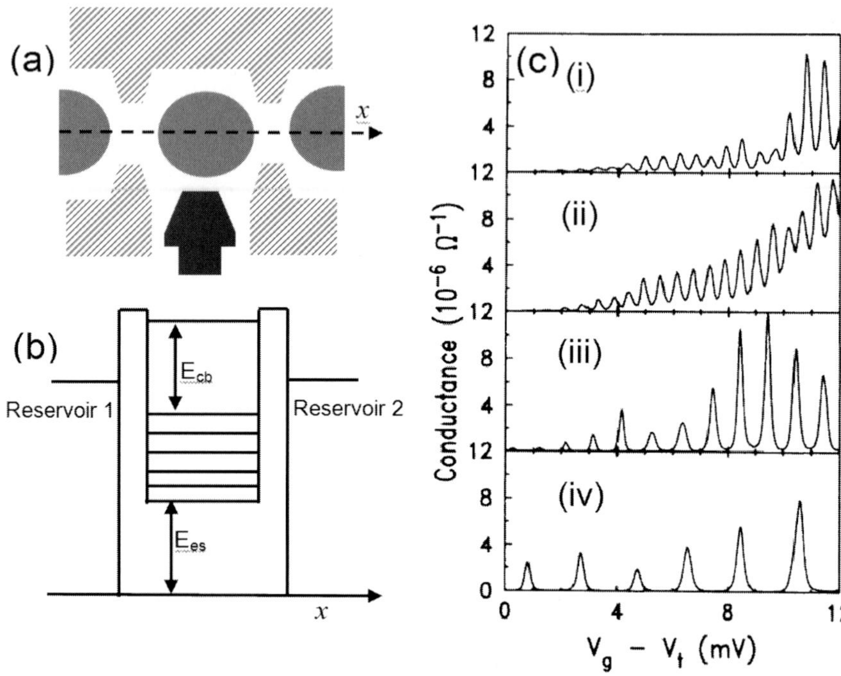

Fig. 4. (a) Schematic top-down view of the gate pattern which can be used to define a zero-dimensional quantum dot in an underlying 2D electron sheet (after Ref. 5). Hatched areas are surface gates used to define the quantum dot, which is shown by the circular gray region. Black region is a plunger surface gate, which can be used to remove single electrons from the dot. Semicircular gray regions on either side of the quantum dot are the 2D reservoirs. (b) Simplified energy diagram for the electron system created underneath the gates is shown in (a). $E_{\rm cb}$ is the Coulomb charging energy, while $E_{\rm es}$ is the electrostatic energy of the dot (set by the plunger gate). 2D reservoirs which can act as both sources or drains for electrons, depending on their potential, are located on either side of the dot. (c) Experimental observation of Coulomb-blockade oscillations in two systems of similar size ((i) and (ii)) and two smaller systems ((iii) and (iv)), taken from Ref. 6.

the current flowing between the reservoirs as a function of the plunger gate potential, as shown in Fig. 4(c). The presence of an impurity atom near the channel of a quantum wire will sometimes cause natural Coulomb-blockade features to appear in the conductance of such a device. The reason for this is that the charged impurity atom modifies the geometry of the 1D channel in such a way as to make an unintentional constriction, similar to that deliberately formed in a quantum-dot system. Such impurity-induced Coulomb blockade could be an important phenomenon to avoid as the dimensions of electronic devices are further reduced.

2.2. *Electron pumps and turnstiles*

The idea of confining electrons in all three dimensions leads naturally to the question of whether single electrons can be moved in a controllable way. A possible way to do this is to regularly perturb as a function of time a system in which single electrons are already confined. Thouless first proposed the idea of generating very accurate single-electron transport using a time-varying potential which moved slowly by comparison with the electron's velocity.[8] Observation of this kind of movement of single electrons has been something of a holy grail for experimentalists in the field. A major reason for this interest in the control of single electrons is that it could give metrologists (scientists who study units of measurement) a worldwide standard of electric current. They seek a highly accurate and stable source of current, whose unit of measurement is the ampere, that can be used to calibrate measurement apparatus, in the same way that they currently maintain standards for resistance and voltage. The present standards of voltage and resistance are maintained using the quantum Hall effect[9] and the Josephson effect,[10] respectively. The ampere is currently represented by combining these voltage and resistance standards. An alternative to this method is sought because it is both difficult to implement and it is indirect, in the sense that it is not an independent representation of the ampere separate from the standards for resistance and voltage. Full details of present methods can be found in Quinn *et al.*[11] The requirements of a current standard set by metrologists are that it should provide an output of at least 1 nA with an accuracy of better than around one part in 10^8, that is, there should be an error of no more than one electron for every one hundred million electrons moved around the circuit. The nanoamp magnitude of the current is important to allow the accurate calibration of secondary standards, which can then be distributed to laboratories and instrument makers around the world. As can be appreciated when one considers the essentially random movement of individual electrons which is a characteristic of the flow of electricity in almost every electronic circuit, this is a very challenging requirement.

The first experiments showing single-electron movement linked to an external frequency signal were performed by Delsing *et al.*[12] with arrays of aluminum/aluminum oxide junctions.[12,13] In these experiments, a series of junctions between which single electrons could move was periodically modulated using a signal source which operated at a few gigahertz. This gave peaks in the resistance of the device. The position of the peaks in resistance corresponded to currents flowing through the device with an approximate magnitude of ef and $2ef$, where f is the (few gigahertz) frequency of the signal source. Around the same time, Geerlings[14] showed that plateaux in the current characteristics could be induced at $I = nef$ in a device formed from four junctions in a row, with the gaps in between forming three islands for single electrons to occupy. The middle island of this device was coupled to a gate-voltage modulated at radio frequencies. The resulting device was termed a turnstile for single electrons since single electrons entered, moved through, and exited the device in a regulated manner. Under suitable voltage-bias conditions the device produced stable current plateaux, with a magnitude given by the applied frequency multiplied by the charge of an electron. A plateaux was a large advance over the peaks in the resistance seen in earlier devices, since it represented more stable single-electron transport. A plateaux in the current characteristic is required for the system to be unaffected by the inevitable small fluctuations in parameters such as temperature, voltage, or amplitude of the modulating applied signal. The flatness of the plateaux degraded when the modulating frequency exceeded 40 MHz however, which limited the magnitude of accurate current transfer to ~ 2 picoamps. This limitation was attributed to the charging time of components in the device. At 50 MHz, for example, the authors calculated that there was a probability of the turnstile failing to transfer an electron once every 10^5 cycles. The problem became worse as the frequency was increased. In 1991, a different device called a single-electron pump was demonstrated by Urbina *et al.*[15] The principle behind the operation of the pump is similar to that of the turnstile, but with some subtle differences. The pump has gates between every junction, which means that there is then no need for an overall DC bias to define the direction of the generated current. The presence of gates at every junction in the current pump improves control over the device optimization relative to the turnstile, while removing the need for an external bias reduces heat dissipation, making it more favorable for high-precision charge transfers. Detailed analysis of pump devices has shown three main errors in the transport. These errors are: co-tunneling, which is an unwanted process whereby multiple electrons move through the device simultaneously; thermally assisted tunneling of electrons through the system; and errors similar to those seen in the turnstile, which are caused by operating the pump at too high a frequency. By 1994 the error rate of five junction pumps under optimum conditions had been reduced to 0.5×10^{-6} per transferred electron.[16] Later results from pumps containing a larger number of controlled junctions showed an even lower error rate per transferred electron $\sim 1 \times 10^{-8}$.[17] It should be noted that the current produced by pump devices is not generally measured since it is so low. Instead, a capacitor is typically charged and the voltage monitored to determine the pumping accuracy. Owing to their high transfer accuracy but small intrinsic current some of the focus of research into pump devices has moved toward their use in a new standard of capacitance; in this context pumps are used to transfer accurately a known number of electrons to a capacitor, and the voltage developed measured accurately which yields a precisely defined capacitance.

2.3. *Surface acoustic wave devices*

An alternative simple way to apply a time-varying external perturbation to a nanoscale system to induce single-electron transport is to use an acoustic wave. A surface acoustic wave is a mechanical wave produced by deforming the lattice of a solid near its surface, similar to an earthquake. Acoustic wavelengths between a few and several hundred microns are excited using an electrode structure on the surface of the crystal, which acts as a transducer for applied high-frequency signals; typically MHz to a few tens of GHz. This high-frequency signal is converted into a surface acoustic wave by the transducer. Such surface acoustic waves, which were first described in a mathematically rigorous way by Lord Rayleigh in the late 1800s, can be propagated on piezoelectric materials. In this case an electrical potential component develops in the substrate whose amplitude varies with the atomic displacement, and which decays into the substrate within around one wavelength. If the piezoelectric material in which the wave propagates is a semiconductor, then the electronic properties of the semiconductor (its conduction and valence bands) will be modulated by the changing electric potential as the wave propagates. If the semiconductor contains an electron sheet held within a heterostructure, such as that described above, then a current is induced in the electron sheet whose magnitude is proportional to the square of the acoustic wave amplitude.[18] A sufficiently powerful wave (with a larger potential than the semiconductor's band gap) can break up the 2D electron sheet into a series of stripes of charge, which will all then move at the acoustic velocity through the semiconductor.[19] To induce single-electron transport, a surface acoustic wave is applied to a quantum wire defined in the heterojunction, as shown in Fig. 5(a).

The first such observations of single-electron transport by a surface acoustic wave were made in 1996 by Shilton and co-workers.[21] Initially,

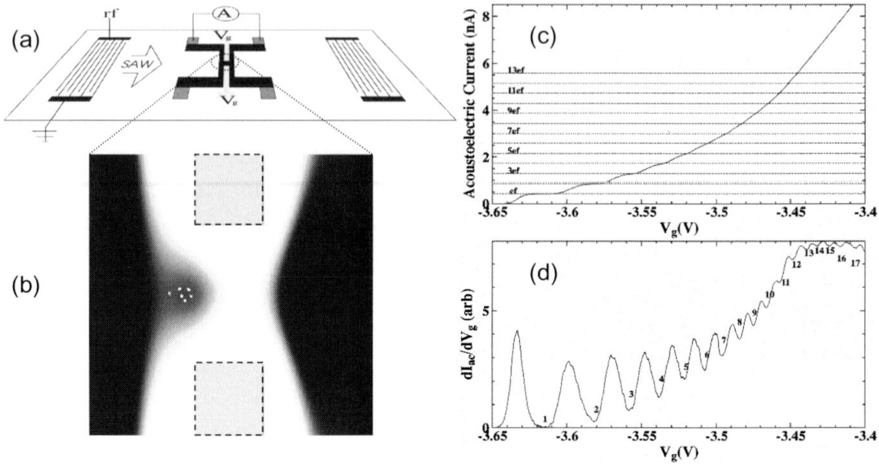

Fig. 5. (a) Schematic of surface acoustic wave device showing 2D electron sheet formed in a GaAs/AlGaAs heterojunction (black region), ohmic contacts to the electron sheet between which current is measured (dark gray squares), and transducers used for surface wave excitation with high-frequency signal applied as shown. Light regions marked V_g are a metal split-gate formed above the heterojunction to which a negative voltage is applied to form a quantum wire. (b) Schematic of the potential profile caused by the acoustic wave as it pushes single electrons (white dots) into the transport channel from the reservoir on the left-hand side (after Ref. 20). Dashed boxes are the split-gate on surface, white areas are at a high relative negative potential. (c) The current generated by a $\sim$3-GHz surface acoustic wave, showing plateaux as a function of spilt-gate voltage which correspond to single-electron transport at $I_{ac} = nef$ (where e is the electron charge, f the acoustic wave frequency, and n is the number of electrons occupying each acoustic wave minima), $T = 1.2\,K$. (d) The derivative of the data shown in part (c), showing the presence of many higher quasi-plateaux at $I_{ac} = nef$.

surface split-gates were used to define the constriction, but later work showed that etched gates could also be used to define a suitable transport channel.[22] The operating regime of single-electron control by the acoustic wave in both systems occurs when the channel is highly resistive. The potential experienced by electrons in the vicinity of the channel is shown in Fig. 5(b).

With no surface acoustic wave perturbation, the 2D regions are separated by a highly resistive (MΩ to GΩ) potential barrier ($\sim$100 s of mV), induced by applying a strong negative potential to the split-gate. The acoustic wave acts first to scoop up in its potential minima a small number of electrons in the entrance (source) region of the channel. The detailed electron dynamics as the minima collect electrons and then propagate through the channel determines the accuracy of the single-electron transport (see Ref. 20). As the well moves into the channel, the increasing electrostatic confinement across the gate causes the electron well to be squeezed, and a traveling quantum-dot begins to form. Any high-energy electrons in the dots are ejected backward toward the source reservoir over the rear barrier of the wave. For a particular range of surface acoustic wave and channel potentials only a small number of electrons are left in the dots by the time the potential minimums reach the far side (drain) of the constriction, where they disperse into the 2D reservoir there. When a small number of electrons is left in each dot then the discrete energy levels of the dot produce stable states in the system which are robust to small variations in temperature, gate voltage or acoustic wave amplitude. A series of plateaux then appear in the current when measured as a function of gate voltage. Each plateaux corresponds to an integer number of electrons held within each packet, at nef (see Fig. 5). The lowest plateaux, which occurs at a current $I = ef$, is always the strongest and most well developed, since the spacing between upper energy levels becomes progressively smaller. The accuracy of the transport, which can be assessed from detailed inspection of the current plateaux, has been measured as at least one part in 10^5, and further improvements seem likely, for example by populating any potential wells with missing electrons using additional surface contacts along the channel.[23] A major advantage of the acoustic wave technique is the high current which is produced relative to the single-electron pumps. As yet, however, neither this nor any other technique fully meets the metrologist's criteria for a quantum standard of DC current.

3. Picosecond Electronics

3.1. *Excitation and detection*

Increasing the operational frequency range of an electronic circuit allows more information to be carried by the circuit in a given time interval. This is important both in the design of digital microelectronic systems, such as a microprocessor, as well as in the design of wireless communications systems. What is the highest frequency electrical signal that it is possible to excite in an electronic circuit? A strongly related, but more useful question to ask experimentally is this: what is the shortest transient voltage pulse that can be propagated some useful distance in a circuit? The present answer to both these questions is produced by the excitation of a semiconductor surface by a pulsed laser beam. Developments in fast pulsed laser systems over the past 20 years have enabled pulses of near-infrared radiation as short as 10 fs to be readily available (provided a particular application can afford it; the price of these systems, though steadily falling, is still in the tens of thousands of dollars bracket). At the Bell Laboratories in 1980, Auston was the first to find that illumination of an amorphous silicon surface could both generate and detect pulses on a picosecond timescale.[24] The experimental arrangement is shown in Fig. 6(a). The idea was as follows: short (sub-picosecond) pulses of optical radiation are focused onto a semiconductor surface, across which an electrical bias is applied by surface electrodes. These laser pulses generate extra carriers in a semiconductor provided the energy associated with the wavelength of the laser exceeds the bandgap of the semiconductor. For the short duration of the additional carriers' lifetime, the semiconductor is rendered conductive. When the semiconductor resistance is low, current flows between electrodes on the semiconductor surface. The current generated is, to the first approximation, proportional both to the intensity of

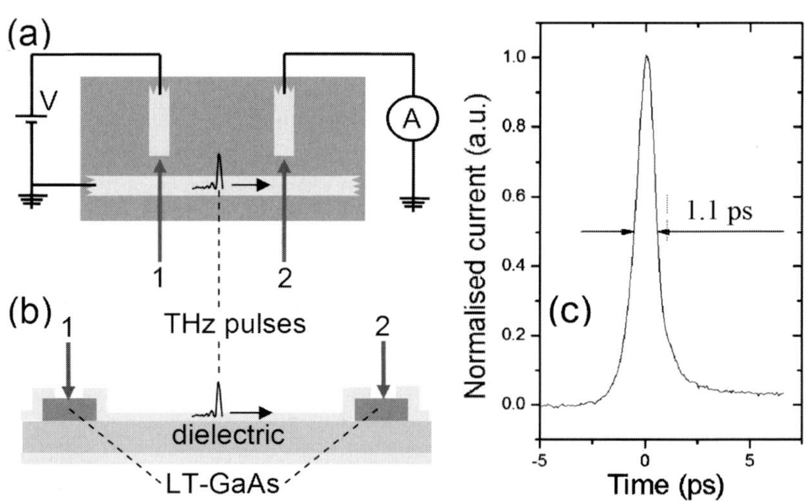

Fig. 6. (a) Schematic of the Auston scheme originally used to generate picosecond duration pulses on the surface of an amorphous semiconductor (plan view of semiconductor wafer surface, after Ref. 24). (b) Schematic cross-section of a recent terahertz microstrip device, with integrated semiconductor regions (LT-GaAs, dark gray areas) used for excitation and detection of picosecond pulses. Near-infrared laser pulses, 1 and 2, are used to generate and detect pulses on the central transmission line, respectively. (c) Typical measurement of a picosecond duration pulse using LT-GaAs emitter and detector integrated into a microstrip circuit (after Ref. 26).

the pulsed laser beam, since a higher-intensity laser beam causes more carriers to be excited, and to the voltage applied across the electrodes. After the laser pulse has finished, and as the optically excited carriers are lost by recombination, the gap becomes resistive once more and the current pulse so generated then propagates along the electrodes. For Auston's original experiment, which used amorphous silicon, the lifetime of the optically induced carriers was tens of picoseconds. More recently, low-temperature-grown GaAs have emerged as a semiconductor which offers shorter carrier lifetime than amorphous silicon. The photocarrier lifetime in GaAs can now be as small as 100 fs when the grown parameters and annealing conditions are suitably optimized.[25] It is therefore possible to obtain pulses of sub-picosecond duration by suitable selection of laser pulse-width and semiconductor carrier lifetime.

To measure the shape of the propagating pulse, the excitation operation is performed in reverse order; the picosecond pulse is made to propagate past another electrode from which it is separated by another region of semiconductor. A portion of the original pulsed laser beam which hit the excitation gap is time delayed and positioned to hit this second semiconductor surface. When the passing picosecond electrical pulse is co-incident at the semiconductor with this laser pulse a portion of its current is transferred across the electrodes. The arrival time of this laser pulse is increased or decreased by adjusting the distance it must propagate before hitting the semiconductor, and the current flowing across the semiconductor gap is measured at the same time. The effect of this is rather similar to a stroboscope; the second laser pulse acts as a window whose opening time is altered until the pulse moving past is co-incident, allowing the pulse through. This process maps out the pulse, which typically has picosecond duration, as shown in Fig. 6, and is called *photoconductive sampling*.

Pulses on a picosecond timescale have corresponding frequency components in the terahertz range (up to a few $\times 10^{12}$ Hz). Variations of the picosecond switch system described above have, over the last decade, become widely used to both create and detect free-space propagating pulses of terahertz electromagnetic radiation (see Ref. 27, for a review). Acceleration of the photo-generated carriers produced by the electrode bias causes emission of free-space photons from the semiconductor surface over a very broad frequency range, centered around terahertz frequencies. The terahertz portion of the electromagnetic spectrum has traditionally been an under-explored regime since, before the recent advances in femtosecond laser systems and short carrier lifetime semiconductors, it was difficult to either detect or create. Photoconductive sampling provides a powerful technique which is capable of detecting not only the amplitude, but also the phase as well as the polarization of terahertz pulses.[28]

3.2. *Transmission of signals*

To use picosecond pulses in a practical circuit, one must propagate the pulses in a geometry where the possible dissipation and dispersion mechanisms which could affect the pulse are minimized. The simple geometry originally used by Auston is not best suited to this propagation, since it is lossy. A better recent alternative has been found in microstrip; which is a highly conductive signal line (typically gold) separated from a ground plane by a low-loss dielectric. Suitable dielectrics for picosecond pulse propagation should be compatible with optical lithography, so that the conductive signal line can be defined on top. They should also have low dielectric constants and small attenuation coefficients in the terahertz range. Such materials include organic polymers such as benzocyclobutene (BCB) (see Ref. 29), and plastic films such as polypropylene and Kapton.[30] Integration of short lifetime LT-GaAs into a device, which then acts as a photoconductive emitter or detector of picosecond pulses, can be achieved in a number of ways. One technique, which is important since it allows lithographic patterning of the picosecond emitter and detector region, is to grow (by MBE) the LT-GaAs on a sacrificial layer of AlAs. Enormously preferential etching of the AlAs layer with respect to the GaAs can be achieved in dilute hydrofluoric acid, which releases the few to 100-nm-thick LT-GaAs layer. Wax layers are typically used to support the thin LT-GaAs, which would otherwise disintegrate by cracking, during its etching. Rather large (cm^2) films of LT-GaAs can be prepared in this way, which can then be attached onto the dielectric surface using a technique called van der Waals' bonding (see Ref. 31). The LT-GaAs layer obtained using this technique is sufficiently robust to allow its patterning by optical lithography, so any required geometry for the picosecond pulse emitter or detector is possible. Further optical lithography is then used to pattern the chip with the gold metallization which forms the signal conductor of the microstrip. The entire structure (LT-GaAs, dielectric, and gold-signal conductor) is fabricated on a gold-coated surface. The substrate for this gold surface is not critical, but silicon wafers are ideal since they provide a very flat surface. Figure 6(b) shows a cross section schematic of one such device. Other systems which have been investigated recently and which allow propagation of picosecond pulses include the radial mode of a free-standing metal wires,[32,33] and cavity modes in parallel plate systems.[34] Both these systems show lower losses than microstrip, and for this reason some future THz systems may employ these guided modes in applications where very low losses are critical.

3.3. *Passive devices, filters, and dielectric loading*

Some of the first terahertz range devices to be investigated using the picosecond pulse technique have been simple filtering structures. We may judge from the development of gigahertz electronic systems that such filters are likely to form a vital component of many terahertz electronic circuits in the future, whenever selection or rejection of some frequency range is required. The design principles of these structures are similar to their microwave counterparts, but extra consideration must be given to the higher dielectric and metallic losses typical at terahertz frequencies, which can lead to additional dispersion and absorption of the radiation. The lithographic dimensions of terahertz devices are smaller than microwave devices, and are therefore less tolerant of any roughness in their geometry; terahertz devices require precision optical lithography to fabricate working devices. A variety of filters have recently been investigated (band-stop, band-pass, and ring resonator structures) in the frequency range 300 GHz to 1.2 THz.[26,35,36] The geometry which comprises the active area of each of these filters is designed to resonate at some frequency in the terahertz range. A typical geometry for one such device, the band-stop filter, is shown in Fig. 7.

An interesting motivation for the study of these passive filtering devices, separate from their use as a simple electrical filter, comes from their extreme sensitivity to overlaid dielectric materials. This sensitivity provides a mechanism to study the dielectric properties of a range of overlaid materials (see Ref. 36). One application is in genetic sensing,[35] where the differences in terahertz range refractive index between double- and single-stranded DNA has been detected using these structures; this difference in terahertz refractive index between two forms of DNA has been attributed to the high density of terahertz frequency vibrational modes predicted to exist in double-stranded DNA.[37]

4. Future Prospects

This chapter has described two of the most exciting developments in modern electron systems; the ability to control single electrons, and to generate,

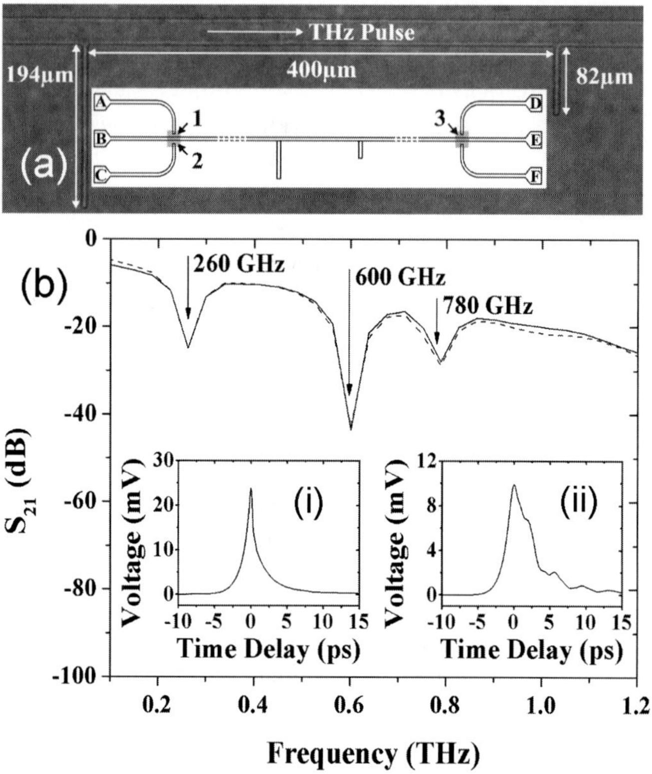

Fig. 7. (a) Photograph (main figure) and schematic diagram (inset) showing an array of two bandstop filters (length 194 μm and 82 μm), connected by microstrip, and fabricated on a low loss dielectric. Numbers 1–3 indicate the position of 100 fs duration laser pulses, used to excite and detect terahertz radiation in the device. (b) Measurement of the signal transmitted (S_{21}) through the two filters in series made using photoconductive sampling, showing drops in transmission near the fundamental mode of each filter (260 GHz and 600 GHz). The position of these drops can be predicted from 3D electromagnetic modeling of the device; such calculations indicate resonances at the positions indicated by arrows (the 780 GHz resonance is a third harmonic of the lower frequency filter). Insets: The picosecond duration pulses measured (i) before and (ii) after traveling through the filter. Notice modification of the pulse shape in (ii) compared with (i), which appears owing to the interaction of the pulse with the filters.

detect, and transfer electronic signals over the shortest timescales. Both these fields are in their relative infancy at present, and major developments are seen in the scientific literature on an almost weekly basis. It is interesting, however, to speculate on some areas in which the two fields are likely to show major advances over the coming years. In single electronics, the search for essentially perfect control over the transfer of single electrons within a circuit has already been achieved by the single-electron pump. Developments must now focus on magnitude of the single-electron currents, and therefore frequency of the clock signals which need to be applied to such systems, for the devices to obtain widespread use in a metrological environment and beyond. Single-electron systems subject to acoustic wave perturbation are one example where frequency limitations are not reached, and rapid developments could be seen here. In the field of picosecond electronics, a wide variety of new devices are now possible which will allow creation of new circuits which will further blur the boundaries between electronics and photonics. It is clear that as the explosion of interest in faster information processing systems continues, it will increase the demand for systems which operate at very high frequencies. The techniques which enable terahertz electronics will then be relevant to an expanding community of technologists faced with practical problems to address.

Acknowledgments

The author gratefully acknowledges funding from the EPSRC, the Royal Society, and the Wolfson Foundation.

References

1. J. Spector, H. L. Störmer, K. W. Baldwin, K. N. Pfeiffer and K. W. West, Electron focusing in two dimensional systems by means of an electrostatic lens, *Appl. Phys. Lett.* **56**, 1290 (1990).
2. C. W. J. Beenaker and H. van Houten, *Quantum Transport in Semiconductor Nanostructures*, Solid State Physics, Vol. 44, eds. H. Ehrenreich and D. Turnbull (Academic Press, 1991).
3. M. A. Reed, J. N. Randall, R. J. Aggarwal, R. J. Matyl and T. M. Moore, Observation of discrete electronic states in a zero-dimensional semiconductor nanostructure, *Phys. Rev. Lett.* **60**, 535 (1998).
4. T. Lundberg, J. E. F. Frost, K. F. Berggren, C. T. Liang, I. M. Castleton, D. A. Ritchie and M. Pepper, Tunneling transmission resonances through a zero-dimensional structure, *Semicond. Sci. Technol.* **12**, 875 (1997).
5. C. J. B. Ford, P. J. Simpson, M. Pepper, D. A. Ritchie, J. E. F. Frost, M. P. Grimshaw and G. A. C. Jones, Coulomb blockade in small quantum dots, *Nanostruct. Mater.* **3**, 283 (1993).
6. U. Meirav, M. A. Kastnew and S. J. Wind, *Phys. Rev. Lett.* **65**, 771 (1990).
7. C. G. Smith, Low-dimensional quantum devices, *Rep. Prog. Phys.* **59**, 235 (1996).

8. D. J. Thouless, Quantization of particle transport, *Phys. Rev. B* **64**, 1812 (1983).
9. K. von Klitzing, G. Dorda and M. Pepper, New method for the high-accuracy determination of the fine-structure constant based on quantized Hall resistance, *Phys. Rev. Lett.* **45**, 494 (1980).
10. B. N. Taylor, W. H. Parker and D. N. Langenberg, Determination of e/h, using macroscopic quantum phase coherence in superconductors: Implications for QED, *Rev. Mod. Phys.* **41**, 375 (1969).
11. T. Quinn and K. Burnett, The fundamental constants of physics, precision measurements and the base units of the SI, *Phil. Trans. R. Soc. A* **363**, 2097–2327 (2005).
12. P. Delsing, K. K. Likharev, L. S. Kuzmin and T. Claeson, Time correlated single-electron tunnelling in one-dimensional arrays of ultrasmall tunnel junctions, *Phys. Rev. Lett.* **63**, 1861 (1989).
13. P. Delsing, T. Claeson, K. K. Likharev and L. S. Kuzmin, Observation of single-electron-tunnelling oscillations, *Phy. Rev. B* **42**, 7439 (1990).
14. L. J. Geerlings, V. F. Anderegg, A. M. Holweg, J. E. Mooij, H. Pothier, D. Esteve, C. Urbina and M. H. Devoret, Frequency-locked turnstile device for single electrons, *Phys. Rev. Lett.* **64**, 2691 (1990).
15. C. Urbina, H. Pothier, P. Lafarge, P. F. Orfila, D. Esteve, M. H. Devoret, L. J. Geerlings, V. F. Andereg, P. A. M. Holweg and J. E. Mooij, Controlled transfer of single charge carriers, *IEEE Trans. Magn.* **27**, 2578 (1991).
16. J. M. Martinis, M. Nahum and H. D. Jensen, Metrological accuracy of the electron pump, *Phys. Rev. Lett.* **72**, 904 (1994).
17. M. W. Keller, J. M. Martinis and R. L. Kautz, Rare errors in a well characterized electron pump: Comparison of experiment and theory, *Phys. Rev. Lett.* **80**, 4530 (1998).
18. F. Maaø and Y. Galperin, Acoustoelectric effect in quantum constrictions, *Phys. Rev. B* **56**, 4028 (1997).
19. M. Rotter, A. V. Kalameitsev, A. O. Govorov, W. Ruile and A. Wixforth, Charge conveyance and nonlinear acoustoelectric phenomena for intense surface acoustic waves on a semiconductor quantum well, *Phys. Rev. Lett.* **82**, 2171 (1999).
20. A. M. Robinson and C. H. W. Barnes, Classical dynamics of electrons in quantized-acoustoelectric-current devices, *Phys. Rev. B* **63**, 165418 (2001).
21. J. M. Shilton, D. R. Mace, V. I. Talyanskii, M. Pepper, D. A. Ritchie, J. E. F. Frost, C. J. B. Ford, C. G. Smith and G. A. C. Jones, High frequency single-electron transport in a quasi-one-dimensional GaAs channel induced by surface acoustic waves, *J. Phys. C* **8**, 531 (1996).
22. J. Cunningham, V. I. Talyanskii, J. Shilton, M. Pepper, P. Kristensen and P. Lindelof, Single-electron acoustic charge transport on shallow-etched channels in a perpendicular magnetic field, *Phys. Rev. B* **62**, 1564 (2000).
23. V. I. Talyanskii, J. M. Shilton, J. Cunningham, M. Pepper, C. J. B. Ford, C. G. Smith, E. H. Linfield, D. A. Ritchie and G. A. C. Jones, Quantized current in a one-dimensional channel by surface acoustic waves, *Physica B* **251**, 140 (1998).

24. D. H. Auston, A. M. Johnson, P. R. Smith and J. C. Bean, Picosecond opto-electronic detection, and correlation measurements in amorphous semiconductors, *Appl. Phys. Lett.* **37**, 371 (1980).
25. I. S. Gregory, C. Baker, W. R. Tribe, M. J. Evans, H. E. Beere, E. H. Linfield, A. G. Davies and M. Missous, 100 fs annealed low-temperature-grown GaAs, *Appl. Phys. Lett.* **83**, 4199 (2003).
26. J. Cunningham, C. D. Wood, A. G. Davies, I. C. Hunter, E. H. Linfield and H. E. Beere, Terahertz frequency range band-stop filters, *Appl. Phys. Lett.* **86**, 213503 (2005).
27. A. G. Davies and E. H. Linfield, Bridging the terahertz gap, *Phys. World* **14**, 37 (2004).
28. E. Castro-Camus, J. Lloyd-Hughes, M. B. Johnston, M. D. Fraser, H. H. Tan and C. Jagadish, Polarization-sensitive terahertz detection by multicontact photoconductive receivers, *Appl. Phys. Lett.* **86**, 254102 (2005).
29. M. Nagel, C. Meyer, H.-M. Heiliger, H. Kurz, R. Hey and K. Ploog, Optical second-harmonic probe for ultra-high frequency on-chip interconnects with benzocyclobutene, *Appl. Phys. Lett.* **72**, 1018 (1998).
30. M. Nagel, T. Dekorsky, M. Brucherseifer, P. H. Bolivar and H. Kurz, Characterization of polypropylene thin-film microstrip at millimetre and submillimetre wavelengths, *Micr. Opt. Tech. Lett.* **29**, 97 (2001).
31. E. Yablonovitch, D. M. Hwang, T. J. Gmitter, L. T. Florez and J. P. Harbinson, Van der Waals bonding of GaAs epitaxial liftoff films onto arbitrary substrates, *Appl. Phys. Lett.* **56**, 2419 (1990).
32. K. Wang and D. M. Mittleman, Metal wires for terahertz wave guiding, *Nature* **432**, 376 (2004).
33. T.-I. Jeon, J. Zhang and D. Grischowsky, THz Sommerfield wave propagation on a single metal wire, *Appl. Phys. Lett.* **86**, 161904 (2005).
34. R. Mendis and D. Grischkowsky, Undistorted guided-wave propagation of subpicosecond pulses, *Opt. Lett.* **26**, 846 (2001).
35. M. Nagel, P. H. Bolivar, M. Brucherseifer, H. Kurz, A. Bosserhoff and R. Büttner, Integrated THz technology for label-free genetic diagnostics, *Appl. Phys. Lett.* **80**, 154 (2002).
36. J. Cunningham, C. D. Wood, A. G. Davies, C. K. Tiang, P. A. Tosch, E. H. Linfield, I. C. Hunter and M. Missous, Multiple-frequency terahertz pulsed sensing of dielectric films, *Appl. Phys. Lett.* **88**, 071112 (2006).
37. P. H. Bolivar, M. Nagel, F. Richter, N. Brucherseifer, H. Kurtz, A. Bosserhoff and R. Büttner, Label-free THz sensing of genetic sequences: Towards "THz biochips", *Phil. Trans. Roy. Soc.* **362**, 1815 (2003).

Dr. John Cunningham started his academic life at University College and Imperial College London, where he obtained a BSc with first class honors and MSc with distinction, respectively. He then took his PhD in Physics at the Cavendish Laboratory, Cambridge, in 2000, subsequently becoming

College Lecturer and Director of Studies in Physics at Selwyn College Cambridge. He moved to Leeds in 2003 to hold first a Royal Academy of Engineering Fellowship, and then an EPSRC Advanced Fellowship. His present research interests encompass terahertz measurement techniques, single electronics and mesoscopic physics. He lives in York with his partner Deborah and his baby daughter Miriam.

CHAPTER 9

ERASABLE ELECTROSTATIC LITHOGRAPHY TO FABRICATE QUANTUM DEVICES

Rolf Crook

Department of Physics, University of Cambridge
Madingley Road, Cambridge
CB3 0HE, UK

A quantum device is an electronic component whose electrical properties can only be understood from quantum mechanical phenomena, such as the wave nature of electrons or the discreet energy levels of a confined electron. A couple of criteria must be met before quantum observations can be made. First, the thermal energy must be less than the electron energy-level spacing which typically requires a combination of temperatures less than 1 K (-272°C) and device dimensions less than one micron (micron = $1\,\mu$m = 10^{-6} m). Second, electron motion must be ballistic, meaning no scattering takes place within the confines of the device, which requires small devices and material of extremely high purity. Quantum devices have been studied intensely for the past 20 years, partly because they provide an ideal arena to investigate diverse quantum-mechanical phenomena, but also to identify and study the next generation of electronic components whose mass production will soon be possible if microchip miniaturization continues at the current rate. The simplest quantum device is a two-dimensional electron gas (2DEG), an electron sheet, where electrons are confined in one dimension and free in two. Confinement by electric fields occurs at the junction between layers of the compound semiconductors GaAs and AlGaAs. Confinement in further dimensions, to create quantum wires and dots for example, is usually achieved using patterned surface electrodes fabricated by electron beam lithography (EBL). During experiments a negative bias is applied to the surface electrodes which expel electrons from the 2DEG lying underneath the electrodes. The quantum wires or dots are created in the 2DEG plane under gaps between the electrodes. Erasable electrostatic lithography (EEL) is a flexible alternative to EBL. Using a charged scanning probe, patterns of negative charge are drawn on a

GaAs sample surface. Electrons are expelled from the 2DEG underneath the surface charge, and the quantum device is created in the 2DEG plane under gaps in the surface charge. EEL has the unique advantages that fabrication takes place in the same low-temperature high-vacuum environment required for measurement and that patterning is erasable. The charge, and therefore the quantum device, can be erased either locally using the scanning probe or globally by illuminating the device with a flash of red light. This means that as part of the experiment, the device geometry can be modified without constraint and the device tuned for optimum performance. With EEL many new fascinating experiments become possible and often with reduced fabrication timescales.

Keywords: Erasable electrostatic lithography, EEL, lithography, quantum device, GaAs surface states, scanning probe.

1. Quantum Devices

A quantum device is an electronic component which exploits quantum mechanical phenomena, such as the wave nature of electrons or the discreet energy levels of a confined electron (see Ref. 1 for a review). They are the subject of intense experimental and theoretical research, partly because they open a window to the fascinating world of quantum mechanics but also because they are likely candidates for the building blocks of the next generation of computers. There are two criteria which must often be met to observe quantum phenomena in quantum devices:

- The thermal energy kT must be less than the electron energy-level spacing to avoid random electron excitation to higher quantum states. This is achieved through a combination of reducing the temperature and reducing the size of the quantum device which increases the energy-level spacing. For example, a device with dimensions of $1\,\mu$m typically requires a temperature less than $1\,$K ($-272°$C) for quantum measurements.
- Electron motion must be ballistic, meaning electrons must not scatter off impurity atoms within the confines of the device. Scattering events randomize the electron phase and energy, making quantum measurements impossible. Ballistic transport is achieved by a combination of reducing the size of the device and decreasing the density of impurity atoms. Indeed, quantum devices only exist because of the availability of high-purity material whose manufacture has received considerable investment and research.

Although quantum mechanics is at the heart of everything, most commercial electronic components cannot be described as quantum devices because quantum phenomena are not observed. Transport is instead described as being diffusive, dominated by scattering events, and can be understood using a semiclassical picture.

Quantum devices are often classified by the number of spatial dimensions in which electrons are free to move. In the remaining spatial dimensions, electrons are restricted to their lowest energy level using strong spatial confinement to take the energy-level spacing greater than the thermal energy. Confinement is attained using a pair of potential barriers through which electron transmission is possible but exceedingly unlikely. For example, in a one-dimensional (1D) electron system, which is also known as a quantum wire, electron motion is free in 1D and confined in two dimensions. The electron energy is a continuum in 1D but restricted to the lowest discreet energy levels in the other two dimensions (2Ds). Electronic properties of quantum devices are further discussed in Sec. 4.

1.1. *Fabrication*

A quantum device is made from a semiconductor wafer, usually GaAs, which is placed in the high-vacuum chamber of a molecular beam epitaxy (MBE) machine. Crucibles containing the compound-semiconductor constituents Ga, As, Al, In, and the dopants Si and Be are heated so that the contents slowly evaporate and shutters control the deposition of these molecular beams onto the wafer surface. Layers of compound-semiconductor are grown on the wafer according to specific recipes, and the finished product is called a heterostructure. Control and measurement in an MBE machine has the finesse to deposit a single atomic layer at a time. The deposited compound-semiconductors are chosen to be lattice matched, meaning the atomic spacing is nearly the same in all the layers, allowing material with almost zero defects to be grown. See Chapter 8 for further details about MBE.

The most important interface in the heterostructure is between GaAs and AlGaAs, due to the alignment of the two band structures. Electronic states in a semiconductor are separated by the bandgap, a region in energy which is devoid of quantum states, into the lower-energy valence band and the higher-energy conduction band. Critically, the bandgap of AlGaAs is larger than the bandgap of GaAs. Occupancy of quantum states is determined by the Fermi level, below which all the states are occupied and above which none are occupied, although there is some statistical leeway as the temperature increases. At low temperature, in pure semiconductors, the Fermi level lies in the bandgap and the material is insulating. For the device to be conductive, and useful, the Fermi level must lie in either the conduction band or the valence band. This is achieved by doping part of the AlGaAs layer with Si during MBE growth. Doping is finely controlled so that the Fermi level lies in the bandgap in the AlGaAs, which is therefore insulating, but lies in the conduction band in the GaAs, which is therefore conductive. This means a AlGaAs/GaAs/AlGaAs sandwich will confine electrons to the GaAs layer forming an electron sheet, which is known as a two-dimensional electron gas (2DEG). In fact, an AlGaAs/GaAs interface is sufficient to create a 2DEG using a triangular, rather than a rectangular, confinement profile.

The quantum device is finished by making electrical contacts to the 2DEG, defining areas of the 2DEG called mesas, and often by confining electrons in further spatial dimensions. Structures larger than 1 μm are created using ultraviolet (UV) photolithography and various chemical processes. One common procedure, which is illustrated in Fig. 1, is the fabrication

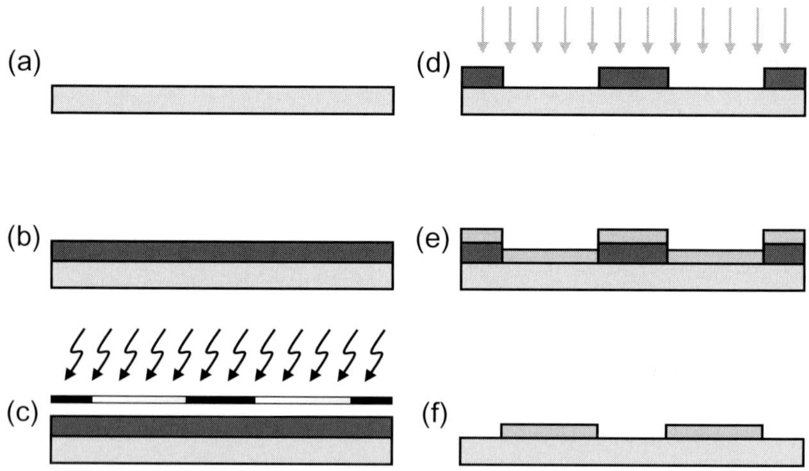

Fig. 1. Device fabrication using UV photolithography with a positive photoresist. (a) The heterostructure surface is thoroughly cleaned. (b) A 0.2-μm layer of positive photoresist is spun on the surface. (c) The resist is selectively exposed to UV radiation using a glass mask. (d) The exposed resist is removed and gold evaporated on the surface. (e) A 0.1-μm layer of gold is evaporated on the device. (f) Lift-off removes the remaining photoresist and the overlying gold, to leave patterned metallic film.

of a patterned thin metal film on the heterostructure surface. After the heterostructure has been thoroughly cleaned, a photoresist is spun on the surface typically forming a 0.2 μm-thick layer. The surface is selectively exposed to UV radiation by placing a mask between the UV source and the photoresist. In exposed regions the photoresist undergoes a chemical change such as cross-linking, degrading, or a molecular rearrangement depending on the choice of resist. During the development stage, the exposed (positive resist) or unexposed (negative resist) regions are washed away leaving windows to the underlying heterostructure. The device is then placed in a high-vacuum chamber with a metal crucible which is heated to evaporate the metal and deposit a thin layer on the device surface. The final step is known as lift-off, where the device is soaked in a solvent to remove the remaining resist and the metal above, leaving a patterned metal film. This procedure has an additional step called *annealing* which is used if electrical contacts to the 2DEG are being made. The device is placed in an oven where the temperature is carefully controlled so that the molten metal spikes down below the surface to contact the 2DEG.

There are many techniques which can be used to create structures smaller than 1 μm, but electron beam lithography (EBL) is the most popular. A positive photoresist made from long-chain polymers is spun on the device forming a layer of the order of 0.05 μm. Under a high vacuum, a beam of electrons is focused on the device surface to locally break up the long-chain polymer making exposed regions more soluble during the subsequent development stage. To expose the desired pattern, the beam is steered by changing perpendicular electric fields. The electron wavelength is thousands of times shorter than the wavelength of UV radiation, which accounts for the improved resolution. However, EBL uses a scanning beam instead of a mask, so the technique is much slower than UV photolithography and is not economic for exposing large features.

Now consider how these technologies are combined to fabricate a complete quantum device. Using UV lithography, two pads of a metal alloy, such as AuGeNi, are patterned on the heterostructure surface and then annealed. These "ohmic contacts" make electrical connections to the 2DEG and are labeled source and drain. Processing could stop here if the desired device is a 2DEG. If a more complex device is required, such as a quantum wire, then thin metal electrodes are fabricated on the surface using UV lithography for large features and EBL for small features. During measurement a negative voltage is applied to the electrodes which expel electrons from the regions of the 2DEG underneath the electrodes. The quantum device forms

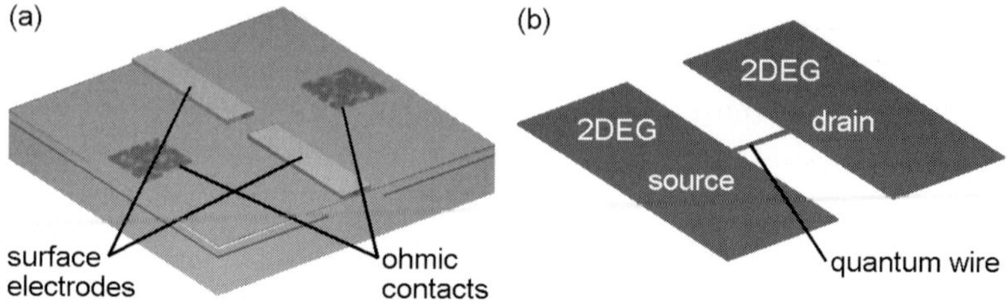

Fig. 2. Quantum wire fabricated using UV lithography and EBL. (a) Schematic showing source and drain ohmic contacts to the 2DEG, and surface electrodes to which a negative bias is applied. The quantum wire forms in the 2DEG plane under the gap between the electrodes. (b) Electron density in the 2DEG plane.

in the 2DEG plane under gaps between electrodes where the electrons are not expelled. This technique is illustrated in Fig. 2 for a quantum wire. Note that no current flows between the electrodes and the 2DEG because a potential barrier, called a Schottky barrier, forms at the metal/GaAs interface.

2. Scanning Probe Lithographic Techniques

Various scanning probe techniques can be used to fabricate the fine structures in a quantum device (see Ref. 2 for a review). Most scanning probe lithographic systems are cheaper than an EBL machine and attain a comparable resolution. Issues restricting the more widespread adoption of scanning probe lithographic systems are poor reproducibility, reliability, flexibility, and productivity. Only the most widely used of these techniques are discussed next.

2.1. *Local anodic oxidation*

For local anodic oxidation (LOA), a negative voltage is applied to the conductive tip of an atomic force microscope (AFM) in a humid environment at room temperature. An electrochemical reaction takes place where the tip is the cathode, the surface is the anode, and water is the electrolyte. In the high electric field between the tip and the surface, water is ionized and the OH^- ions are drawn toward the surface where they locally oxidize the surface underneath the tip. The process requires an electrical current, hence both the tip and sample must be somewhat conductive. As the oxide becomes thicker, the electric field between the tip and surface becomes less, so the reaction is self-limiting.

The LAO technique for fabricating quantum devices was first demonstrated by oxidizing lines in a thin titanium film on top of a GaAs heterostructure.[3] The oxide lines divided the titanium film into surface gates to which a bias was subsequently applied to define the quantum device from a subsurface 2DEG at 1.6 K. Direct oxidation of GaAs, which is illustrated in Fig. 3, was later demonstrated[4,5] and has become the preferred method. Oxidized regions effectively bring the sample surface closer to the 2DEG which causes the local depletion of electrons from the 2DEG. Due to the self-limiting nature of LAO, the technique is only suitable for fabricating devices containing a 2DEG closer than 40 nm to the device surface. Nevertheless, LAO has been refined by several groups, and is now a viable alternative to EBL for many types of devices.

2.2. *Scribing*

It is also possible to increase the force between the tip and sample then scribe, or plough, a line out of the device. The tips clearly need to be harder than GaAs, and are usually made of Si_3N_4 or diamond and the forces applied are several orders of magnitude larger than those used for contact-mode AFM imaging. Although a high lithographic resolution can be obtained using this technique, the depth of the grooves is often limited to a few nanometers, to avoid excess tip damage. This limitation can be overcome by scribing only a thin titanium,[6] or different semiconductor,[7]

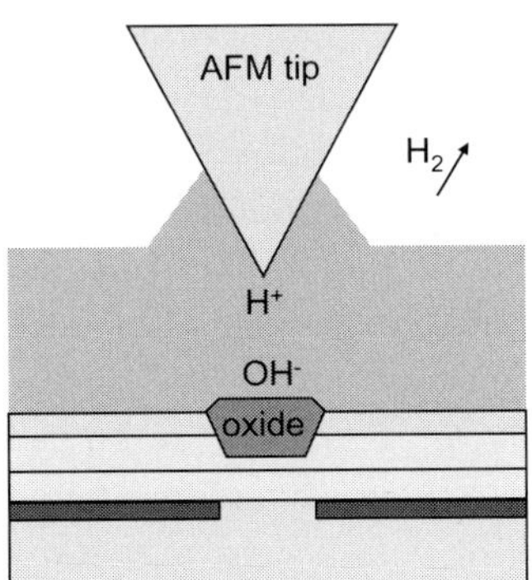

Fig. 3. LAO using an AFM. OH^- ions are drawn to the surface where they locally oxidize the GaAs and AlGaAs. H^+ ions combine at the tip and are released as H_2 gas.

surface layer to provide a mask for the subsequent chemical processes, which fabricate the quantum device.

2.3. *Atomic manipulation*

A scanning tunneling microscope (STM) can be used in a remarkable way to move individual atoms to specific lattice locations on a crystal surface. Pioneering research used STM to move Xe atoms on a crystalline Ni surface.[8] More recently an elliptical quantum corral was drawn using Cu atoms on a crystalline Cu surface, and interference effects beautifully demonstrating the wave nature of electrons were observed in subsequent STM images.[9] The corral also acted as a quantum mechanical resonator, and projected the image of a magnetic Co atom from one focus to the other. Using STM, the atoms are simply moved by approaching the tip above the target atom, dragging the atom to the desired location, then retracting the tip. These experiments require a low-temperature ultra-high-vacuum environment for the STM. Finding material systems, which can be integrated into an electrical circuit, and making electrical connections to such tiny devices are the next challenges for atomic manipulation experiments.

3. Erasable Electrostatic Lithography

Erasable electrostatic lithography (EEL) is a low-temperature scanning-probe lithographic technique. The essential component is a conductive tip with a method to determine when the tip contacts the device surface. Surface detection in the very first scanning probe microscope, STM, measured the current between the tip and a necessarily conductive sample.[10] For EEL the surface charge must be static, otherwise the charge patterns would not be stable, hence STM is unusable and AFM is used instead.[11] AFM measures the force between the tip and surface by measuring the deflection of a flexible beam, called a cantilever, which supports the tip at one end and is secured to the scanning mechanism at the other. Most room-temperature AFMs measure the cantilever deflection by bouncing a laser beam off the cantilever back and detecting the change in angle or phase of the reflected beam. However, this scheme is impractical at low temperatures as thermal movements complicate optical alignment, and any stray laser light would disrupt the EEL process and the electrical properties of the sample. Piezoresistive cantilevers,[12] where the cantilever resistance is proportional to the cantilever deflection, require no alignment and are ideal for EEL. Piezoresistive materials change their resistance under mechanical stress but, unlike piezoelectric materials, they do not generate a charge. Cantilevers used for EEL are typically 0.2 mm long, and the tip is typically 2 μm tall with a radius at the point of 0.05 μm. A further development, which has enabled the recent increase in low-temperature scanning probe experiments, is the availability of reliable low-temperature micropositioners. A micropositioner is required to bring the sample in range of the tip, and the most successful designs use piezoelectric actuators to jolt the sample along in many small steps.

To draw a pattern of surface charge using EEL, the scanning-probe tip is taken to − 6 V and brought into contact with the GaAs device surface as illustrated in Fig. 4(a). The force between the tip and the surface is kept smaller than that required to break molecular bonds, hence the surface is not damaged. A current momentarily flows between the tip and the surface which locally charges the device surface states to the same negative voltage as the tip. Surface states are available quantum states associated with dangling bonds on the material surface. If the tip is then retracted a spot of charge is left on the surface, or the tip can be moved while maintaining contact with the surface to draw a line of charge. The surface charge density is sufficiently high to locally expel the electrons from the underlying 2DEG, so the charge pattern is projected onto the 2DEG as a pattern of depletion and a quantum device is created. If a less negative tip voltage is used during EEL, the 2DEG electron density is locally reduced without

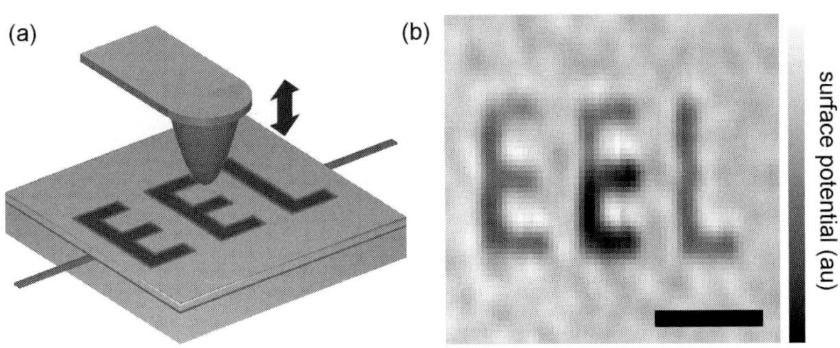

Fig. 4. (a) Illustration of EEL operation. Dark regions show the surface charge which locally depletes the underlying 2DEG. (b) KPM image showing surface charge previously drawn using EEL. Scale bar 1 μm. Parts of this figure were originally published in Ref. 13.

expelling all the electrons. GaAs surface charge is stable at temperatures less than approximately 100 K (− 173°C) which sets the maximum temperature for EEL. Above this temperature thermal energy is sufficient to redistribute the charge between the surface, the 2DEG, and the donor layer, which supplies the electrons for the 2DEG. A technique similar to EEL has also been reported where tip-induced charging of the surface, or the donor layer locally perturbed the 2DEG which was detected by imaging charge transparency.[14]

If the tip is returned to a point on the surface where charge has been previously drawn, the surface is simply charged to the new voltage on the tip. Therefore the charge pattern, and hence the quantum device, can be locally erased using the tip biased to 0 V. Alternatively, all the charge can be erased by raising the temperature above 100 K, or by illuminating the device for a few seconds with red light with the convenience of maintaining a low temperature. Both heating and illumination redistribute the surface charge between the surface states, 2DEG and the donor layer, in a process known as Fermi level pinning.

Uniquely among scanning-probe lithographies, EEL takes place in the same low-temperature high-vacuum environment required for taking measurements, which has a number of advantages:

- The layout for a new quantum device often needs to be modified when initial measurements fail. Such modifications take at least a week using conventional lithographic techniques as a new sample must be fabricated and cooled to cryogenic temperatures. Using EEL there is no need to replace the sample as the charge can be globally erased, then the new design drawn, and the cycle typically takes a few hours.
- Due to inherent disorder in material, such as the precise configuration of ionized donor atoms, two quantum devices fabricated to exactly the same design invariably exhibit slightly different behavior. This problem is usually overcome using surface electrodes which are biased to compensate for disorder during the experiment, but the number of electrodes becomes unmanagable when arrays of quantum devices are required. Using EEL, the local disorder can be measured then the device geometry tuned with additional spots of charge to compensate for disorder, without using surface electrodes.
- EEL provides a new freedom to vary device geometry as part of the experiment. This is particularly useful for experiments investigating the interaction between electrons, as interaction depends on the separation

between electrons which in turn can depend on geometry. For example, the length of a quantum wire could be incrementally increased between measurements. Other lithographic technologies could in principle achieve similar results with a series of devices, but this becomes time-consuming and disorder leads to unintentional variation between devices which can mask the intended change in geometry.

3.1. *Characterizing erasable electrostatic lithography*

To image charge patterns on the device surface, a different scanning probe technique called Kelvin probe microscopy (KPM) is used.[15] KPM can be easily integrated into an EEL system, requiring only a modification of room temperature wiring to switch between the techniques. For KPM, the cantilever is driven to oscillate at or near its resonant frequency. The electric interaction force between the tip and the sample draws the tip slightly closer to the surface which modifies the resonant properties of the cantilever. By recording the cantilever deflection amplitude and phase shift, the interaction force and therefore the surface potential can be deduced. A popular KPM configuration uses negative feedback to maintain a constant cantilever resonant frequency by applying a feedback bias to the tip. The feedback bias is recorded to provide the image-contrast and single-electron charge resolution can be attained. However, without a precise knowledge of the tip geometry it is impossible to calibrate the KPM measurement accurately which accounts for the frequent use of arbitrary units (a.u.). KPM is ideal for imaging surface charge drawn by EEL, as is evident in the KPM image presented in Fig. 4(b).

Not only do the KPM images provide information on EEL device geometry, but they can also be used to characterize the EEL technique and help deduce the mechanism of charge storage.[13] Figure 5(a) presents a KPM image of 12 spots of EEL charge drawn with tip voltages from -6 V to $+3$ V. The electric potential of each charge spot, albeit in arbitrary units, was recorded and is plotted in Fig. 5(b). When the tip voltage is near -4.5 V, the plateau seen in spot potential is due to the complete removal of 2DEG electrons under the tip, which is known as depletion. During fabrication with a tip voltage less than -4.5 V, when the tip is in contact with the surface, the capacitance between the charge spot and the 2DEG is reduced so less charge is drawn. Between EEL tip voltages of -1 and $+1$ V the spot potential is suppressed which indicates the presence of a potential barrier, perhaps due to a thin oxide on the silicon tip, which must

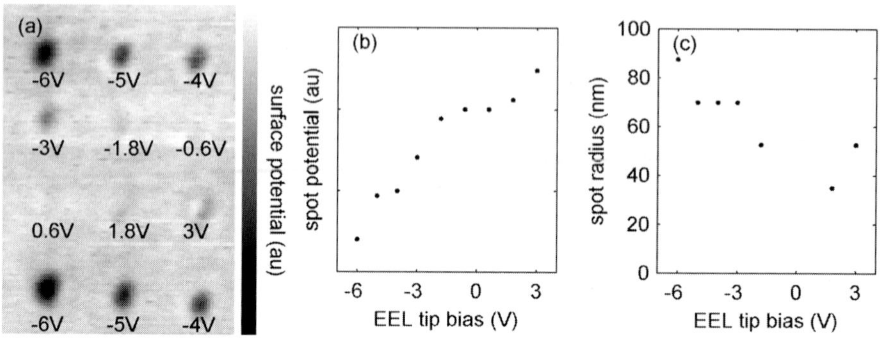

Fig. 5. (a) KPM image of 12 spots of charge, drawn using EEL with the tip biases shown. (b) Plot of spot potential as a function of EEL tip bias. (c) Plot of spot radius as a function of EEL tip bias. This figure was originally published in Ref. 13.

be overcome before current can momentarily flow between the tip and the surface. To investigate EEL resolution, the radius at half-maximum of each spot is plotted in Fig. 5(c) as a function of EEL tip voltage. This measurement is not possible between -1 and $+1$ V. To be useful as a lithographic technique the 2DEG must be depleted, and at a sufficient EEL tip bias the radius is approximately 90 nm ($1\,\text{nm} = 10^{-9}\,\text{m}$) which is in rough agreement with other indirect measurements.[16] Without depletion a radius of 50 nm can be attained with an EEL tip bias of -2 V. Of course, the resolution depends on the contact area of the tip, but this image suggests charge spreading, rather than contact area, is setting the resolution limit.

Global erasure by illumination has also been investigated using KPM. Figures 6(a)–6(d) present a sequence of images showing the same charge spots after various durations of illumination. In these experiments, the photon flux at the sample is of the order $10^6\,\mu\text{m}^{-2}\,\text{s}^{-1}$ which corresponds to 2.5×10^4 photons s^{-1} over a charge spot with radius 90 nm. From the images, the peak potential of the -6, -5, and -4 V spots is plotted in Fig. 6(e) as a function of illumination duration where a logarithmic-linear axis provides a straight line fit. It has been calculated that a -6 V spot contains about 240 electrons, and the probability is small so that an incident photon removes an electron from the charge spot. The exponential, rather than linear, decay seen in Fig. 6(e) suggests that the photon capture area is small compared to the charge spot size. With reference to Figs. 6(a)–6(d), it is apparent that the spot radius is not changed by illumination, which rules out the possibility that during illumination the surface becomes conductive allowing the charge to diffuse then leak to surface electrodes.

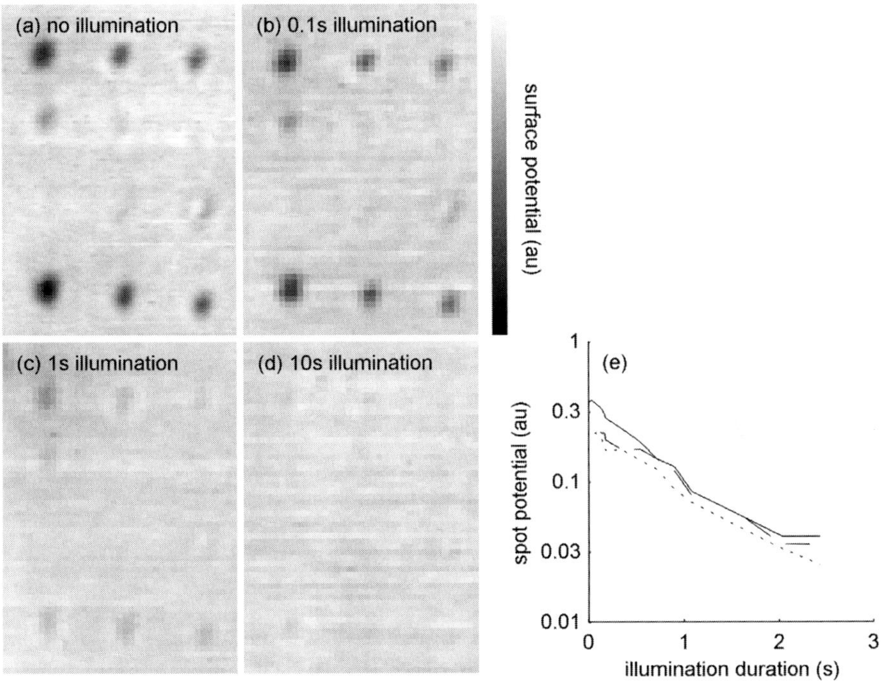

Fig. 6. ((a)–(d)) Sequence of KPM images of EEL charge spots where the device has been illuminated with red light between each image. (e) Plot of spot potential as a function of time. Solid, dashed, and dotted lines correspond to − 6, − 5, and − 4 V applied to tip during EEL. This figure was originally published in Ref. 13.

These experiments provide useful guidance for the future fabrication of quantum devices using EEL. Depletion occurs with an EEL tip bias near − 4.5 V, although this may be reduced in magnitude when a line, or several spots in close proximity, is drawn. Resolution is limited by the requirement to have sufficient surface charge to deplete the 2DEG. The measurements suggest that a resolution of 50 nm can be attained by using a shallower 2DEG where depletion occurs with a smaller bias. With this apparatus, illumination of 10 s is sufficient to reduce charge patterns by two orders of magnitude.

3.2. *Future developments*

Above 100 K, thermal energy is sufficient to overcome the confinement energy which traps charge in GaAs surface states, and the EEL technique fails.[17] If the confinement energy can be raised, by using a surface treatment, a surface layer, or a different material system, then room-temperature operation is possible. Charge can be stored in an Si_3N_4 layer at room temperature and sourced from a conductive AFM tip.[18] Charge storage has also been demonstrated in a polymer, which is used in semiconductor processing, called PMMA using a conductive rubber stamp to deposit a pattern of charge.[19] So long as the charge density is sufficient to repel electrons from the 2DEG, a thin layer of these materials should allow room-temperature operation of EEL.

EEL provides the fascinating possibility of locally reducing electrical noise and disorder which often hinders quantum measurements. Using a proposed scheme called defect compensation, KPM first locates the impurity complex responsible for generating device noise, then a spot of EEL charge eliminates the noise source by raising its energy above that of the transport electrons. A similar scheme can be used to locate the ionized donors responsible for device disorder, then compensate with an appropriate pattern of surface charge.

4. Quantum Devices and Scanning Probes

Scanned gate microscopy (SGM) is another low-temperature scanning-probe technique which is easily integrated in an EEL system. Unlike EEL and KPM which are tools for fabrication and imaging, SGM is a technique for studying the quantum properties of the finished device. A variety of quantum systems have been studied using SGM, including backscattering from quantum wires,[20,21] defects in carbon nanotubes,[22] and electron wave functions in quantum billiards.[23] For SGM, the tip is taken to a fixed voltage and scanned over a rectangular region a constant height above the device surface, without ever touching the surface. While scanning proceeds, the device resistance is recorded to set the color of an image pixel corresponding to the tip's spatial coordinates. SGM images reveal information about electron flow and electron density within the device, but are sometimes difficult to fully understand due to device disorder.

4.1. *Quantum wires*

The first quantum device to be fabricated using EEL or LAO was a quantum wire. The quantum wire is a 1D electron system, meaning the electrons are free in one spatial dimension and confined in two, and has become the archetypal quantum device.[24,25] It is straightforward to fabricate compared to other quantum devices, and it exhibits an unambiguous quantum behavior which is readily observable. The device conductance (one divided by resistance) is independent of electron energy due to a fortuitous cancellation, in 1D electron systems, of the energy dependence in electron velocity and electron energy density. This results in a device conductance of $2e^2/h$ which is independent of wire width, length, and electron spatial density. In practice, higher 1D energy-levels can become transmissive as the wire width, which is set by the voltage on the tip or surface gates, is increased. Each new energy level contributes a further $2e^2/h$ to the device conductance, so a series of steps in conductance, called plateaus, quantized to integer multiples of $2e^2/h$ are observed. For further discussion about quantum wires see Chapter 8.

The first EEL quantum wire was created in the 2DEG under a small gap in a line of negative charge, as illustrated in Fig. 7(a). The lines of charge extend to surface electrodes to which a negative voltage is applied sufficient to deplete the underlying 2DEG. If the gap were not present, there would be a continuous route of 2DEG depletion across the entire sample, which would electrically isolate the source and drain 2DEG regions giving a device conductance of zero. The width of the wire is controlled by the voltage applied to the tip which is positioned locally. Taking the tip bias less negative locally reduces the extent of 2DEG depletion, which increases the width of the quantum wire, as is evident in the plot shown in Fig. 7(b).

An SGM image is presented in Fig. 7(b) (lower inset) which was made by scanning the tip over the quantum wire 50 nm above the device surface, while the device conductance was recorded to generate the image. The tip voltage is negative, so as the tip approaches the quantum wire, the wire width and therefore conductance both decrease. So the image reveals a circular region of reduced conductance centred over the quantum wire, which demonstrates how SGM can be used to locate a quantum device precisely. The device conductance transition from $4e^2/h$ to $2e^2/h$ is seen in the SGM image.

Quantum wires have also been fabricated using LAO by oxidizing a Ti film to define surface gates,[3] and by oxidizing the GaAs surface to define gates from the 2DEG which are called in-plane gates.[4,5] By applying a DC bias between the source and drain contacts, the 1D energy-level separation was reported as 5.5 meV which is much larger than typical GaAs quantum wires fabricated using other techniques.[5] This is caused by a strong lateral

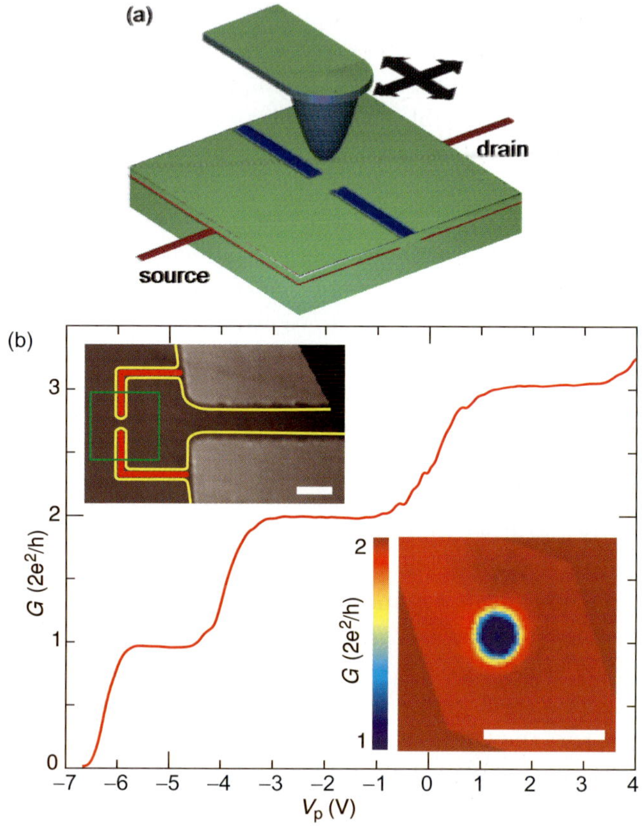

Fig. 7. (a) Experimental setup for the study of a quantum wire fabricated by EEL. Negative surface charge is shown in blue. The quantum wire is defined in the subsurface 2DEG, below the gap in the line of surface charge. (b) Plot of device conductance as a function of tip bias and therefore wire width. Plateaus are observed at integer multiples of $2e^2/h$. Upper inset: AFM image of surface electrodes superimposed with surface charge and depletion outlined in yellow. Lower inset: SGM image of the quantum wire made over the region indicated in the upper inset. Scale bar 1 μm. Parts of this figure were originally published in Ref. 16.

confinement from the in-plane gates, and meant plateaus in conductance were observed up to 20 K ($-253°$C).

If the tip is positioned over a 2DEG region, away from the quantum wire, and a sufficiently negative bias applied to the tip, it is possible to deplete electrons from the 2DEG over a small spot immediately under the tip. The spot of depletion can be considered as a potential hill which is impervious to electrons traveling in the 2DEG. Electrons approaching the potential hill head on, will be scattered back from where they came in a process known as *backscattering*. The backscattering process can be demonstrated experimentally by scanning a biased tip over a 2DEG region adjacent to a quantum wire. A small fraction of the electrons transmitted through the wire are backscattered by the tip-induced potential hill and travel back through the quantum wire, which reduces the recorded device current and therefore conductance. The backscattered current depends on the tip position, and the resulting image is related to the flow of electrons emanating from the quantum wire into the 2DEG. Backscattering experiments were pioneered by Topinka *et al.*[20,21] producing wonderfully detailed SGM images, one of which is reproduced in Fig. 8, where the quantum wire is fabricated using EBL. Superimposed on the filaments of electron flow are oscillations with a period that is half the electron wavelength. This is an interference effect, demonstrating the wave nature of electrons, where electrons bounce between the potential hill and the quantum wire. The interference is understood to be regenerated at points called cusps, which accounts for the observation of oscillations further from the quantum wire than is expected. Note that the inset SGM image presented in Fig. 7(b) does not show backscattering because the tip bias is insufficiently negative to create a potential hill capable of backscattering electrons.

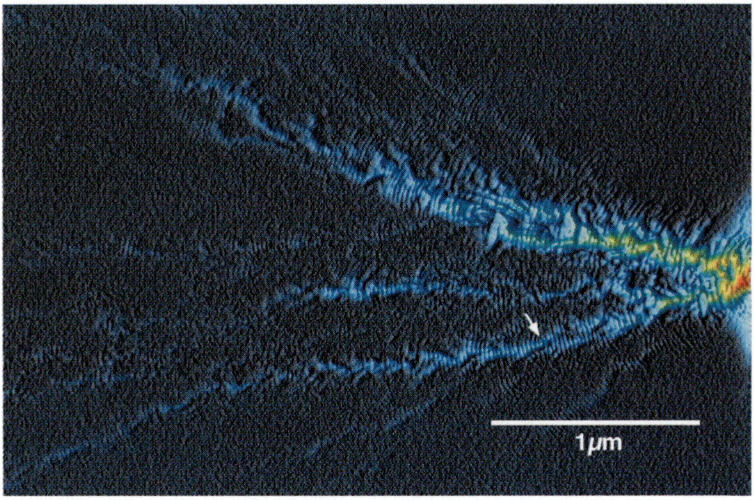

Fig. 8. SGM image made over a region to the left of a quantum wire. The arrow indicates a cusp where electron interference is regenerated. Scale bar 1 μm. Figure originally published in Ref. 21.

4.2. *Quantum billiards*

Another device which has been fabricated using EEL is a quantum billiard.[23] Imagine shrinking a billiard table, found in a bar, one million times and replacing the billiard balls with electrons. The pockets in a quantum billiard, of which there are only two, are quantum wires. Electrons enter the quantum billiard by either quantum wire, then scatter off the billiard boundaries a number of times. If they exit by the wire through which they entered, then the electron is backscattered and the conductance will be reduced, whereas if they exit by the opposite hole, then the conductance is maintained. Therefore, measuring conductance provides information about the electron trajectories inside the billiard. If the billiard boundaries are somewhat curved and the billiard is within certain size and temperature limits, then electron dynamics become chaotic. This means a small perturbation, such as a small change in magnetic field or tip voltage, leads to a large change in conductance. As a function of magnetic field or tip bias, the device conductance is described as being fractal where structure is present on, and proportional to, all length-scales. Fractals are often associated with self-similar structure, but note that although self-similar structure is fractal, fractals are not necessarily self-similar. With one exception,[26] fractal structure from quantum billiards is not found to be self-similar. A quantum billiard can exhibit fractal structure over several orders of magnitude in magnetic field (e.g., see Ref. 27), and has consequentially received much theoretical attention (e.g., see Ref. 28).

The EEL quantum billiard was fabricated by drawing a series of charge lines, as illustrated in Fig. 9(a), which deplete the underlying 2DEG to create the billiard walls. The lithographic dimensions of the billiard were $2\,\mu$m by $3.5\,\mu$m, but EEL linewidth and lateral depletion reduce the actual billiard dimensions to $1.4\,\mu$m by $2.9\,\mu$m. Increased surface charge density at the billiard corners means the actual billiard corners will be rounded, and inherent material disorder means the billiard is unlikely to be perfectly symmetric. Both of the quantum wires, the billiard entrance and exit, were tuned to transmit only the first 1D energy level by drawing additional spots of surface charge. This maximizes the chaotic behavior of a quantum billiard.

Figure 9(b) plots conductance against the external perpendicular magnetic field, which is called *magnetoconductance*, and the plot is fractal because the irregularity is proportional to the length-scale. The average magnetoconductance is $0.5 \times 2e^2/h$ because, averaged over all magnetic

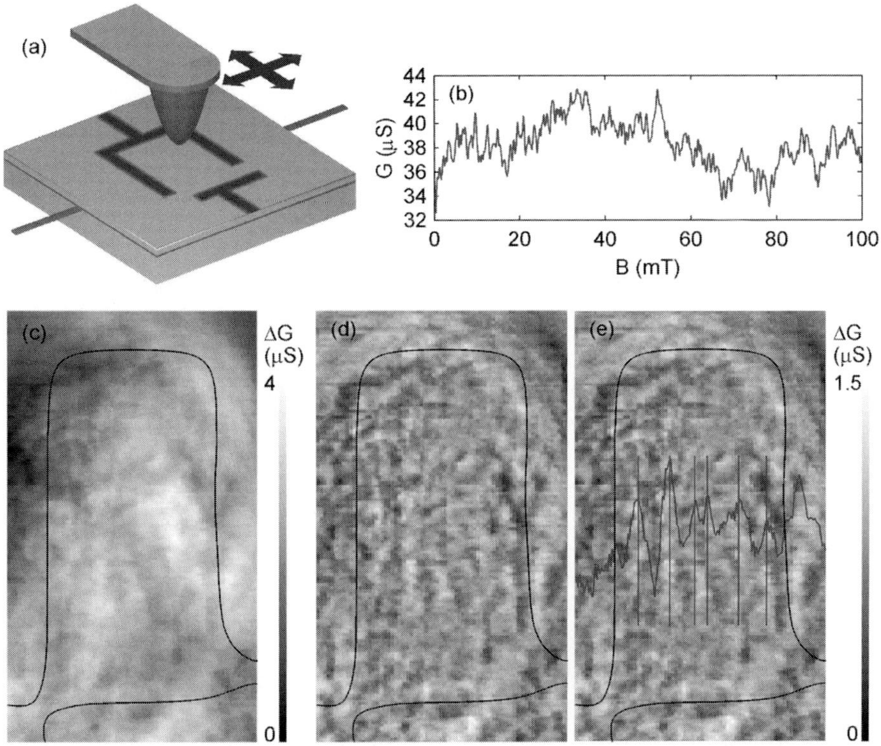

Fig. 9. (a) Experimental setup for the study of a quantum billiard fabricated by EEL. (b) Plot of device magnetoconductance made at 20 mK. (c) SGM image of the quantum billiard made in a 39.1 mT perpendicular magnetic field. (d) Enhanced version of (c) made by subtracting low-frequency fractal structure. (e) Annotated version of (d) to highlight lines parallel to the billiard long-axis, including a plot of average conductance across the width of the image. Parts of this figure were originally published in Ref. 23.

fields, half of the electrons are transmitted through the billiard and half are backscattered. The power spectrum, or frequency spectrum, of the magnetoconductance reveals the presence of a strong periodicity with frequency $0.33\,\mathrm{mT}^{-1}$. Such periodicities are a consequence of simple closed-loop orbits. An electron will follow an orbit several times, interfering with itself, and because the phase shift is proportional to the magnetic field, oscillations between constructive and destructive interference are observed when the magnetic field changes.

SGM is a valuable tool to study quantum billiards as it provides spatial information. To generate the SGM image presented in Fig. 9(c), the tip was scanned over the quantum billiard, 50 nm above the device surface, while the device conductance was recorded. Structure, which is entirely reproducible, is visible on many length scales which, not surprisingly, turns out to be fractal. In fact, the low-frequency fractal structure is obscuring additional regular structure which is revealed in Fig. 9(d) by subtracting the fractal background. Close inspection reveals nodes, boundaries, and lines parallel to the billiard long-axis which are highlighted in Fig. 9(e). These features are understood to indicate the presence of regular closed-loop orbits which are more stable and robust than the mass of chaotic orbits. The quantum-mechanical equivalent of such an orbit is called a *scarred wave function*, of which these same lines, boundaries, and nodes are characteristic. Confirmation of the observation of scarred wave functions is provided by the periodicity of the images in magnetic field, which is equal to the periodicity seen in magnetoconductance.

4.3. *Quantum rings*

One of the most intricate quantum devices fabricated by scanning probe lithography is the quantum ring, which exhibits fascinating quantum properties. Figure 10(a) presents an AFM image of the device which was fabricated by LAO at ETH Zurich.[29] The ring is connected to the source and drain 2DEG regions using quantum wires. Six in-plane gates are used to tune the width of the quantum wires and the size of the quantum ring. A quantum ring can be studied in two regimes (see Ref. 1):

- The device is described as being "open" when the quantum wires are tuned to transmit one or more 1D energy levels. Conductance is determined by the Aharonov–Bohm (AB) effect where each electron travels both ways round the ring (clockwise and anticlockwise) to interfere with itself at the other side. Interference is either constructive or destructive depending upon the relative phase shift around either side of the ring. The relative phase shift is a linear function of the magnetic field, leading to characteristic AB oscillations.
- The device is described as being "closed" when the quantum wires are tuned to be potential barriers and transmission is only possible by quantum mechanical tunneling. Conductance is determined by the Coulomb blockade (CB) effect where the tiny capacitance of the device adds a dominating contribution to the energy-level spectra. When the 2DEG electron energy falls between two energy levels, transmission is forbidden. By varying the size of the device, energy levels are brought into

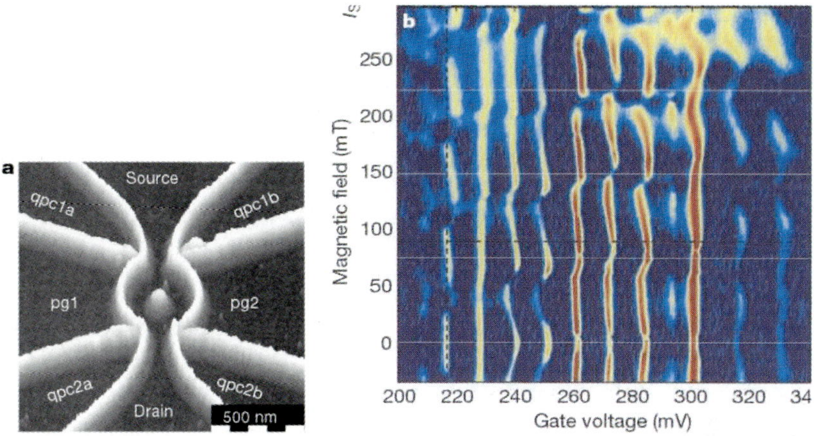

Fig. 10. (a) AFM image of a quantum ring fabricated using LAO. The in-plane gates defined by LAO (qpc1a, qpc1b, qpc2a, qpc2b, pg1, and pg2) are used to tune the size of the quantum wires and the quantum ring. (b) Plot of device conductance as a function of the size of the ring (voltage to pg1 and pg2) and the magnetic field. Coulomb blockade oscillations are seen along the x-axis, and AB-like oscillations along the y-axis. Figure originally published in Ref. 29.

resonance with the 2DEG energy, and transmission is allowed, leading to characteristic CB oscillations.

The quantum billiard discussed in the previous section operated as an open device, whereas this quantum ring operates as a closed device. Figure 10(b) plots the ring conductance as a function of both the gate voltage and the magnetic field. Along the x-axis CB oscillations are clearly visible whose amplitude and spacing provides information about the discreet electron energy levels within the ring. Along the y-axis AB-like oscillations are visible despite the system being closed. The cylindrical symmetry of this ring, in contrast to the quantum billiard, means the electron motion is regular, rather than chaotic.

4.4. *Future devices*

Quantum devices suitable for quantum-information processing are receiving much interest in current research. This is partly driven by the somewhat distant prospect of making a quantum computer where processing is dictated by quantum, rather than classical, mechanics. Such a machine promises unprecedented computation power, albeit only for certain algorithms. Building a quantum computer using today's technology is simply impossible, so instead research is focused on building and understanding the essential component which is called *a qubit*. A classical bit, meaning a memory element storing either a one or a zero, is an isolated entity. If a classical computer consists of n classical bits, then the information capacity is n. However, information stored in an array of n qubits cannot be identified with individual qubits, but with ensembles of quantum states formed by a phenomenon called entanglement. This means the information capacity of a quantum computer is 2^n because there are 2^n ways of entangling n qubits. Pioneering research created a single qubit device from a double-quantum dot, meaning a pair of coupled 0D-quantum systems.[30] In this embodiment classical information is stored by a single electron being in either dot. However, the quantum information, which is a superposition of an electron in both dots, was lost within about 1 ns, so it was not possible to connect more than one qubit. Several hundreds of nearly identical qubits communicating in a quantum coherent fashion are required to build a useful quantum computer.

Future scanning probe lithographies, perhaps similar to EEL, may be suitable for the fabrication of a quantum computer. The ability to modify device geometry during a measurement is vital, as this will allow the fabrication of an array of nearly identical qubits. Once a qubit has been fabricated, an *in situ* measurement is made, then additional charge drawn to slightly modify the qubit geometry and precisely tune the qubit quantum properties. The scanning probe then moves onto the next qubit.

Acknowledgments

The Erasable Electrostatic Lithography research was funded by the EPSRC UK. I thank M. A. Topinka and A. Fuhrer for their generous permission to use Figs. 8 and 10, respectively.

References

1. C. G. Smith, Low-dimensional quantum devices, *Rep. Prog. Phys.* **59**, 235–282 (1996).
2. A. A. Tseng, A. Notargiacomo and T. P. Chen, Nanofabrication by scanning probe microscope lithography: A review, *J. Vac. Sci. Technol. B* **23**, 877–894 (2005).

3. R. Held, T. Heinzel, P. Studerus, K. Ensslin and M. Holland, Semiconductor quantum point contact fabricated by lithography with an atomic force microscope, *Appl. Phys. Lett.* **71**, 2689–2691 (1997).
4. R. Held, T. Vancura, T. Heinzel, K. Ensslin, M. Holland and W. Wegscheider, In-plane gates and nanostructures fabricated by direct oxidation of semiconductor heterostructures with an atomic force microscope, *Appl. Phys. Lett.* **73**, 262–264 (1998).
5. N. J. Curson, R. Nemutudi, N. J. Appleyard, M. Pepper, D. A. Ritchie and G. A. C. Jones, Ballistic transport in a GaAs/AlGaAs one-dimensional channel fabricated using an atomic force microscope, *Appl. Phys. Lett.* **78**, 3466–3468 (2001).
6. S. Hu, A. Hamidi, S. Altmeyer, T. Koster, B. Spangenberg and H. Kurz, Fabrication of silicon and metal nanowires and dots using mechanical atomic force lithography, *J. Vac. Sci. Technol. B* **16**, 2822–2824 (1998).
7. R. Magno and B. R. Bennett, Nanostructure patterns written in III–V semiconductors by an atomic force microscope, *Appl. Phys. Lett.* **70**, 1855–1857 (1997).
8. D. M. Eigler and E. K. Schweizer, Positioning single atoms with a scanning tunneling microscope, *Nature* **344**, 524–526 (1990).
9. H. C. Maroharan, C. P. Lutz and D. M. Eigler, Quantum mirages formed by coherent projection of electronic structure, *Nature* **403**, 512 (2000).
10. G. Binnig, H. Rohrer, Ch. Gerber and E. Weibel, Surface studies by scanning tunneling microscopy, *Phys. Rev. Lett.* **49**, 57–61 (1982).
11. G. Binnig, C. F. Quate and Ch. Gerber, Atomic force microscope, *Phys. Rev. Lett.* **56**, 930–933 (1986).
12. M. Tortonese, R. C. Barret and C. F. Quate, Atomic resolution with an atomic force microscope using piezoresistive detection, *Appl. Phys. Lett.* **62**, 834–836 (1993).
13. R. Crook, C. G. Smith, S. Chorley, I. Farrer, H. E. Beere and D. A. Ritchie, Kelvin probe microscopy to image and characterize erasable electrostatic lithography, *Physica E* **34**, 686–688 (2006).
14. S. H. Tessmer, P. I. Glicofridis, R. C. Ashoori, L. S. Levitov and M. R. Melloch, Subsurface charge accumulation imaging of a quantum Hall liquid, *Nature* **392**, 51–54 (1998).
15. C. Schonenberger and S. F. Alvarado, Observation of single charge-carriers by force microscopy, *Phys. Rev. Lett.* **65**, 3162–3164 (1990).
16. R. Crook, A. C. Graham, C. G. Smith, I. Farrer, H. E. Beere and D. A. Ritchie, Erasable electrostatic lithography for quantum components, *Nature* **424**, 751–754 (2003).
17. M. G. Pala, G. Iannaccone, S. Kaiser, A. Schliemann, L. Worschech and A. Forchel, Extraction of parameters of surface states from experimental test structures, *Nanotechnology* **13**, 373–377 (2002).
18. B. D. Terris, R. C. Barrett and H. J. Mamin, *SPIE 1855 Scanning Probe Microscopies II* (1993), p. 195.
19. H. O. Jacobs and G. M. Whitesides, Submicrometer patterning of charge in thin-film electrets, *Science* **291**, 1763–1766 (2001).

20. M. A. Topinka, J. B. LeRoy, S. E. J. Shaw, E. J. Heller, R. M. Westervelt, K. D. Maranowski and A. C. Gossard, Imaging coherent branched flow from a quantum point contact, *Science* **289**, 2323–2326 (2000).
21. M. A. Topinka, J. B. LeRoy, R. M. Westervelt, S. E. J. Shaw, R. Fleischmann, E. J. Heller, K. D. Maranowski and A. C. Gossard, Coherent branched flow in a two-dimensional electron gas, *Nature* **410**, 183–186 (2001).
22. M. T. Woodside and P. L. McEuen, Scanned probe imaging of single-electron charge states in nanotube quantum dots, *Science* **296**, 1098–1101 (2002).
23. R. Crook, C. G. Smith, A. C. Graham, I. Farrer, H. E. Beere and D. A. Ritchie, Imaging fractal conductance fluctuations and scarred wave functions in a quantum billiard, *Phys. Rev. Lett.* **91**, 246803 (2003).
24. D. A. Wharam, T. J. Thornton, R. Newbury, M. Pepper, H. Ahmed, J. E. F. Frost, D. G. Hasko, D. C. Peacock, D. A. Ritchie and G. A. C. Jones, One-dimensional transport and the quantization of the ballistic resistance, *J. Phys. C* **21**, L209–L214 (1988).
25. B. J. vanWees, H. vanHouten, C. W. J. Beenakker, J. G. Williamson, L. P. Kouwenhoven, D. Vandermarel and C. T. Foxon, Quantized conductance of point contacts in a two-dimensional electron-gas, *Phys. Rev. Lett.* **60**, 848–850 (1988).
26. R. P. Taylor, R. Newbury, A. S. Sachrajda, Y. Feng, P. T. Coleridge, C. Dettmann, N. J. Zhu, H. Guo, A. Delange, P. J. Kelly and Z. Wasilewski, Self-similar magnetoresistance of a semiconductor Sinia billiard, *Phys. Rev. Lett.* **78**, 1952–1955 (1997).
27. A. P. Micholich, R. P. Taylor, A. G. Davies, J. P. Bird, R. Newbury, T. M. Fromhold, A. Ehlert, H. Linke, L. D. Macks, W. R. Tribe, W. R. Linfield, D. A. Ritchie, J. Cooper, Y. Aoyagi and P. B. Wilkinson, Evolution of fractal patterns during a classical-quantum transition, *Phys. Rev. Lett.* **87**, 036802 (2001).
28. R. Akis, D. K. Ferry and J. P. Bird, Wave function scarring effects in open stadium shaped quantum dots, *Phys. Rev. Lett.* **79**, 123–126 (1997).
29. A. Fuhrer, S. Luscher, T. Ihn, T. Heinzel, K. Ensslin, W. Wegscheider and M. Bichler, Energy spectra of quantum rings, *Nature* **413**, 822–825 (2001).
30. T. Hayashi, T. Fujisawa, H. D. Cheong, Y. H. Jeong and Y. Hirayama, Coherent manipulation of electronic states in a double quantum dot, *Phys. Rev. Lett.* **91**, 226804 (2003).

Rolf Crook studied physics and electronics engineering at Lancaster University, graduated with first class honors in 1996. He obtained his PhD in 2000 from the Department of Physics at the University of Cambridge for the study of electron transport in low-dimensional quantum devices. In 2007 he moved to the Energy and Resources Research Institute at the University of Leeds to set up a new research group studying a range of nanostructured photovoltaic devices using novel electro-optical low-temperature scanning-probe techniques. Rolf is a RCUK Academic Fellow. His other interests include installing low-carbon buildings technology and competing in orienteering events.

CHAPTER 10

ULTRAFAST NANOMAGNETS: SEEING DATA STORAGE IN A NEW LIGHT

Robert J. Hicken

School of Physics, University of Exeter
Stocker Road, Exeter EX4 4QL, UK

Magnetic materials provide the most important form of erasable data storage for information technology today. The demand for increased storage capacity has caused the size of the region used to represent a "1" or "0" of binary data and features of the read–write transducers to be reduced to the nanometer scale. However, increased storage capacity is useful only if there is a commensurate reduction in the time taken to read and write the data. In this chapter the basic principles that determine the behavior of nanomagnetic materials are introduced and their use in data storage systems is described. Particular attention is paid to processes that limit the speed of operation of the data storage system. It is shown that optical pump–probe experiments may be used to characterize dynamic magnetic processes with femtosecond temporal resolution. The macroscopic magnetization of a ferromagnet can be made to precess in response to an optically triggered magnetic field pulse, leading to reduced switching times. Alternatively, an ultrashort laser pulse may be used to manipulate the magnitude of the magnetization on femtosecond time scales, leading to an ultrafast demagnetization in certain ferromagnets, and providing new insight into magnetotransport phenomena. Finally, the outlook for increased record and replay rates is assessed and the prospect of further use of optical techniques within magnetic data storage technology is discussed.

Keywords: Magnetism, magneto-optics, ultrafast, data storage.

1. Introduction

The field of nanotechnology has caused excitement within many of the traditional scientific disciplines. By manipulating structure on atomic and nanometer-length scales new properties may emerge. Nanometer scale (nanoscale) building blocks may then be used to construct larger structures that perform multiple functions with increased speed. The latest generation of magnetic materials provide an example of this process of development. Nanostructuring has allowed researchers to enhance the properties of bulk materials, providing stronger magnets that remain magnetized for longer, and discover new phenomena, such as giant magnetoresistance, that has led to a new generation of improved magnetic field sensors. Today magnetic materials are used in a multitude of devices from electric motors and actuators to the devices that store data in computers. Indeed let us consider the hard disk drive in a desk or lap-top computer. The storage capacity doubles approximately every year. This has been possible because the size of the region used to store each bit ("1" or "0") of binary data has continued to decrease deep into the nanoscale regime. Since we expect the stored information to be almost instantly available, reduced bit sizes have been accompanied by reduced record and replay times.

In this chapter the speed limits that constrain the development of magnetics-based technology will be considered. We will consider why familiar materials such as iron are magnetic, and how manipulation of atomic and nanoscale structure may alter their magnetic properties. Some common formats used in magnetic data storage will be examined and their speed of operation discussed. It will become apparent that many important processes have characteristic time scales shorter than 1 ns and so call for specialized measurement techniques. In particular it will be shown that ultrashort optical pulses can be used to probe and perhaps manipulate the magnetic state. We will see that magnets may be made to reorient or have their magnetism temporarily suppressed. Finally we will explore how these processes might be exploited in future technologies and used to extend our existing understanding of magnetic materials.

2. What Makes a Magnet?

Before attempting to understand the properties of nanomagnetic materials, we must first understand the origin of magnetism in familiar materials such as iron. Our first experience of magnetism is gained from permanent bar magnets and loops of wire that carry a current. It is tempting to believe that the magnetism has a common source. We know that the atoms of a magnetic material contain shells of orbiting electrons, each of which can be considered as a tiny current loop. Each electron has orbital angular momentum and an associated magnetic moment. We might obtain a macroscopic magnetic moment if the electrons could be persuaded to orbit in the same sense. However this is not the dominant source of magnetism in iron. Electrons are indistinguishable particles that satisfy the Pauli exclusion principle. This means that only one electron can occupy each quantum state defined by a unique set of "quantum numbers". They therefore fill the available energy states within metallic iron, so that those in the highest energy states, near the Fermi level, have a large kinetic energy. This corresponds to a velocity of about 1% of the speed of light, and suggests that relativistic effects must be considered.

The quantum mechanics of relativistic electrons are described by the Dirac equation, from which it is found that electrons have an additional degree of freedom known as "spin". Spin is a form of angular momentum, which is sometimes visualized as the electron rotating about an axis through its center. Each electron has a magnetic moment associated with its spin. In most metals the electron spins are randomly oriented and no macroscopic magnetic moment is observed. For the Pauli exclusion principle to be satisfied the many body wave function of the electrons must change sign (be antisymmetric) when the coordinates of any pair of electrons are interchanged. The overall symmetry is the product of that of the spatial and spin parts of the wave function. In a ferromagnetic metal, due to the electrostatic "exchange" interaction between electrons, an anti-symmetric spatial wave function has the lowest energy. The spin wave function of the lowest energy state must therefore be symmetric, requiring the spins of different electrons to be aligned, and leading to the magnetization observed in Fe, Ni, and Co, the ferromagnetic transition metals. The situation in a real metal is complicated by the fact that electrons fill a number of different energy bands. These may experience a different exchange interaction, and the number of available states at a particular energy $N(E)$ may contain many fine features. However, we may consider an idealized case, shown in Fig. 1, where electrons occupy a single band. Spins may occupy either up or down states and the exchange interaction leads to an excess of up spins.

Many objects formed from magnetic metals exhibit no net magnetization unless they are subject to an external magnetic field. Let us consider a

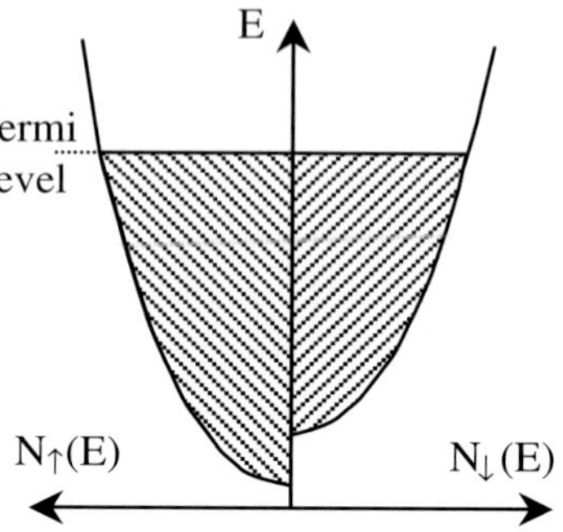

Fig. 1. The number of spin up $N_\uparrow(E)$ and spin down $N_\downarrow(E)$ electrons in an idealized ferromagnetic metal is shown in schematic form.

thin rectangular element of magnetic material. When fully magnetized the element generates a magnetic field, as shown in Fig. 2(a), in which energy is stored. This energy depends on the orientation of the magnetization and gives rise to "shape anisotropy". The energy is minimum when the magnetization lies parallel to the long side of the element, and maximum when the magnetization lies perpendicular to the plane of the element. However a lower energy state may be obtained by the formation of "domains" that generate no external field. The transitions between domains are not abrupt because the exchange interaction favors parallel alignment of neighboring spins. Instead a domain wall forms across which the spins rotate gradually. The width of the wall is limited by magnetocrystalline anisotropy. This means that it is energetically favorable to align the magnetization in certain directions relative to the crystallographic axes of the element. This anisotropy arises because the spin and orbital angular momentum of an

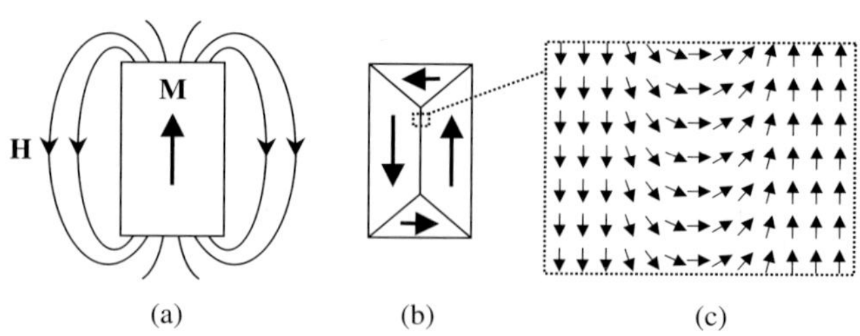

Fig. 2. (a) The stray field **H** generated by the magnetization **M** of a uniformly magnetized thin film element is shown. (b) A typical closure domain structure formed in the absence of an applied field is shown. No external field is generated in this case. (c) A Néel-type domain wall separating two anti-parallel domains is shown.

individual electron interact. In this way the spin senses the anisotropy of the orbital angular momentum that occurs because the spatial wave function must have the symmetry of the crystal lattice. The structure of a wall in a real material can be complicated. However if the element in Fig. 2 is made sufficiently thin then the spins in the domain wall rotate within the plane of the element to form a Néel wall. When a magnetic field is applied to the element in Fig. 2(b) the domain walls move so that the domain in which the magnetization is parallel to the applied field grows until the element is magnetized as in Fig. 2(a).

3. How Are Nanomagnets Different?

Let us define a nanomagnet as an object with at least one nanoscale dimension. Continuous films of nanometer thickness may be fabricated by vapor deposition. Electron beam lithography and focused ion beam etching can then be used to define nanoparticles or nanowires of a chosen size and shape. These techniques are rather slow and unsuitable for fabricating the large quantities of material needed for data storage. However, through careful control of the vapor deposition process alloys may be fabricated that self-segregate into a monolayer of weakly interacting nanoparticles. This is the method used to fabricate the coatings on magnetic hard disk drives, and much effort is needed to control the distribution of grain sizes and the strength of intergranular interactions. New methods, such as chemical and biological self-assembly and nanosphere lithography, have recently attracted interest as new means by which to achieve this aim.

Nanomagnets may behave very differently to their bulk counterparts. This may be because new crystallographic structures can be stabilized in a nanomagnet, or because the atoms at its surfaces and edges represent a larger fraction of the total volume. These atoms have fewer nearest neighbors and experience electric and magnetic fields of different symmetry to those within the interior of the crystal. They may possess a different magnetic moment and magneto-crystalline anisotropy that significantly modifies the overall behavior of the nanoparticle. Finally, as the size of the particle becomes comparable to the domain wall width, it is no longer possible to accommodate a domain structure. The response of the magnetization to an external field is then qualitatively different. The magnetization remains nearly uniform and rotates toward an applied field as the strength of the field is increased.

Completely new properties may be obtained when composition is modulated at the nanoscale. The spin-split band structure of a ferromagnetic metal, shown schematically in Fig. 1, underlies a class of giant magnetoresistance (GMR) effects.[1,2] Electrons close to the Fermi level are responsible for electrical conduction. The resistance increases with the rate of electronic scattering and hence with the number of vacant states available. Assuming that the electron spin is unaffected when the electron is scattered, electrons with spin parallel (spin up) or anti-parallel (spin down) to the majority spin direction will experience different resistance due to the different density of states at the Fermi level. When an ultrathin layer of nonmagnetic metal is used to separate two ferromagnetic metals, forming a "spin-valve" structure, electrons passing through the structure experience a resistance that depends on the local orientation of the magnetization. Conduction can be thought to occur through two parallel spin channels and the total resistance depends on whether the layer magnetizations are parallel or anti-parallel as shown in Fig. 3. Changes in resistance of tens of percent may be observed when a magnetic field causes the relative alignment of magnetizations to change. However the size of the GMR effect depends on the spin-dependent scattering time and also the longer spin relaxation time. These quantities are poorly known and difficult to measure.

Let us instead assume that the ferromagnetic layers are separated by an insulating layer as shown in Fig. 4. A potential energy barrier prevents the flow of current between the layers. However electrons may pass through a sufficiently thin barrier by quantum mechanical tunneling. The tunnel current is proportional to the number of electrons in the left-hand electrode and the number of vacant states in the right-hand electrode. Assuming that

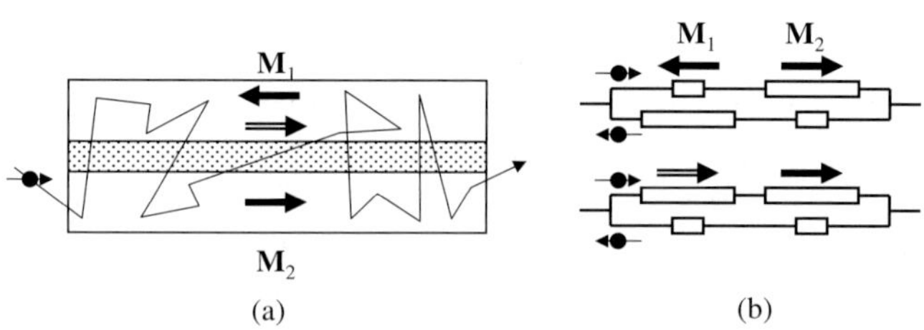

Fig. 3. (a) The trajectory of an electron within a spin-valve device is shown. (b) The effective resistances seen by the two spin types are shown for parallel and anti-parallel alignment of the magnetization of the two layers. Resistances are understood to be proportional to the length of the symbols used to represent them within the circuit diagram. Hence the parallel arrangement has lower resistance.

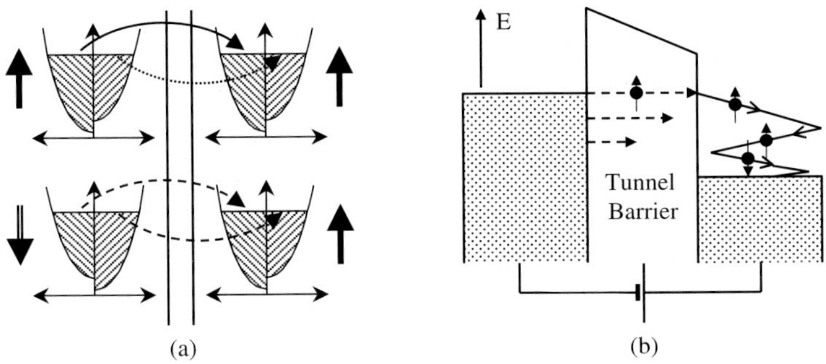

Fig. 4. (a) The tunnel process is shown schematically for parallel and anti-parallel alignment of the magnetizations of the magnetic layers. Since the spin is unchanged during tunneling, the process indicated by the solid arrow has greatest probability and so resistance is smallest for the parallel case where the magnetizations are parallel. (b) Injection and scattering of hot electrons within the second layer is shown.

spin is unchanged during tunneling, electrons of up and down spin tunnel in parallel. Changing the relative alignment of the layer magnetizations results in a tunnel magnetoresistance (TMR)[3] as shown in Fig. 4(a). A larger bias voltage allows electrons to be injected into the right-hand electrode at energies far above the Fermi level as shown in Fig. 4(b). These "hot" electrons are strongly scattered and lose their excess energy within tens of femtoseconds. However the spin dependence of the energy relaxation and the time taken for the spin of the hot electron to relax are not well known.

Tunnel barriers have the important property that they allow spin-polarized electric currents to be transferred between two materials of greatly different conductivity. They are one of the building blocks of a new technology known as "spintronics" in which the flow of charge and spin current is controlled by using applied electric and magnetic fields to act on both the charge *and* the spin of the conduction electrons. While semiconducting materials are usually preferred for the construction of conventional electronic devices, the development of room temperature ferromagnetic semiconductors is still at an early stage. Therefore tunnel junctions between metallic ferromagnet and semiconductor electrodes are expected to play an important role as sources and detectors of spin-polarized current.

Let us now consider one further fascinating feature of the trilayer structures that exhibit GMR and TMR. The conduction electrons within each electrode carry spin angular momentum that has quantization axis parallel to the local magnetization. When electrons are injected from the first electrode into the second electrode, the injected electrons must rapidly lose the spin polarization characteristic of the first electrode and adopt that characteristic of the second electrode. There is hence a transfer of angular momentum to the second electrode. Furthermore, this angular momentum can be transferred to the macroscopic magnetization of the second electrode.[4,5] In general the vector magnetizations of the first and second electrodes, $\mathbf{M}_1$ and $\mathbf{M}_2$, respectively, are not parallel and then it is helpful to consider separately the components of the transferred angular momentum that lie parallel and perpendicular to $\mathbf{M}_2$. The parallel component causes the magnitude of $\mathbf{M}_2$ to be slightly modified and is generally of less interest. In contrast, the perpendicular component can cause $\mathbf{M}_2$ to rotate and switch its orientation. The perpendicular component lies in the direction $\hat{\mathbf{M}}_1 - (\hat{\mathbf{M}}_1 \cdot \hat{\mathbf{M}}_2)\hat{\mathbf{M}}_2$, where $\hat{\mathbf{M}}_1$ and $\hat{\mathbf{M}}_2$ are unit vectors parallel to $\mathbf{M}_1$ and $\mathbf{M}_2$, respectively, which can be written as $-\hat{\mathbf{M}}_2 \times (\hat{\mathbf{M}}_2 \times \hat{\mathbf{M}}_1)$ through the use of a standard vector identity. Let us assume that, on average, each electron transfers angular momentum $\eta\hbar/2$, where η is an efficiency factor with value between 0 and 1. The rate at which angular momentum is transferred to unit volume of the second electrode is then equal to $\eta\hbar J/(2ed_2)$, in which J is the injected current density and d_2 is the thickness of the second electrode. Hence, the torque acting on the magnetization of the second electrode is equal to the rate of transfer of the perpendicular component of angular momentum and is given by

$$-\frac{\eta\hbar J}{2ed_2}\hat{\mathbf{M}}_2 \times \left(\hat{\mathbf{M}}_2 \times \hat{\mathbf{M}}_1\right). \tag{1}$$

This is the most commonly assumed form of the spin-transfer torque although additional terms are sometimes added on detailed consideration of microscopic models.[6] The value of η generally depends on the spin polarization of each electrode and the spin-dependent transmission coefficients of the different interfaces within the structure. Finally it should be mentioned that the injected current generates an additional classical magnetic field, as described by Ampère's law, which also exerts a torque on $\mathbf{M}_2$. This azimuthal magnetic field is spatially inhomogeneous and greatly complicates the resulting dynamical response. For a uniform current density, the magnitude of the classical field increases linearly with distance from the center of the structure. For this reason the structure should be no more than a few hundred nanometers in diameter if the spin-transfer torque is to dominate.[4]

4. Recording Technology and Speed Bottlenecks

The use of nanomagnets in data storage technology has been reviewed by Cowburn[7] and Kirk.[8] In this section we focus on the factors that limit speed of operation. Let us return to the magnetic hard disk drive.[9] It consists of the storage medium and a recording head containing a miniature electromagnet and a spin-valve sensor, as shown in Fig. 5(a). Fabrication of the recording head requires multiple stages of thin film deposition and photolithography. The pole pieces of the electromagnet generate the magnetic field that realigns the magnetization of the storage medium. The magnetization of the pole pieces switches by a domain wall process. Since wall velocities are of the order of 100 m/s, tens of nanoseconds may be required for a wall to propagate across a pole piece that extends tens of microns from the surface of the storage medium. Data is read out from the storage medium by using the stray field from the bit transitions to reorient the magnetization in one of the layers of a spin-valve sensor. The sensor normally has a built in biasing magnetic field so that the magnetization rotates in response to the stray field from the medium. The uniformity of this reorientation is highly important if the maximum GMR response is to be obtained.

Each "bit" of stored data is less than 100 nm long and is comprised of a large number of grains, with diameter of order 10 nm, that behave as single domain particles. These particles possess a strong uniaxial anisotropy, so that the magnetization occupies one of two stable states, parallel or

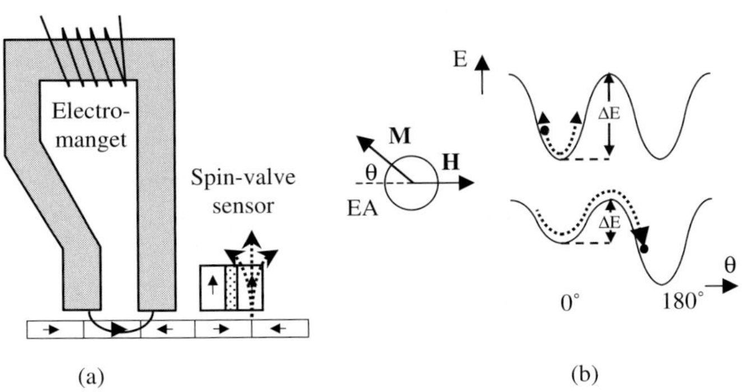

Fig. 5. (a) The essential components of a hard disk drive system are shown schematically. (b) A single grain from the storage medium is shown. The dependence of the magnetic free energy on the orientation of the magnetization both with (bottom) and without (top) an applied magnetic field is shown.

anti-parallel to the easy axis (EA) as shown in Fig. 5(b). The energy barrier between these two states must be sufficiently high that the magnetic moment remains in the desired state in spite of thermal energy fluctuations of order k_BT, where k_B is Boltzmann's constant and T is the temperature. The average time τ taken for the magnetization to move between the two states is roughly equal to $\tau_0 \exp(\Delta E/k_BT)$ where ΔE is the height of the energy barrier. Here τ_0 is the characteristic time with which the system oscillates about the minimum energy position. Its value is proportional to the anisotropy and is typically tens of picoseconds. For data to be stable for 10 years, ΔE must be greater than about 40 k_BT. During the write process an applied field reduces the height of the barrier so that the system may cross to the other minimum with the assistance of the thermal fluctuations. The height of the barrier is proportional to both the anisotropy and the particle volume. Although τ values of just a few nanoseconds are commonly achieved, these values still greatly exceed τ_0. On shorter time scales problems are encountered due to the distribution of grain sizes present, and because the nature of the oscillatory motion of the magnetization must be more carefully considered. As the grain size is reduced, the anisotropy must be made stronger if thermal stability is to be maintained. Larger anisotropy then requires larger fields to be applied during writing. Hard disk technology now faces a major difficulty because the write field is limited by the finite magnetization of the pole piece material. Despite extensive efforts, little progress has been made in developing materials with higher magnetization than the standard Fe–Co alloys.

Magneto-optical (MO) recording uses a different approach to read and write data as shown in Fig. 6.[10] Again data is stored on a disk that is moved beneath a stationary recording head. The disk coating consists of an amorphous rare earth–transition metal (RE–TM) alloy that has magnetization perpendicular to the plane. Bits are written by heating the coating with a focused laser beam in the presence of a reversed magnetic field. Heating causes the magnitude of the magnetization to be reduced and allows it to reverse more easily in the applied field. Data may be recorded by modulating either the laser beam or the magnetic field. The speed of writing is limited by the time taken to demagnetize the storage medium. This is achieved within a few nanoseconds using a low-power semiconductor laser. Further time may be required to modulate the applied magnetic field, as in a hard disk drive, and for motion of the domain wall that marks the edge of the bit. The latter effect is not yet a limiting factor for submicron bit sizes. Data is read out with the same laser by means of the magneto-optical

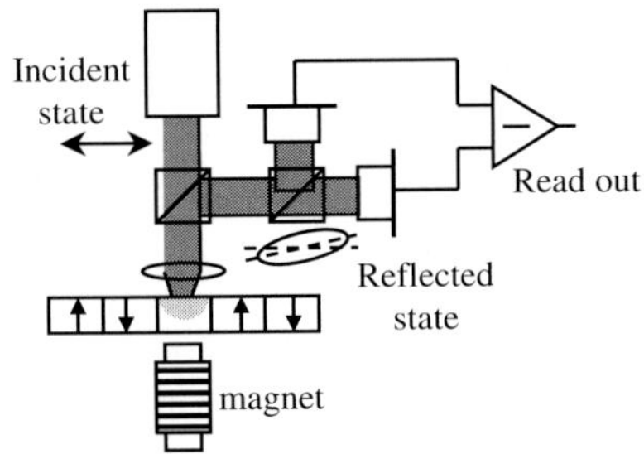

Fig. 6. The essential components of a magneto-optical recording system are shown. The same laser is used to heat the surface of the storage layer during writing and then read the written data.

Kerr effect (MOKE). Here the optical electric field transfers momentum to electrons in the surface of the metal film, which are then deflected due to the spin–orbit interaction before re-radiating light of modified polarization. If the incident light is plane polarized, the reflected light generally has a small ellipticity and has its principal plane of polarization rotated slightly relative to that of the incident beam. This Kerr rotation is proportional to the magnetization of the storage medium, and changes by about 1° as the orientation of the magnetization changes between in and out of the plane of the disk. Rotations of this magnitude are readily measured using an optical bridge arrangement as shown in Fig. 6. The reflected beam is passed through a beam splitting polarizer so that near equal intensity falls on the two photodiode detectors. The difference of their outputs is then proportional to the Kerr rotation. The time of interaction of the light with the storage medium is extremely short, a few femtoseconds only, and so the speed of the read out process is limited by other factors such as the bandwidth of the detector electronics.

Much effort has recently been devoted to the development of magnetic random access memory (MRAM) chips. Spin-valves and MTJs may possess the property of bistability, whereby parallel and anti-parallel alignment of the magnetizations gives two stable states. Arrays of these devices may be used to store large quantities of binary data that remains present when the power to the device is switched off. A matrix of address lines is used to read out the state of a selected element. These lines may also be used to propagate large current pulses that generate the magnetic field that switches the storage element between its stable states. Alternatively, the crossed address lines can be used to inject current through nanoscale elements so that the spin-transfer torque induces magnetic switching. In either case the speed of operation of MRAM is constrained by similar principles to those affecting the spin-valve sensor in a hard disk system.

5. Observing Ultrafast Magnetization Dynamics

The study of magnetic processes in microscopic samples on sub-nanosecond time scales has required the development of new experimental techniques. Optical "pump–probe" techniques are strongly favored because a focused laser beam can provide submicron spatial resolution, while the recent development of commercial ultrafast laser systems provides unrivaled temporal resolution. A pump–probe apparatus is shown in schematic form in Fig. 7(a). A Ti:sapphire laser produces 100-fs pulses at a repetition rate of 80 MHz. Each pulse is split into an intense pump and a weaker probe part. The pump is used to stimulate the sample while the probe is delayed in time and used to determine the instantaneous state of the sample at a later time. A retro-reflector on a translation stage is used to delay the probe pulse with a time resolution of about 1 fs. By measuring the response of the system at a series of different time delays, the dynamic behavior can be mapped out. The magnetic state of the sample is determined with a MOKE measurement using an optical bridge similar to that employed in an MO recording drive.

The pump pulse may be used to directly excite electrons within the surface of the sample. However, another way to excite a magnetic sample is to apply a magnetic field pulse generated by a device such as that shown in Fig. 7(b). A transmission line is deposited on a semi-insulating GaAs

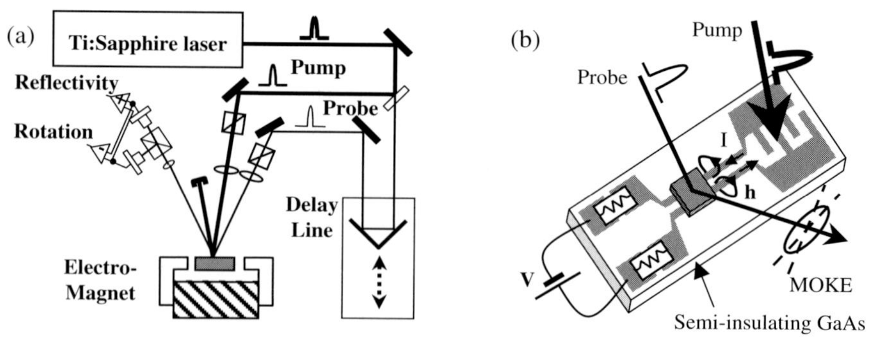

Fig. 7. (a) The optical pump–probe apparatus is shown in schematic format. (b) A device used to supply an optically triggered magnetic field pulse to the sample is shown.

substrate and a bias voltage applied. When the pump pulse is directed onto the exposed GaAs between the interlocking fingers at one end of the device, a photoconductive switch, electrons are excited to the conduction band of the GaAs, causing it to conduct. A current pulse then propagates along the transmission line, generating a magnetic field that interacts with the sample.[11] The pulsed field has a rise time of tens of picoseconds, and a nanosecond decay time that is related to the time taken for carriers to recombine within the GaAs. Fields of tens of gauss may be generated when the width of the tracks is reduced to a few tens of microns. This measurement technique is entirely free of electronic jitter. However if a temporal resolution of 1–10 ps is sufficient, the amplitude and profile of the exciting waveform can be varied more widely by replacing the photoconductive switch with an optically triggered pulse generator or microwave synthesizer.[12]

6. Harnessing Precession

From the previous discussion it was found that domain wall motion in micron scale elements, and thermally activated switching of single domain particles are essentially nanosecond processes. Let us now consider what may happen when a magnetic moment is stimulated by a magnetic field pulse with sub-nanosecond rise time. The response of the magnetization vector $\mathbf{M}$ is described by the Landau–Lifshitz–Gilbert (LLG) equation:

$$\frac{1}{|\gamma|}\frac{\partial \mathbf{M}}{\partial t} = -\mathbf{M} \times \mathbf{H}_{\text{eff}} + \frac{\alpha}{|\gamma| M}\left(\mathbf{M} \times \frac{\partial \mathbf{M}}{\partial t}\right), \tag{2}$$

in which the material parameters γ and α are the gyromagnetic ratio and Gilbert damping parameter, respectively. An effective field $\mathbf{H}_{\text{eff}}$ is used to describe the effect of the applied field and other contributions to the free energy of the system. When $\mathbf{H}_{\text{eff}}$ is independent of time, the first term in Eq. (2) causes the magnetization vector to sweep out a cone with axis parallel to $\mathbf{H}_{\text{eff}}$. This is referred to as magnetization "precession". The second term damps the motion and causes the angle of the cone to decrease with time. For the trilayer structures described previously, the expression (1) may be added to the right-hand side of Eq. (2) to describe the effect of the spin-transfer torque on $\mathbf{M}_2$. The spin-transfer torque acts like an effective magnetic field in stimulating the precessional motion.

Figure 8 shows the result of a measurement[13] made on a circular element of $Ni_{81}Fe_{19}$ with diameter of 30 μm and thickness of 50 nm. The element

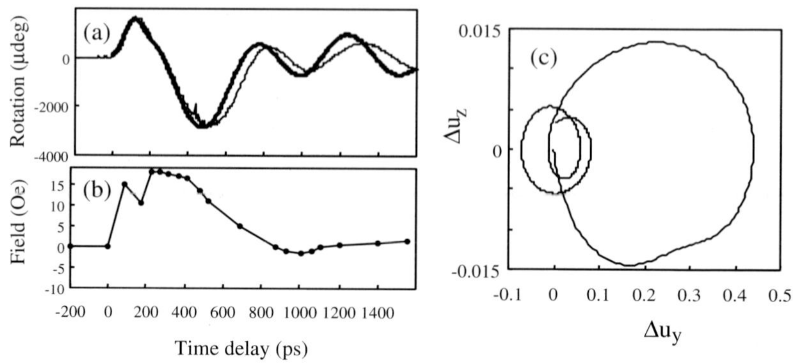

Fig. 8. (a) The measured (fine) and calculated (bold) time dependent Kerr rotation from a $Ni_{81}Fe_{19}$ dot of 30-μm diameter are shown. (b) The form of the pulsed field is shown. (c) The calculated trajectory of the normalized magnetization vector $\mathbf{u} = \mathbf{M}/M$ is shown. The z-axis lies normal to the plane of the dot.

was placed on a track of a transmission line, similar to that in Fig. 7, so that the pulsed field lay in the plane of the element, 140° from a static field of 54 Oe. A focused probe spot with diameter of about 15 μm was placed at the center of the element. The time variation of the pulsed field was determined by a magneto-optical sampling technique and is shown in Fig. 8(b). The magnetization initially lies close to the static field but is then deflected by the pulsed field as shown in Fig. 8(c). The trajectory is very flat because the shape anisotropy of the element prevents the magnetization tipping very far out of the film plane. A maximum in-plane deflection of about 30° occurs after 350 ps, demonstrating that large-angle reorientation is possible on sub-nanosecond time scales. However the precession is lightly damped and continues for a number of nanoseconds afterward.

A static field was applied in the experiment to reset the magnetization to its initial value before the arrival of the next pump pulse. However, the LLG equation can be used to predict the trajectory in the absence of a static field as in Fig. 9. The magnetization, $\mathbf{M}$, initially lies parallel to the in-plane uniaxial anisotropy axis. The anisotropy field is equal to 12 Oe (as in Fig. 8). When a pulsed field of 10 Oe is applied along the y-axis for 700 ps, the magnetization begins by moving out of the film plane, parallel to the z-axis. Due to the shape anisotropy of the thin film, a large out-of-plane contribution to $\mathbf{H}_{\text{eff}}$ develops that deflects $\mathbf{M}$ toward the y-axis. The angular deflection of $\mathbf{M}$ exceeds 90° after about 570 ps and then spirals down toward the anisotropy axis in the reverse direction. The first observations

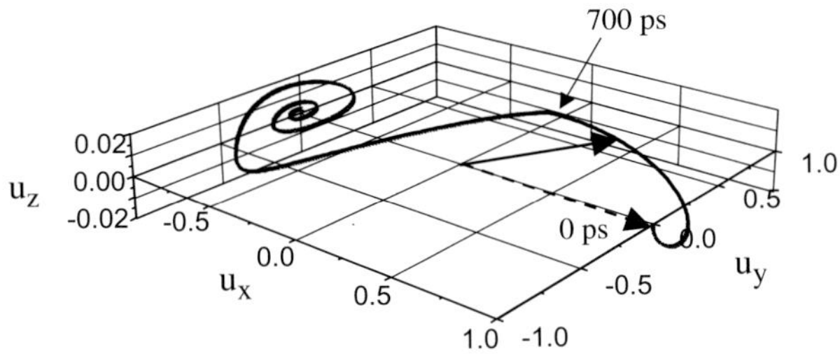

Fig. 9. A simulation of magnetization reversal of the element in Fig. 8 is shown. There is no static field and the trajectory is plotted over a period of 4000 ps.

of full precessional switching were recently reported.[14,15] The reversal time can be further reduced by increasing the anisotropy field of the element and the magnitude of the pulsed field. So far we have assumed that the magnetization of the element rotates uniformly. By focusing the probe to a submicron spot we can check the validity of this assumption. The time delay between pump and probe is fixed and the sample moved under the stationary probe spot, using a computer-controlled piezoelectric translation stage, to acquire a dynamic image. Images of the out-of-plane component of the magnetization within a 10 μm square element of $Ni_{81}Fe_{19}$, of 150 nm thickness, are shown in Fig. 10.[16] The static field of 288 Oe and the pulsed field, with peak value of 27 Oe, were applied in the plane of the element at right angles to one another. The magnetization quickly becomes nonuniform and the images suggest that standing waves are generated with wave vector parallel to the static field. In general, in microscale elements,

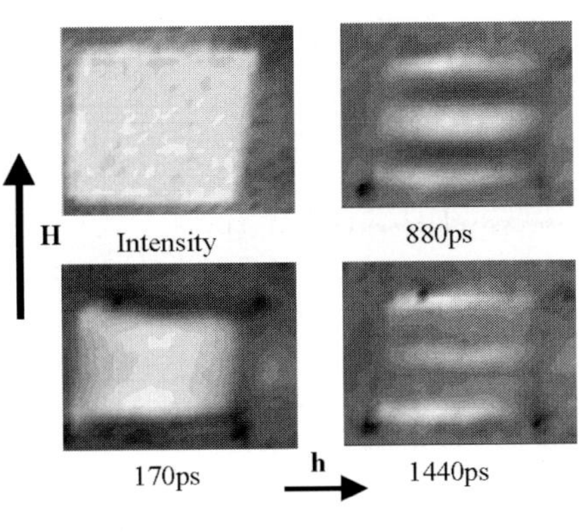

Fig. 10. Images of the static reflectivity, and the out-of-plane component of magnetization at different time delays are shown for a 10-μm square element.

these spin wave excitations are confined within either the center region of the element or near the edges that lie perpendicular to the static applied field.[17] Recently, the first measurements made on elements within the deep nanoscale regime showed that edge modes dominate the response of a square element to a pulsed field when the element is less than 200 nm in length.[18] This observation has serious implications for future recording technology where a spatially uniform magnetic response is normally required.

7. Optical Modification of the Spontaneous Magnetization

While a pulsed field may modify the orientation of the local magnetization, it has negligible effect on its magnitude. However the writing in MO recording requires a reduction of the magnetization. The pulses from a typical ultrafast laser oscillator have modest energy, of order 10 nJ, yet very high peak intensity, of order 100 kW. When the laser pulse is incident on a metallic surface, it transfers a large amount of energy to the conduction electrons in a very short time. These electrons become very hot while the lattice temperature is almost unchanged. The effect of pumping a nickel thin film with ultrashort pulses of different helicity is shown in Fig. 11(a).[19] The *change* in the Kerr rotation of the probe pulse is plotted, so that the step observed with a linearly polarized pump beam corresponds to a *reduction* of the magnetization of about 10%. The peak demagnetization signal is obtained after about 200 fs. Experiments with 20-fs laser pulses have shown that demagnetization may be achieved in less than 50 fs[20] in a $CoPt_3$ alloy, while theoretical investigations[21] suggest an intrinsic time scale as short as 10 fs.

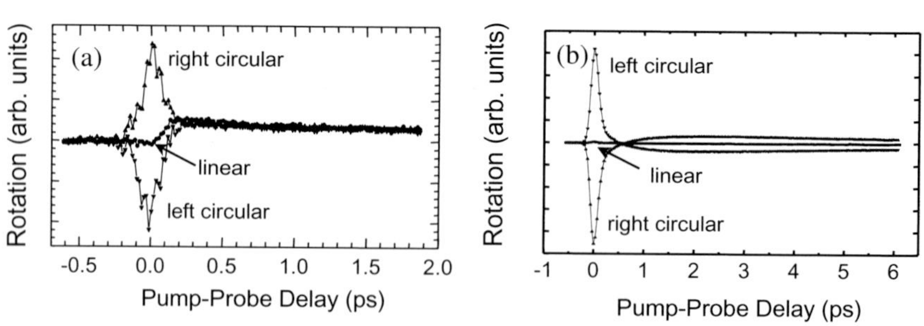

Fig. 11. The transient Kerr rotation obtained from (a) an Ni thin film and (b) an intrinsic GaAs wafer is shown. The curves are labeled according to the helicity of the pump pulse.

When the pump beam is circularly polarized, a peak is observed at zero time delay with width limited by the laser pulse width. The circularly polarized light transfers angular momentum to the electrons in the metal and so induces a magnetic moment. The electrons are scattered on time scales of the order of 1 fs so that the orbital magnetic moment is soon destroyed. However in some circumstances spin–orbit coupling can cause the conduction electrons to obtain a net spin polarization that is much longer lived. This may be seen in the measurements made on GaAs in Fig. 11(b) where a long tail in the Kerr signal is observed. The decay of this tail allows the spin relaxation time of electrons in the conduction band to be deduced. Having created a nonequilibrium spin polarization, it is then possible to manipulate its orientation through precession in an applied magnetic field, or, with the electric field from a second intense optical pulse, by means of the optical Stark effect.[22] Optical manipulation of spin hence provides another approach in the development of spintronic technology.

Optical experiments may potentially give access to a variety of other electronic relaxation times. The time resolved two photon photoemission (TRTPPE) technique is shown schematically in Fig. 12. The pump pulse is used to excite electrons to vacant states above the Fermi level, after which the excited electrons cascade back toward the Fermi level within a few hundred femtoseconds. The probe pulse may be used to promote some of the hot electrons to above the vacuum level before they relax. They are then emitted from the sample surface where an electric field is used to select electrons of a specific energy. Using a spin-sensitive electron detector the spin dependence of the energy relaxation process may be deduced. Measurements made on Co (Ref. 23) have shown that at 1 eV above the Fermi level, minority spin electrons relax at about twice the rate of those with majority spin.

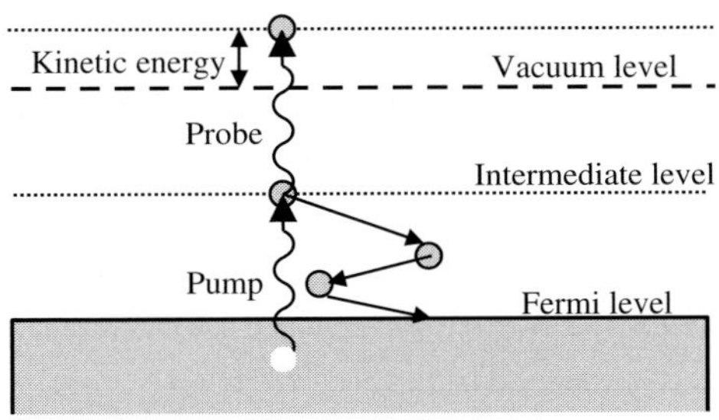

Fig. 12. The principle of the TRTPPE technique is illustrated.

8. Future Trends

The examples given in the previous two sections suggest that we are some way from the fundamental speed limits relevant to magnetic data storage technology. Magnetic switching of the transducers in a hard disk drive can clearly be pushed into the picosecond regime if precessional motion is utilized instead of domain wall motion. However this may require some redesign of existing heads since precession is most easily stimulated when the pulsed field is applied perpendicular rather than anti-parallel to the magnetic moment that is to be switched. It is also important to curtail precession after switching has been achieved. Otherwise the moment might continue to rotate through a full 360° and so regain its initial orientation. Such control may be achieved by tailoring the pulse duration to the period of precession.[24,25] Further increases in switching speeds require the frequency of precession to increase through the use of larger anisotropies and stronger pulsed fields.

There is certainly scope for further reduction of the time, τ, taken for thermally activated switching of storage media. Media with larger anisotropy are needed to maintain thermal stability at higher storage densities and so the value of τ_0, the period of precession, will continue to decrease. Tighter control of grain size distributions may also allow τ to be reduced to a smaller fraction of τ_0. Ultimately the precessional character of the motion will have to be taken into account, with further implications for the design of recording transducers. Since generation of the necessary write fields may be difficult to achieve, serious consideration is being given to mechanisms that may assist switching. Heating the storage medium during the write process can enhance the effect of thermal activation and near-field optical sources[26] could be incorporated into the recording head so that the heating is spatially localized. Very recently there have also been the first reports that circularly polarized light can directly exert a torque on the magnetization within certain magnetic materials.[27] An additional effective magnetic field can be generated by an optically triggered "exchange-bias" mechanism. Optical heating of an FeRh underlayer is known to induce an ultrafast phase transition[28,29] from an antiferromagnetic to a ferromagnetic state that can exert a torque on the magnetization of the storage layer.[30] Alternatively, spin-transfer torque can generate precession in the GMR sensor leading to the emission of microwaves that could in turn stimulate precession of the magnetization of the storage medium. If the associated tribological problems could be solved then it might be possible to inject current from the recording head into the storage medium so that switching could be induced directly by means of the spin-transfer torque. Finally, a further possibility might be to dispense with the traditional writer structure altogether. The transmission line structure used in pump–probe experiments has a bandwidth of tens of gigahertz, but so far the maximum field has been limited to about 1 kOe. The field is inversely proportional to the linear dimensions of the transmission line, and so much larger fields might be generated by a much smaller structure designed to switch a single nanoscale bit.

It is now clear that intense laser pulses can be used to change the spin polarization of certain materials on femtosecond time scales. Laser technology continues to evolve rapidly and battery powered femtosecond lasers have been demonstrated.[31] We can even imagine ultrafast lasers being used in future recording systems. Studies of ultrafast demagnetization have already shown that erasure times in MO recording could be reduced to tens of femtoseconds. More serious constraints are introduced by the attendant electronics and the need to switch an applied magnetic field on comparable time scales. Consequently, for speed of operation, an all-optical technology will always be favored. If circularly polarized light could be used to reset the spin polarization of the demagnetized medium, then the entire write process might be completed within 1 ps. There is also no reason why such a process should be confined to rotating disks systems. Future architectures might involve steerable beam access, or the incorporation of magnetic materials into photonic chips where they could provide on-board memory function. Indeed all-optical magnetic recording would be an important step toward the further integration of computer and telecommunications technology.

Ultrafast optical techniques will continue to be used to increase our understanding of the processes that underlie electronic device operation, and magneto-optical probes are of particular relevance to spintronics. While spin relaxation times have been determined in semiconductors, there is a pressing need to apply similar techniques to the metals from which existing GMR devices are constructed. It will also be necessary to determine the characteristic time scales for spin-dependent momentum and energy relaxation, which may range from 1 to 100 fs. Optical techniques seem like the most viable candidate for such studies as laser pulse widths are reduced to the attosecond regime. The TRRPE technique has provided an indication of what may be possible. The challenge now is to develop methods that allow the important relaxation processes to be investigated within more complicated structures such as spin valves. Recent developments in laser technology promise to make this possible. Ultrafast lasers are now available that allow the photon wavelength to be tuned through the entire visible spectrum. By tuning to optical resonances in the electronic band structure of the constituent materials it may be possible to characterize spin-dependent processes within different regions of a heterogenous nanoscale structure.

In conclusion, there is an urgent need to reduce read and write times within magnetic data storage technology in line with increased storage densities. Existing systems operate on nanometer length scales, where new phenomena such as single domain switching, GMR and spin-transfer torque have emerged, and on nanosecond time scales. The fundamental limits on speed of operation have been assessed and we have seen that optical measurement techniques allow dynamic magnetic processes to be investigated with femtosecond temporal resolution. While the speed of recording systems will undoubtedly increase, the nanosecond barrier represents a watershed beyond which the precessional nature of magnetization reversal must be taken into account. Ultimately magneto-optical technology may offer the highest possible record and replay rates. Indeed magnetic data storage is far from the limits of what may be achieved with this new light.

Acknowledgment

The financial support of the Engineering and Physical Sciences Research Council is gratefully acknowledged.

References

1. M. N. Baibich, J. M. Broto, A. Fert, F. Nguyen Van Dau, F. Petroff, P. Etienne, G. Creuzet, A. Friederich and J. Chazelas, Giant magnetoresistance of (001)Fe/(001)Cr magnetic superlattices, *Phys. Rev. Lett.* **61**, 2472 (1988).
2. G. Binasch, P. Grunberg, F. Saurenbach and W. Zinn, Enhanced magnetoresistance in layered magnetic structures with antiferromagnetic interlayer exchange, *Phys. Rev. B* **39**, 4828 (1989).
3. M. Julliere, Tunneling between ferromagnetic films, *Phys. Lett. A* **54**, 225 (1975).
4. J. C. Slonczewski, Current-driven excitation of magnetic multilayers, *J. Magn. Magn. Mater.* **159**, L1 (1996).
5. L. Berger, Emission of spin waves by a magnetic multilayer traversed by a current, *Phys. Rev. B* **54**, 9353 (1996).
6. J. Xiao, A. Zangwill and M. D. Stiles, Macrospin models of spin transfer dynamics, *Phys. Rev. B* **72**, 014446 (2005).

7. R. P. Cowburn, The attractions of magnetism for nanoscale data storage, *Phil. Trans. R. Soc. Lond. A* **358**, 281 (2000).
8. K. J. Kirk, Nanomagnets for sensors and data storage, *Contemp. Phys.* **41**, 61 (2000).
9. D. Weller and A. Moser, Thermal effect limits in ultrahigh-density magnetic recording, *IEEE Trans. Magn.* **35**, 4423 (1999).
10. S. Tsunashima, Magneto-optical recording, *J. Phys. D: Appl. Phys.* **34**, R87 (2001).
11. M. R. Freeman, R. R. Ruf and R. J. Gambino, Picosecond pulsed magnetic fields for studies of ultrafast magnetic phenomena, *IEEE Trans. Magn.* **27**, 4840 (1991).
12. S. Tamaru, J. A. Bain, R. J. M. van de Veerdonk, T. M. Crawford, M. Covington and M. H. Kryder, Imaging of quantized magnetostatic modes using spatially resolved ferromagnetic resonance, *J. Appl. Phys.* **91**, 8034 (2002).
13. J. Wu, D. S. Schmool, N. D. Hughes, J. R. Moore and R. J. Hicken, Picosecond large angle reorientation of the magnetisation in $Ni_{81}Fe_{19}$ circular thin film elements, *J. Appl. Phys.* **91**, 278 (2002).
14. Th. Gerrits, H. A. M. van den Berg, J. Hohlfeld, L. Bar and Th. Rasing, Ultrafast precessional magnetisation reversal by picosecond magnetic field pulse shaping, *Nature* **418**, 509 (2002).
15. H. W. Schumacher, C. Chappert, R. C. Sousa, P. P. Freitas and J. Miltat, Quasiballistic magnetisation reversal, *Phys. Rev. Lett.* **90**, 017204 (2003).
16. R. J. Hicken, A. Barman, V. V. Kruglyak and S. Ladak, Optical ferromagnetic resonance studies of thin film magnetic structures, *J. Phys. D: Appl. Phys.* **36**, 2183 (2003).
17. J. Jorzick, S. O. Demokritov, B. Hillebrands, M. Bailleul, C. Fermon, K. Y. Guslienko, A. N. Slavin, D. V. Berkov and N. L. Gorn, Spin wave wells in nonellipsoidal micrometer size magnetic elements, *Phys. Rev. Lett.* **88**, 47204 (2002).
18. V. V. Kruglyak, A. Barman, R. J. Hicken, J. R. Childress and J. A. Katine, Picosecond magnetization dynamics in nanomagnets: Crossover to nonuniform precession, *Phys. Rev. B* **71**, 220409R (2005).
19. R. Wilks, N. D. Hughes and R. J. Hicken, Investigation of transient linear and circular birefringence in metallic thin films, *J. Phys.: Condens. Matter* **15**, 5129 (2003).
20. L. Guidoni, E. Beaurepaire and J.-Y. Bigot, Magneto-optics in the ultrafast regime: Thermalization of spin populations in ferromagnetic films, *Phys. Rev. Lett.* **89**, 17401 (2002).
21. G. P. Zhang and W. Hübner, Femtosecond spin dynamics in the time domain, *J. Appl. Phys.* **85**, 5657 (1999).
22. J. A. Gupta, R. Knobel, N. Samarth and D. D. Awschalom, Ultrafast manipulation of electron spin coherence, *Science* **292**, 2458 (2001).
23. M. E. Aeschlimann, M. Bauer, S. Pawlik, W. Weber, R. Burgermeister, D. Oberli and H. C. Siegmann, Ultrafast spin-dependent electron dynamics in fcc Co, *Phys. Rev. Lett.* **79**, 5158 (1997).

24. M. Bauer, R. Lopusnik, J. Fassbender and B. Hillebrands, Suppression of magnetic-field pulse-induced magnetization precession by pulse tailoring, *Appl. Phys. Lett.* **76**, 2758 (2000).
25. T. M. Crawford, P. Kabos and T. J. Silva, Coherent control of precessional dynamics in thin film permalloy, *Appl. Phys. Lett.* **76**, 2113 (2000).
26. T. W. McDaniel and W. A. Challener, Issues in the design of media for hybrid recording, *Trans. Mag. Soc. Jpn.* **2**, 316 (2002).
27. A. V. Kimel, A. Kirilyuk, P. A. Usachev, R. V. Pisarev, A. M. Balbashov and Th. Rasing, Ultrafast non-thermal control of magnetization by instantaneous photomagnetic pulses, *Nature* **435**, 655 (2005).
28. G. Ju, J. Hohlfeld, B. Bergman, R. J. M. van deVeerdonk, O. N. Mryasov, J.-Y. Kim, X. Wu, D. Weller and B. Koopmans, Ultrafast generation of ferromagnetic order via a laser-induced phase transformation in FeRh thin films, *Phys. Rev. Lett.* **93**, 197403 (2004).
29. J.-U. Thiele, M. Buess and C. H. Back, Spin dynamics of the antiferromagnetic-to-ferromagnetic phase transition in FeRh on a sub-picosecond time scale, *Appl. Phys. Lett.* **85**, 2857 (2004).
30. J.-U. Thiele, S. Maat and E. E. Fullerton, Exchange spring films for thermally assisted magnetic recording media, *Appl. Phys. Lett.* **82**, 2859 (2003).
31. B. Agate, A. J. Kemp, C. T. A. Brown and W. Sibbett, Efficient, high repetition-rate femtosecond blue source using a compact Cr:LiSAF laser, *Opt. Exp.* **10**, 824 (2002).

Robert J. Hicken was born in Nantwich, Cheshire in 1965. He read Physics at Brasenose College, Oxford, graduating in 1986, and obtained his PhD in Physics from the Johns Hopkins University in 1991. Following a postdoctoral appointment at Cambridge University, he became a Lecturer at Exeter University in 1996, was awarded an Engineering and Physical Sciences Research Council (EPSRC) Advanced Research Fellowship in 1999, and became Professor of Condensed Matter Physics in 2006.

CHAPTER 11

NEAR-FIELD MICROSCOPY: THROWING LIGHT ON THE NANOWORLD

David Richards

Department of Physics, King's College London
Strand, London WC2R 2LS, UK

Optical imaging with nanoscale resolution, beyond that possible with conventional diffraction-limited microscopy, may be achieved by scanning a nano-antenna in close proximity to a sample surface. This review will first aim to provide an overview of the basic principles of the technique of scanning near-field optical microscopy, before moving on to consider its most widely implemented form, in which the sample is illuminated through a small aperture held less than 10 nm to the sample surface, for optical imaging with a resolution of about 50 nm. The exciting new possibilities for high-resolution optical imaging and spectroscopy promised by "apertureless" near-field optical microscopy are then considered. Such techniques may involve local scattering of light from a sample surface by a tip, local enhancement of an optical signal by a metal tip, or the use of a fluorescent molecule or nanoparticle attached to a tip as a local optical probe of a surface. These new optical nanoprobes offer the promise of optical microscopy with true nanometer spatial resolution.

Keywords: SNOM, NSOM, nano-optics, optical microscopy.

1. Introduction

1.1. *The need for nanoscale resolution optical microscopy*

There is an ever-increasing need for optical imaging techniques for nanometer-scale measurements and analysis. That is, optical techniques with the ability to resolve features with sizes in the range of 1 nm to 1 μm, the length-scales relevant for biological cells, large biological molecules and complexes, and for the rapidly developing class of materials known as nanostructures — artificial materials and molecular assemblies with sub-micrometer dimensions, fabricated for an ever-increasing range of applications. For example, techniques which enable the study of processes in cells, or even allow the investigation of a single molecule or nanostructure, will provide essential information for drug discovery, medicine, and nanotechnology, while the development of new optical and opto-electronic devices for optical communications technology requires techniques which are able to map the flow of light and capture localized optical fields on the nanometer scale.

There exists already a range of well-developed techniques which provide high resolving power for the visualization of materials. These include electron microscopy, which employs a beam of electrons to probe a material, and atomic force microscopy (AFM), which maps out the relief of a surface with a sharp needle. However, the use of light for microscopy offers many distinct advantages over other approaches. It is nondestructive and can operate under a wide range of conditions unlike, for example, electron microscopy. But in particular, *optical* microscopy offers access to unique information. We can study the structural and dynamic properties of matter through our knowledge of the way light interacts with it. We can make use of different wavelengths of light for spectroscopy to enable spectral contrast in imaging, with the chemical specificity that this brings. We can also exploit another dimension, time, to monitor processes of change as they happen, down to femtosecond temporal resolutions. To take a specific but increasingly important example, it has now become standard practice in cell and molecular biology to use fluorescent tags. These tags are molecules which emit light of a well-defined wavelength range when illuminated and can be used as labels for particular biological molecules, or to flag the occurrence of specific processes. The ability to identify, using spectral contrast, these fluorescent molecules with high spatial resolution is essential.

1.2. *Breaking the diffraction limit*

So the need for high-resolution microscopy is clear. The problem is that conventional optical microscopy, in which a sample is imaged with a lens, is unable to provide the resolution necessary to access nanometer length-scales. In fact the spatial resolution attainable, the size Δx of the smallest object we can visualize, is fundamentally limited by diffraction to

$$\Delta x = \frac{\lambda}{2NA}, \tag{1}$$

where NA is the numerical aperture of the lens. We can see from the above expression that the resolution is therefore limited to the order of the wavelength λ of light, which is in the range 0.4–0.8 μm in the visible part of the electromagnetic spectrum. This diffraction limit can be attained using *confocal* techniques, in which a point source or laser beam is used for excitation. A high numerical aperture lens focuses the light onto the sample and then collects the emitted or scattered light, which is focused onto a pinhole in the image plane of the lens. An image is formed by raster scanning the focal point across the sample, enabling the achievement of resolutions of ~200 nm with the use of oil immersion objective lenses. So if we are to probe optically molecules and nanostructures with true nanometer resolution, somehow the diffraction limit must be overcome.

By exploiting nonlinear optical properties of fluorescent labels, three-dimensional (3D) imaging with 30 nm resolution has in fact been achieved using lens-based techniques, such as the stimulated emission depletion (STED) microscopy of Hell and co-workers.[1] Alternatively, the diffraction limit may be circumvented altogether by performing measurements with the optical source or detector held much closer to the sample than the wavelength of the light. In this *near-field* regime the attainable resolution is now determined by the size of the source or the detector and not by the wavelength of the light. One way in which this may be achieved combines optical spectroscopy with scanning probe microscopy: the family of techniques which includes AFM and scanning tunneling microscopy (STM).

1.3. *Scanning near-field optical microscopy*

The new technique, scanning near-field optical microscopy (SNOM), the subject of the present chapter, exploits the optical interaction between a probe and the sample to investigate surfaces with a resolution far higher than that of traditional optical microscopy techniques.[2] In the most widely implemented form of this microscopy (to which we will refer as aperture-SNOM) a small aperture is held in close proximity (<10 nm) to the sample surface (see Fig. 1) and raster scanned across the sample. The aperture can either operate as an emitter (i.e., as a nanoscopic light source to illuminate the sample) or as a receiver (to collect light from the sample), or both. An optical signal is detected at a number of points in each line of the raster-scan, to form an optical image. This form of microscopy was in fact first proposed in 1928 by Synge.[3] However, it was not until the late 1980s that it was actually realized for the first time, following the invention of STM

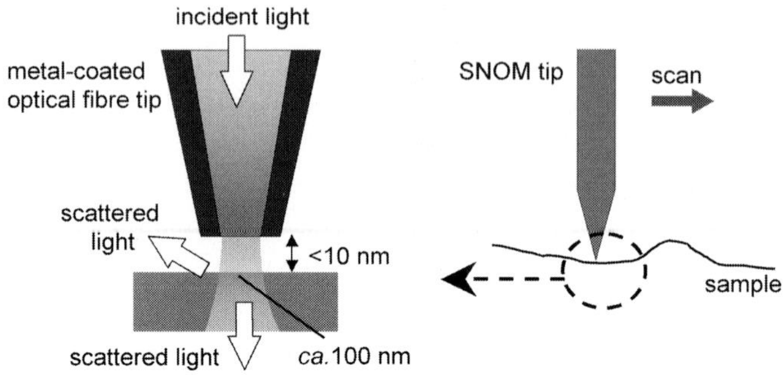

Fig. 1. A schematic illustration of aperture-SNOM. An optical fiber tip is scanned across a sample surface (right) to form an image. The tip is coated with metal everywhere except at the apex, where a small aperture has been created. It is held within a few nanometers of the sample surface so that a region $< 100\,\text{nm}$ across is illuminated (left).

which provided the technology to position a tip within a few nanometers of a sample surface and scan with sub-nanometer resolution. Images with a resolution of about 50 nm can be obtained with aperture-SNOM, which is just now coming of age with reliable systems appearing in the market. Further details about the implemetation of aperture-SNOM and an example of its application are provided in Sec. 2.

1.4. *Nano-optics: The path toward nanometer optical resolution*

The subsequent progress in near-field microscopy has lead to development of the new area of research of "nano-optics", concerned with the interaction of light and matter on the nanoscale, the manipulation of light in sub-wavelength dimensions and nanolocal spectroscopy. In particular, this has led to exciting new possibilities for high-resolution optical imaging, promised by "apertureless"-SNOM, so-called because it does not rely on the illumination of the sample through a small aperture. Apertureless approaches may involve local scattering of light from a sample surface, local enhancement of an optical signal, or local fluorescence (e.g., light emission from a single molecule attached to a scanning tip). A scheme for the realization of apertureless-SNOM is illustrated schematically in Fig. 2. Apertureless-SNOM is considered in greater detail in Secs. 3 and 4.

As well as providing nanometer resolution optical imaging of surfaces, a powerful further feature of near-field optical microscopy results from the requirement for the near-field probe (whether this is an aperture or otherwise) to be held a short distance above the sample surface at all times.

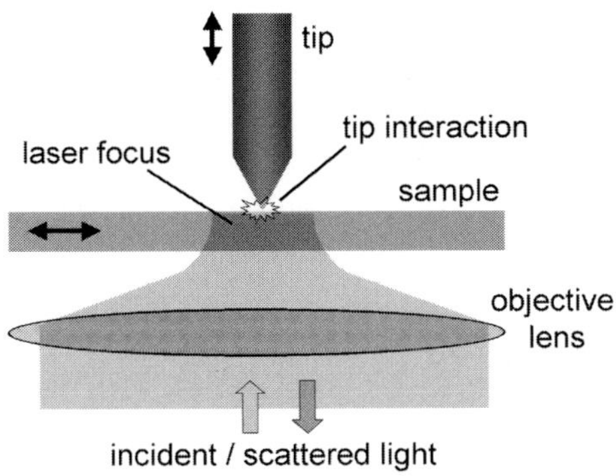

Fig. 2. In apertureless-SNOM high-resolution optical information may be provided by the scattering of light or signal enhancement by a very sharp metal tip, or excitation of a nanoscopic light source on the tip. The sample is scanned beneath the tip to provide an image.

In this way sample topography is mapped simultaneously with the formation of the optical image of the sample as the tip is scanned over the sample surface, allowing a correlation between optical properties and topography and enabling a comparison with the more mature technique of AFM. Furthermore, although a surface-sensitive technique, being optical in nature near-field microscopy, and aperture-SNOM in particular, is able to probe systems beneath the surface of a transparent material.[4]

Near-field optical microscopy has been demonstrated to have great potential for application to the study of nanostructured semiconductor, ferromagnetic, photonic, biological, liquid crystal, and single molecular systems and this powerful family of techniques clearly has great potential for application in chemistry, physics, materials, and the life sciences (see, e.g., a collection of reviews edited by Metiu[5] and by Richards and Zayats[6]). We now stand at the dawn of an exciting new era in optical microscopy, with the development of new probes with great optical and chemical specificity and resolutions to <10 nm looking an attainable goal.

2. Aperture-SNOM

2.1. *Implementation*

In the most common implementation of aperture-SNOM, an aperture of 50–100 nm diameter is made at the apex of a metallized optical fiber tip. Light can then be easily sent to the aperture through the optical fiber to illuminate the sample, with scattered light or fluorescence from the sample collected using a microscope objective lens and then spectrally analyzed, using filters or a spectrometer, before detection with a sensitive detector. This implementation for SNOM is illustrated schematically in Fig. 1.

The tip of an SNOM probe may be prepared by etching a single-mode optical fiber in hydrofluoric acid, or pulling the fiber to a tip. The optical fiber tip is then coated with aluminum and an aperture of size $<100\,\text{nm}$ may be created by shadowing the tip apex from the evaporation source when coating with aluminum, or using a focused-ion-beam to produce a well-defined aperture in the metal coating (see Fig. 3(a)).[7]

The tip–sample distance for such optical fiber probes is controlled via shear-force feedback. This is achieved by laterally dithering the tip with respect to the sample and detecting the shear-force damping of this motion due to the interation between tip and sample. Detection of the tip dither amplitude is commonly performed using a small quartz tuning-fork to which the probe is attached.[8] Figure 3(b) shows an optical fiber glued to the side of one prong of a quartz tuning fork. By keeping this shear-force damping constant, it is possible to hold the tip just a few nanometers above a sample surface. For measurements under ambient conditions it is likely that the shear-force results from the interaction of the tip with a thin film of water on the surface.[9]

2.2. *Near-field fluorescence microscopy of light-emitting polymer blends*

As an example of the application of aperture-SNOM, one class of materials we have investigated with this technique is that of thin film blends

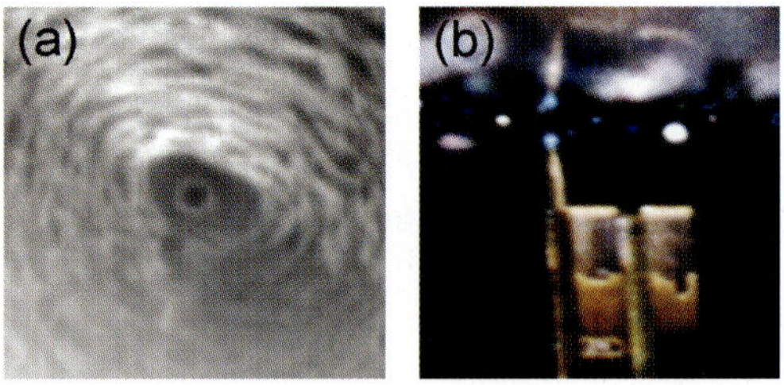

Fig. 3. (a) End-on view of an aperture-SNOM tip. The black circle in the center is an aperture of 100 nm diameter which has been drilled in the aluminum coating over the optical fiber tip using a focused beam of gallium ions. (b) The blue-green emission (wavelength 488 nm) from an optical fiber SNOM tip, and its reflection in the sample above, can be seen in the figure. The optical fiber is glued along one arm of a quartz tuning fork (of width 0.5 mm) which is used for the control of the tip–sample distance.

of light-emitting polymers, which have a wide range of application from solar cells[10] to large area displays.[11] A polymer blend is a thin film containing a mixture of two or more polymers, which may be selected as the optimum materials for each relevant process in an emissive device (charge transport, charge injection, luminescence) or for optimal charge generation and transport in a photocell. The different components of a polymer blend, a mixture of two or more polymers, will often separate into different phases on a length-scale in the range of tens to a few hundreds of nanometers. In this way a large interfacial area is created between the different polymers to optimize charge and energy transfer between them for further enhanced device performance.

Although the morphology and phase separation properties of these systems have been studied by various techniques, such as AFM and transmission electron microscopy (TEM), near-field optical microscopy is required for the study of the optical properties of the blends on the length-scale of the phase separation, which are beyond the resolution limit of a conventional optical microscope. Furthermore, TEM involves staining of one of the phases which may be impossible or undesirable. SNOM has the advantage over AFM of allowing the immediate discrimination of different materials on the basis of their spectral properties — analyzing the color of the fluorescence enables the unambiguous identification of the different phases present in the blends. Moreover, near-field optical microscopy provides information about morphology even in the absence of topography.

We have used aperture-SNOM to investigate blends of luminescent derivatives of poly(*p*-phenylenevinylene), PPV, and inert matrices[12] and binary blends of polyfluorene-derivative polymers.[7,13] We have also demonstrated near-field photocurrent measurements of such systems, exploiting the purely local excitation which SNOM provides at a surface, with an evanescent decay into the material.[14] An example of an SNOM image of a blend of two polyfluorene-based polymers is presented in Fig. 4, indicating the presence of two submicron-scale phases. The image was obtained by illuminating the sample through the SNOM probe and, while scanning the tip over the sample surface, collecting fluorescence from the sample. The excitation wavelength (488 nm) was chosen such that absorption occurred in only one of the constituent polymers, poly(9, 9′-dioctylfluorene-alt-benzothiadiazole), F8BT. By using filters to select only fluorescence from this polymer (rejecting scattered light at the excitation wavelength and any other emission from the film), the measurement is sensitive only to the presence of F8BT in the blend.

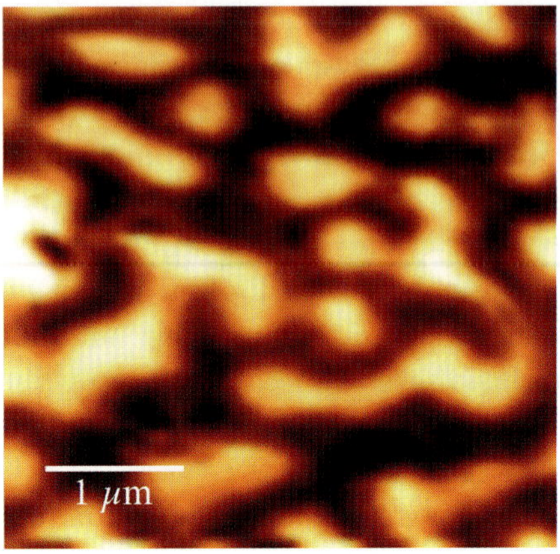

Fig. 4. An image of the fluorescence emitted from a blend of two light-emitting polymers, measured using aperture-SNOM. By exciting with carefully chosen wavelengths of light, only one absorbs and then re-emits light.

SNOM images from a binary blend chosen for its suitability for photovoltaic applications for solar cells are presented in Fig. 5.[13] The blend contained 10% by weight F8BT and 90% by weight poly(9, 9-dioctylfluorene-co-bis-N, N'-(4-butylphenyl)-bis-N, N'-phenyl-1, 4-phenylenediamine), PFB. As above, the fluorescence detected in the SNOM measurement results only from the presence of F8BT. The photovoltaic efficiency of devices fabricated from these films is governed by the extent of the interfacial area between these two constituent polymers and it is expected that a

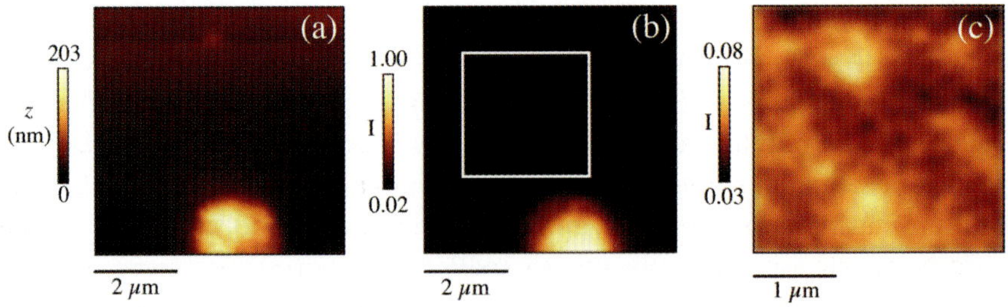

Fig. 5. SNOM images of a conjugated polymer blend film containing 10% by weight F8BT and 90% by weight PFB. Intensity (I) is in arbitrary units, and height (z) is in nm. (a) Topographic image. (b) The corresponding fluorescence image for (a). The topographic and fluorescence images were measured simultaneously. (c) An enlarged fluorescence image, taken from the white box shown in (b); the intensity scale is the same as for (b).

pre-requisite for highly efficient photovoltaic devices based on conjugated polymers is phase separation on a length-scale comparable to, or smaller than $\sim$10 nm.[10] Yet photovoltaic devices fabricated from films with apparently large-scale phase separation, as indicated from AFM measurements, were found to exhibit high efficiencies.[15] Indeed, the images in Fig. 5 show the presence of micron-size regions protruding from the surface of the film, which are unambiguously identified from their fluorescence as F8BT rich. However, comparatively strong and locally varying fluorescence from F8BT is also detected in the PFB-rich phase, which is topographically flat, indicating the presence of further phase separation at much smaller length-scales, of benefit for high photovoltaic efficiency.

2.3. *Beware of artifacts*

No discussion of SNOM would be complete without the issue of a warning for measurement artifacts,[16] which plagued early work in the field, leading to many false claims for high resolution. Owing to the evanescent (rapidly decaying) nature of the electric field intensity beneath the near-field probe, relatively small changes in the separation of tip and sample can lead to a large change in the intensity of the light at the sample surface. This results in variations in fluorescence intensity caused by the probe following the topography of the surface, and not by changes in the type of material illuminated. The corollary of this is that aperture-SNOM is an unsuitable technique for very rough surfaces as such topographic artifacts are likely to dominate any measurement.

The most likely scenario is for the tip motion to underestimate variations in the sample topography, particularly given the blunt nature of SNOM probes (see Fig. 3(a)). In this case, for a sample which should produce a uniform optical signal, protruding areas will appear more intense as the separation between tip and sample will be less than when the tip is over a valley in the sample surface. In such a case there should be perfect registration between optical and topographic images.

In fact, one can often expect to see a shift of 50 nm or more between topographic and optical images in aperture-SNOM. This can be understood when one considers the nature of the probe itself; an aperture defined in an aluminum coating at the end of an optical fiber tip. The "optical probe" is effectively defined as the center of the aperture. However, the "topographic probe" could be a protrusion in the aluminum coating defining the aperture, which could well be $>$50 nm from the aperture center. Thus, observation of a well-defined displacement between topographic and optical images can serve to validate a measurement.

3. Apertureless Near-Field Microscopy: The Promise of True Nanometer-Resolution Optical Imaging

The resolution of aperture-SNOM is limited to ~50 nm, determined by limitations on the reproducible fabrication of optical fiber tips with smaller aperture diameters and, fundamentally, by the penetration of the incident light into the metal used to create the aperture. Also, importantly, the light throughput decreases rapidly with decreasing aperture size rendering the detection of many optical signals for apertures smaller than 50 nm very difficult. However, exciting recent advances give the promise of new optical nanoprobes which offer the promise of optical microscopy with true nanometer resolution. Such techniques will allow individual nanostructures and molecules to be addressed optically, for the determination of their identity, structure, and function.

There are three main approaches to such "apertureless"-SNOM techniques (see Fig. 2), based on local scattering, local fluorescence, and local field enhancement. However, for convenience we will split apertureless-SNOM techniques into two main categories, based on the type of probe: (i) metallic or dielectric tips and (ii) fluorescent probes.

3.1. *Near-field optical microscopy with a metal or dielectric tip*

Initial reports of apertureless-SNOM involved the detection of light elastically scattered by a metal tip of atomic dimensions in close proximity with the sample surface; the tip locally perturbs the fields at the sample surface and the response to this pertubation is detected in the far field.[17,18] By applying interferometric signal detection techniques, such scattering SNOM has been demonstrated to provide information about the refractive absorptive properties of surfaces in the visible and infra-red with 10-nm resolution.[19]

Apertureless near-field optical microscopy may also be achieved using a true local excitation source, for the measurement of the spectroscopic response of a system, by exploiting the highly confined enhancement of electric field in the vicinity of a metal tip.[20] This may simply be the "lightning rod" effect, the large electric field enhancement that can occur near sharp geometrical structures.[21,22] Further significant local electromagnetic field enhancement may also be achieved through the resonant excitation of local fields at the apex of a silver or gold tip.

Using this approach, one route to the achievement of high-resolution fluorescence imaging is with two-photon absorption. In two-photon absorption processes, a molecule absorbs simultaneously two photons, each of half the energy (corresponding to double the wavelength) to that required by the molecule for absorption of a single photon. So, for example, molecules which absorb incident light of 400 nm wavelength will absorb 800 nm light through two-photon absorption. However, whereas normal absorption varies linearly with incident power, two-photon absorption displays a quadratic intensity dependence. The enhanced electric fields in the vicinity of a sharp tip and this nonlinear intensity dependence of two-photon absorption effectively result in a localized excitation source for molecular fluorescence. Such two-photon apertureless-SNOM has been demonstrated to give 30-nm resolution tip-enhanced fluorescence images.[23]

Tip enhancement through the resonant excitation of local fields in the tip has particular application for nanometer-resolution Raman and fluorescence microscopy, and these will be considered in greater detail in Sec. 4.

3.2. *"Single-molecule" fluorescent probes for SNOM*

Conceptually the simplest technique is fluorescence-based near-field microscopy in which a nanoscopic fluorescent light source, located at the end of a tip, is scanned over a sample (see Fig. 6(a)). This nanoscopic fluorescent source could be just a single molecule[24] or, for example, a single color center in an irradiated diamond microparticle attached to a tip.[25]

Within this category of SNOM using a nanoscopic light source, we can also consider another approach for the realization of <10 nm resolution which exploits Förster resonant energy transfer (FRET) between tip and sample.[26,27] FRET involves the radiation-free transfer of energy from excited "donor" fluorophores to "acceptor" fluorophores through the near-field components of the emission and absorption dipole moments (the Förster mechanism). For this to occur the emission spectrum of the donor must overlap the absorption spectrum of the acceptor. The energy transfer rate for a given FRET pair (of donor and acceptor fluorphores) depends inversely on the sixth power of their separation, with typical length-scales (characterized by the 50% transfer distance) of 1–10 nm. The excitation

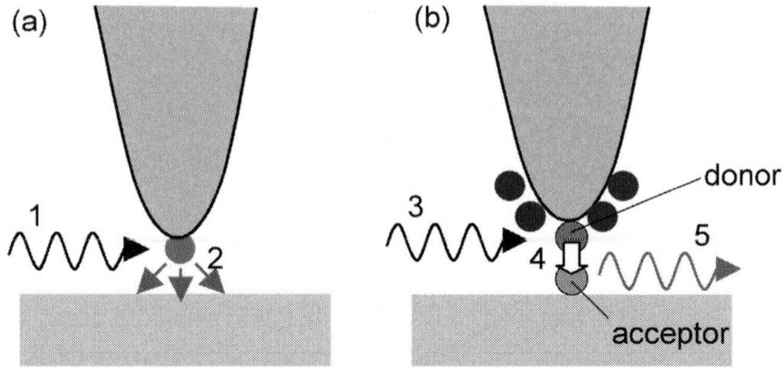

Fig. 6. Apertureless-SNOM using a fluorescent probe. The shaded circles represent fluorescent molecules or nanoparticles. (a) Incident light (1) excites a single flurophore on the tip (2), the fluorescence from which provides a local optical probe of the sample. (b) Scanning FRET imaging. Incident light (3) excites a donor fluorophore on the tip. Förster energy transfer (4) occurs and the acceptor molecule in the sample emits fluorescence (5) with a dimming of the fluorescence from the donor.

wavelength is chosen such that only the donor fluorophore absorbs and so emits fluorescence, while the acceptor fluorophore remains unexcited. However, when the two FRET-complementary fluorophores come in close proximity the emission from the donor fluorophore will dim as resonant energy transfer to the acceptor occurs, which in turn emits fluorescence. This local optical interaction offers itself for exploitation for the realization of optical microscopy with molecular resolution, as illustrated in Fig. 6(b).

4. Tip-Enhanced Spectroscopy

4.1. *Tip-enhanced Raman scattering*

Raman spectroscopy is able to measure the frequencies of molecular and crystal vibrations, providing a high degree of chemical specificity. It is well established that Raman scattering is greatly enhanced in the vicinity of small particles or rough surfaces of noble metals such as silver; the widely accepted cause of this effect is that the external electromagnetic field of the incident light couples to surface plasmons (local field modes) of metal particles with dimensions less than the wavelength of light, and strong fields associated with these plasmon resonances build up at the particle surface. Such surface enhanced Raman scattering (SERS) exploits the fact that both excitation and emission in the Raman process are enhanced and

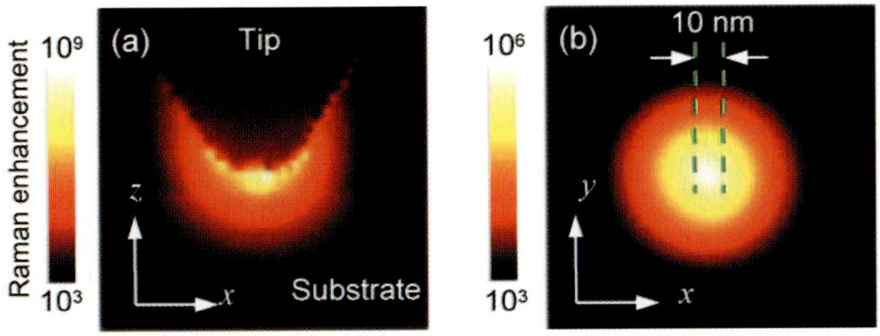

Fig. 7. Calculated enhancement (logarithmic scale) in intensity of Raman scattering in the vicinity of a 20-nm radius metal tip held 4 nm above a glass surface. (a) Cross-section through the tip, along the tip axis. (b) Enhancement in the plane of the sample. The wavelength of illumination is such that a localized mode, strongly confined at the tip apex, is resonantly excited, leading to strong electric field enhancement and hence, Raman scattering.

are proportional to the electric field intensity, so that the Raman signal is increased by the fourth power of the local electric field enhancement.

A silver or gold tip in close proximity to the sample surface can give rise to similar effects — tip-enhanced Raman scattering (TERS). The results of a calculation of the Raman enhancement beneath a resonantly excited metal tip are presented in Fig. 7. Over a narrow range of wavelengths of the incident light, local fields are resonantly excited in the tip, producing a strong, highly localized electric field enhancement, such that a Raman scattering intensity of the order of 10^7 can be expected on the sample surface beneath the tip, over an area of diameter ∼7 nm.[28] Thus it seems that a scheme for nano-optical microscopy exploiting TERS holds much promise for Raman spectroscopy with <10 nm spatial resolution, to provide high-resolution Raman images of systems. TERS enhancements up to 10^6 have been observed[29] and tip-enhanced near-field Raman imaging with a resolution of 25 nm has been demonstrated from measurements of isolated single-walled carbon nanotubes.[30]

4.2. *Tip-enhanced fluorescence*

The enhancement of the local electromagnetic field can also lead to enhanced fluorescence emission. However, placing a sharp metallic tip in the vicinity of a fluorescence molecule also leads to a modification of both the radiative and the nonradiative rates for a molecule, inducing changes in both the fluorescence life-time and the emission intensity.[31] The enhanced local field and radiative decay rate tend to increase the fluorescence intensity, while the increased nonradiative decay rate, which results from energy transfer from the fluorophore to metal, will diminish the fluorescence intensity dramatically. The net enhancement of fluorescence results from a competition between this enhancement and quenching, and whether enhanced or diminished fluorescence intensity is observed depends strongly on the particular experimental conditions. Indeed, both fluorescence intensity enhancement[32,33] and fluorescence intensity quenching[34] have been reported experimentally, paving the way for nanoscale fluorescence microscopy.[33,35]

For example, a tip-enhanced fluorescence image of an isolated cluster of CdSe "quantum dot" nanoparticles is presented in Fig. 8, obtained using a configuration similar to that illustrated in Fig. 2.[33] Confocal fluorescence imaging leads to the observation of a resolution of 200 nm, consistent with the size of the diffraction-limited focus. When a sharp gold tip is brought within a few nanometers from the sample surface, the resulting enhancement in quantum dot fluorescence in the vicinity of the tip leads to a resolution of about 60 nm. The observed four-fold enhancement of the fluorescence is consistent with the value expected from the competition between fluorescence quenching and electromagnetic field enhancement.

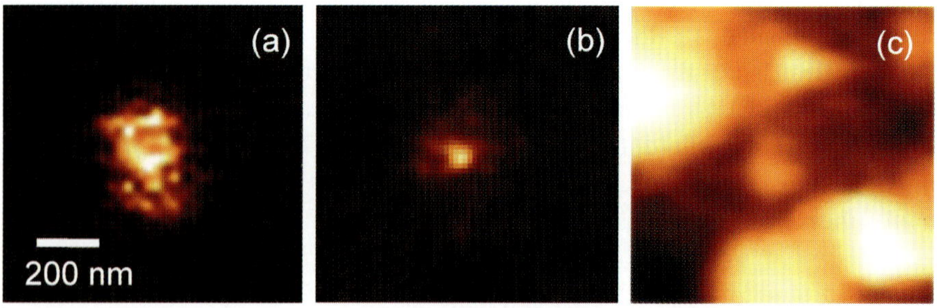

Fig. 8. 1 μm × 1 μm (a) fluorescence confocal (no tip) image (the fluctuations result from intermittent "blinking" of the quantum dot fluorescence), (b) fluorescence apertureless-SNOM image, and (c) topographic image of a small quantum dot cluster. (b) and (c) obtained simultaneously with a sharp gold tip close at the center of the laser focus.

5. Future Developments

Since its first demonstration the research field of scanning near-field microscopy has moved through periods of promise followed by disappointment, often by the huge technical challenges presented by these techniques. Yet the need for optical microscopy with nanometer-spatial resolution is strong, to address new challenges in the life sciences and nanotechnology, and I believe we are at a turning point.

Aperture-based SNOM is now finally reaching maturity and, in the same way that AFM has become ubiquitous with the development of reproducible micro-fabricated probes, the same is likely to be the case for aperture-SNOM, to replace the fragile optical fiber tips used predominantly at present. Although aperture-SNOM is never going to provide resolutions much better than 50 nm, the availability of robust, reproducible probes and the complemetarity of the technique with AFM will mean that the technique will become well established. Recent advances have enabled reliable imaging of biological cells,[36] for which SNOM offers significant potential for understanding protein events in the cell membrane.

For the dream of optical microscopy with true nanometer resolution we must turn to apertureless techniques. The chemical specificity of fluorescence and Raman spectroscopy with the high spatial resolution afforded by tip-enhanced techniques will enable the isolation and differentiation of single molecules, through their Raman "fingerprint" or fluorescence spectrum and lifetime, as well as providing important information about molecular, biological, and nanostructured systems on nanometer length-scales. Exciting recent advances in the fabrication of custom-designed nano-antenna[37] with well-defined optical resonances, or the development of "tip-on-aperture" probes combining the best of aperture- and apertureless-SNOM,[38] offer much promise for the implementation of such techniques for nanoscale imaging of materials.

Acknowledgments

The author would like to acknowledge and thank A. L. Demming, F. Festy, F. M. Huang, R. G. Milner, R. Riehn, and R. Stevenson, who performed the SNOM measurements and theoretical work presented in this chapter, and F. Cacialli for many helpful discussions. He thanks the UK Engineering and Physical Sciences Research Council and the Royal Society for support.

References

1. S. W. Hell, Towards fluorescence nanoscopy, *Nature Biotechnology* **21**, 1347 (2003).
2. M. A. Paesler and P. J. Moyer, *Near-Field Optics: Theory, Instrumentation and Applications* (Wiley, New York, 1996).
3. E. H. Synge, A suggested method for extending microscopic resolution into the ultra-microscopic region, *Phil. Mag.* **6**, 356–362 (1928).
4. R. Stevenson and D. Richards, The use of a near-field probe for the study of semiconductor heterostructures, *Semicond. Sci. Technol.* **13**, 882–886 (1998).
5. H. Metiu (ed.), Near-field microscopy and spectroscopy, *J. Chem. Phys.* **112**, 7761–7821 (2000).
6. D. Richards and A. Zayats (eds.), Nano-optics and near-field microscopy, *Phil. Trans. Roy. Soc. A* **362**, Issue 1817 (2004) and references therein.
7. R. Stevenson, R. Riehn, R. G. Milner, D. Richards, E. Moons, D.-J. Kang, M. Blamire, J. Morgado and F. Cacialli, Ultraviolet-visible near-field microscopy of phase-separated blends of polyfluorene-based conjugated semiconductors, *Appl. Phys. Lett.* **79**, 833–835 (2001).
8. K. Karrai and R. D. Grober, Piezoelectric tip–sample distance control for near-field optical microscopes, *Appl. Phys. Lett.* **66**, 1842–1844 (1995).
9. S. Davy, M. Spajer and D. Courjon, Influence of the water layer on the shear force damping in near-field microscopy, *Appl. Phys. Lett.* **73**, 2594–2596 (1998).
10. J. J. M. Halls, C. A. Walsh, N. C. Greenham, E. A. Marseglia, R. H. Friend, S. C. Moratti and A. B. Holmes, Efficient photodiodes from interpenetrating polymer networks, *Nature* **376**, 498–500 (1995).
11. R. H. Friend *et al.*, Electroluminescence in conjugated polymers, *Nature* **397**, 121–127 (1999).
12. R. Stevenson, M. Granström and D. Richards, Fluorescence scanning near-field optical microscopy of conjugated polymer blends, *Appl. Phys. Lett.* **75**, 1574–1576 (1999).
13. R. Stevenson, R. G. Milner, D. Richards, A. C. Arias, J. D. Mackenzie, J. J. M. Halls, R. H. Friend, D.-J. Kang and M. Blamire, Fluorescence scanning near-field optical microscopy of polyfluorene composites, *J. Microscopy* **202**, 433–438 (2001).
14. R. Riehn, R. Stevenson, D. Richards, D.-J. Kang, M. Blamire, A. Downes and F. Cacialli, Local probing of photocurrent and photoluminescence in a phase-separated conjugated polymer blend by means of near-field excitation, *Adv. Funct. Mater.* **16**, 469–476 (2006).
15. A. C. Arias, J. D. MacKenzie, R. Stevenson, J. J. M. Halls, M. Inbasekaran, E. P. Woo, D. Richards and R. H. Friend, Microscale to nanoscale morphology control of polyfluorene blends: A combined microscopic and photovoltaic investigation, *Macromolecules* **34**, 6005–6013 (2001).
16. B. Hecht, H. Bielefeldt, Y. Inouye, D. W. Pohl and L. Novotny, Facts and artifacts in near-field optical microscopy, *J. Appl. Phys.* **81**, 2492–2498 (1997).

17. F. Zenhausern, Y. Martin and H. K. Wickramasinghe, Scanning interferometric apertureless microscopy — Optical imaging at 10 Angstrom resolution, *Science* **269**, 1083–1085 (1995).
18. R. Bachelot, P. Gleyzes and A. C. Boccara, Near-field optical microscopy by local perturbation of a difraction spot, *Microsc. Microanal. Microstruct.* **5**, 389–397 (1994).
19. F. Keilmann and R. Hillenbrand, Near-field microscopy by elastic light scattering from a tip, *Phil. Trans. Roy. Soc. A* **362**, 787–805 (2004).
20. J. Wessel, Surface-enhanced optical microscopy, *J. Opt. Soc. Am. B* **2**, 1538–1541 (1985).
21. L. Novotny, R. X. Bian and X. S. Xie, Theory of nanometric optical tweezers, *Phys. Rev. Lett.* **79**, 645–648 (1997).
22. Y. Kawata, C. Xu and W. Denk, Feasibility of molecular-resolution fluorescence near-field microscopy using multi-photon absorption and field enhancement near a sharp tip, *J. Appl. Phys.* **85**, 1294–1301 (1999).
23. E. J. Sánchez, L. Novotny and X. S. Xie, Near-field fluorescence microscopy based on two-photon excitation with metal tips, *Phys. Rev. Lett.* **82**, 4014–4017 (1999).
24. J. Michaelis, C. Hettich, J. Mlynek and V. Sandoghdar, Optical microscopy using a single-molecule light source, *Nature* **405**, 325–327 (2000).
25. S. Kühn, C. Hettich, C. Schmitt, J.-Ph. Poizat and V. Sandoghdar, Diamond colour centres as a nanoscopic light source for scanning near-field optical microscopy, *J. Microscopy* **202**, 2–6 (2001).
26. G. T. Shubeita, S. K. Sekatskii, M. Chergui, G. Dietler and V. S. Letokhov, Investigation of nanolocal fluorescence resonance energy transfer for scanning probe microscopy, *Appl. Phys. Lett.* **74**, 3453–3455 (1999).
27. S. A. Vickery and R. C. Dunn, Scanning near-field fluorescence resonance energy transfer microscopy, *Biophys. J.* **76**, 1812–1818 (1999).
28. F. Festy, A. Demming and D. Richards, Resonant excitation of tip plasmons for tip-enhanced Raman SNOM, *Ultramicroscopy* **100**, 437441 (2004).
29. B. Pettinger, B. Ren, G. Picardi, R. Schuster and G. Ertl, Nanoscale probing of adsorbed species by tip-enhanced Raman spectroscopy, *Phys. Rev. Lett.* **92**, 096101 (2004).
30. A. Hartschuh, E. J. Sánchez, X. S. Xie and L. Novotny, High-resolution near-field Raman microscopy of single-walled carbon nanotubes, *Phys. Rev. Lett.* **90**, 095503 (2003).
31. W. L. Barnes, Fluorescence near interfaces: The role of photonic mode density, *J. Mod. Opt.* **45**, 661–669 (1998).
32. A. Kramer, W. Trabesinger, B. Hecht and U. P. Wild, Optical near-field enhancement at a metal tip probed by a single fluorophore, *Appl. Phys. Lett.* **80**, 1652–1654 (2002).
33. F. M. Huang, F. Festy and D. Richards, Tip-enhanced fluorescence imaging of quantum dots, *Appl. Phys. Lett.* **87**, 183101 (2005).
34. T. J. Yang, G. A. Lessard and S. R. Quake, An apertureless near-field microscope for fluorescence imaging, *Appl. Phys. Lett.* **76**, 378–380 (2000).

35. J. M. Gerton, L. A. Wade, G. A. Lessard, Z. Ma and S. R. Quake, Tip-enhanced fluorescence microscopy at 10 nanometer resolution, *Phys. Rev. Lett.* **93**, 180801 (2004).
36. M. Koopman, A. Cambi, B. I. de Bakker, B. Joosten, C. G. Figdor, N. F. van Hulst and M. F. Garcia-Parajo, Near-field scanning optical microscopy in liquid for high resolution single molecule detection on dendritic cells, *FEBS Lett.* **573**, 6–10 (2004).
37. J. N. Farahani, D. W. Pohl, H. J. Eisler and B. Hecht, Single quantum dot coupled to a scanning optical antenna: A tunable superemitter, *Phys. Rev. Lett.* **95**, 017402 (2005).
38. H. G. Frey, S. Witt, K. Felderer and R. Guckenberger, High-resolution imaging of single fluorescent molecules with the optical near-field of a metal tip, *Phys. Rev. Lett.* **93**, 200801 (2004).

David Richards studied Natural Sciences at St. John's College Cambridge, graduating in Physics and Theoretical Physics in 1987. He then began research at the Cavendish Laboratory in Cambridge on the electronic optical properties of low-dimensional semiconductor systems, obtaining his PhD in 1991. He worked as a PostDoctoral Guest Scientist at the Fraunhofer Institute for Applied Solid State Physics in Freiburg, before returning to the Cavendish Laboratory where he held first a research fellowship at St. John's College, then a Lloyds's of London Tercentenary Foundation Fellowship and, from 1995, a Royal Society University Research Fellowship. In 2000 he moved to King's College London where he was appointed Reader in Physics in 2003. David's scientific interests are now principally concerned with nano- and bio-photonics. These include the development of scanning near-field optical microscopy techniques, surface-enhanced Raman and fluorescence spectroscopy, and biological cell imaging.

CHAPTER 12

SMALL THINGS BRIGHT AND BEAUTIFUL: SINGLE MOLECULE FLUORESCENCE DETECTION

Mark A. Osborne

Department of Chemistry, School of Life Sciences
University of Sussex, Falmer, Brighton
BN1 9QJ, UK

Single molecule detection has become instrumental in our understanding of the molecular world. The ability to perform experiments molecule by molecule and probe directly the static and dynamic disorder in a system has revealed new insights into a range of fundamental problems in physics, chemistry, and biology. These insights are now laying the foundations for engineering devices such as optical switches and photonic wires at the molecular level. Of the many single molecule techniques available, fluorescence microscopy has proved an invaluable tool by virtue of its high sensitivity, time-resolution, noninvasive nature, and compatibility with other spectroscopic techniques. The development of specific detection modalities has allowed a multitude of previously impracticable experiments to be realized, from photon antibunching to single particle tracking in living cells. Advances in fluorescent probe design, detectors, modes of acquisition and data analysis will undoubtedly push single molecule techniques to new limits of sensitivity and into new areas of application.

Keywords: Single molecule, fluorescence microscopy, time-resolved spectroscopy, anisotropy, dipole orientation, spectral diffusion, FCS, FRET, localization, biomolecular dynamics.

1. Introduction

The ability to detect, identify, modify, and manipulate a single molecule is having a profound impact on both our understanding of matter and our capacity to engineer materials at the molecular level. Until recently our models of the molecular world, whether from a physical, chemical, or biological perspective have been derived from observations made on populations of molecules where only ensemble averaged measurements can be made. Molecular structures, transformations, processes, and pathways examined by conventional techniques generally represent only the mean of what is often a highly heterogeneous distribution. In contrast, single molecule techniques allow individual members of an ensemble to be interrogated and the distribution of a particular property to be both spatially and temporally resolved. Fluctuations of a system can be monitored without the need for the nigh-impossible synchronization of a molecular population while the molecular surroundings can be spatially mapped using environmentally sensitive probes.

A multitude of ultra-sensitive techniques for detecting and manipulating single molecules have been developed in recent years. These include, in various guises, atomic force microscopy (AFM),[1] scanning near-field optical microscopy (SNOM),[2] and surface enhanced Raman spectroscopy (SERS).[3] Amongst the most popular and widely accessible techniques is fluorescence microscopy. The high signal to noise ratio afforded by fluorescence detection allows single molecule dynamics to be studied with a temporal resolution spanning 12 decades,[4] from nanoseconds to minutes, and a spatial accuracy as high as 1 nm.[5] Moreover the photons emitted by a single florescent molecule contain a wealth of information about the molecule's identity, the state it is in, its relative motion and any interaction it may have with its environment. Together these factors make fluorescence microscopy one of the most versatile, data-rich tools for single molecule investigation. Since the earliest observations of fluorescence from a single molecule, albeit a protein labeled with up to 100 fluorescent tags[6] and more recently the first demonstration of single fluorophore detection at room temperature,[7] there have been a number of major advances in single molecule instrumentation, techniques, and applications. Aspects of these advances have been reviewed elsewhere.[8–10] In this chapter, we explore the underlying principles of single molecule detection, the tools of the trade and the remarkable range of methods that have been developed to harness the full information content from single molecule fluorescence across the entire dynamic range in time.

1.1. *Principles*

Single molecule fluorescence microscopy relies on the detection of photons emitted during the spontaneous radiative relaxation of a molecule from an excited electronic state to its ground state. Laser-induced fluorescence is typically used to drive the molecule through the absorption/emission cycle which is schematized in Fig. 1. The excitation rate is proportional to the laser intensity I_L, but the emission rate is ultimately limited to a maximum I_∞, by the finite lifetime of the excited state. For the three-level system illustrated the fluorescence rate is given by[11]:

$$I_F = \frac{I_\infty}{1 + I_S/I_L},$$

where I_S is the saturation intensity at which $I_F = I_\infty/2$. For a typical fluorophore such as tetramethylrhodamine (TMR), the fluorescence rate only reaches around 10^6 photons/s, even at saturation. Under these conditions the photon flux through the cross-section of a tightly focused laser can approach $10^{14}\,\mathrm{s}^{-1}$, a level where without adequate spectral filtering, Rayleigh (elastic) scattering from optical surfaces and particulates can contribute to a significant background signal of "chemical noise". Efficient detection of fluorescence relies on the shift of the emission to the

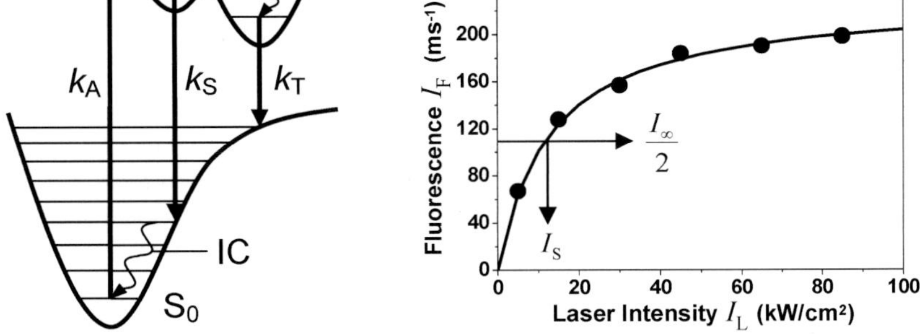

Fig. 1. (a) Schematic of the fluorescence cycle between the ground electronic state S_0 and excited state S_1 and phosphorescence from triplet state T_1. Internal conversion (IC) rates are fast ($10^{12}\,\mathrm{s}^{-1}$) compared to the other processes ($10^6 - 10^9\,\mathrm{s}^{-1}$). (b) The rate equations from which the steady-state fluorescence intensity I_F is determined (proportional to the fraction of the cycle spent in S_1). (c) Graph showing the saturation of fluorescence at increasing excitation intensities.

$$\dot{S}_0 = k_S S_1 - k_A S_0 + k_T T_1$$
$$\dot{S}_1 = k_A S_0 - k_S S_1 - k_{ISC} S_1$$
$$\dot{T}_1 = k_{ISC} S_1 - k_T T_1$$

red of the absorbing transition, the Stokes-shift, and the use of two or more high-quality longpass or bandpass filters with $>80\%$ transmission and $>10^6$ times attenuation at the fluorescence and excitation wavelengths, respectively.

As well as Rayleigh scatter, another source of chemical noise is Raman (inelastic) scatter of the excitation source from the host solvent or matrix. For example, the bending mode of water at $1645\,\mathrm{cm}^{-1}$ will inelastically scatter the 532-nm line of a frequency-doubled Nd:YAG laser that might typically be used to excite our TMR molecule, at around 583 nm, a wavelength close to the TMR fluorescence peak.[12] Although Raman scattering cross-sections are some 10^{14} times smaller than the absorption cross-sections of most single molecule fluorescent probes,[3] it is essential to constrain the excitation zone to minimize the number of host molecules and hence reduce the Raman contribution to the background. For sub-picoliter volumes containing $<10^{12}$ host molecules the signal to background ratio can exceed 100. As well as spectrally and spatially filtering Rayleigh and Raman noise, time gated detection can also be applied where the fluorescence lifetime is long compared with what are essentially instantaneous scattering processes.[7] Autofluorescence that is often encountered in cellular imaging can, in some circumstances, be dealt with in a similar manner if the lifetime of the autofluorescent components are markedly different from that of the single fluorophore.[13] However, fluorescent probes are now being developed with near infrared emission,[14,15] a spectral region where autofluorescence is significantly reduced.

In addition to saturation, the total number of detectable photons emitted by a single molecule is limited by the photochemical lifetime of the fluorophore. The abrupt, one-step photobleaching of a fluorophore after $\sim 10^6$ cycles is a classic (though not a definitive) indicator of singleness and results from the irreversible photoinduced chemical transformation of the molecule to a nonfluorescent product. Singlet oxygen and hydroxyl radicals are generally considered the reactive species. One pathway to singlet oxygen formation is, paradoxically, through the efficient annihilation of the triplet state of the fluorophore by ground state triplet oxygen. Thus, while purging oxygen from the host can improve the longevity of the fluorophore, the photon yield is not generally increased since triplet build-up can be significant without the addition of substitute triplet quenchers.[16] The most common way to scavenge oxygen is enzymatically with a glucose oxidase and catalase mix.[17] Small molecule scavengers such as mercaptoethanol, propyl gallate,[18] and antioxidants such as Trolox (a synthetic

Vitamin E) and ascorbic acid have proved useful in extending single molecule observation times, while mercaptoethylamine and diazabicylco-octane appear to improve fluorescence detection by efficiently quenching the triplet state.[16]

1.2. *Probes*

Fluorescent probes that are suitable for single molecule detection are characterized by high extinction coefficients (probability of excitation), high quantum yields (number of photons emitted per quanta absorbed) and high photostability. Engineered fluorophores with enhanced properties have led to a number of interesting new single molecule fluorescent probes. These include the Alexa series of sulfonated coumarins and rhodamines[19] and the Atto series of amine-bridged benzopyrylium- and carbopyronin-based dyes with steric rigidity that improves photostability.[20] Bifunctional rhodamine now allows conjugation at both ends of the fluorophore, enabling a well-defined orientation of the probe to be fixed in the frame of a host protein or complex.[21] A recent addition is a new class of fluorophore consisting of an amine donor, dicyanodihydrofuran, and conjugated bridge that promises flexibility in tuning the environmental sensitivity of the probe for particular applications.[22] Another promising avenue to improving properties such as photostability, is the encapsulation of fluorophores in the supramolecular cavities of cyclodextrins, rotaxanes, and dendrimers.[23] The catalog of probes now extends to the visible fluorescent proteins (VFPs)[24] which includes the ubiquitous green fluorescent protein (GFP)[25] and its variants such as yellow (YFP);[26] a monomeric red fluorescent protein (DS-RED);[27] photoactive GFP (PA-GFP);[28] and a new fast photo-switching protein, Dronpa.[29]

Inorganic nanocrystals or quantum dots (QDs) form another category of single fluorescent probes. These clusters of a few to a few thousand atoms come in a variety of elemental flavors, the most widely studied being ZnS capped CdSe nanocrystals.[30,31] Their attraction stems from the ability to tune their emission wavelength from 470 to 620 nm simply with the size of dot from 2.3 to 5.5 nm. Moreover a broad absorption profile allows excitation of all colors by a single source, usually the 488-nm line of an argon ion laser. Tuning emission from blue[32] to the near infrared[33] has been achieved by changing the semi-metal composition of the core and/or shell. Highly fluorescent, Au particles containing only 5, 8, 13, 23, and 31 atoms that have emission profiles in the UV, blue, green, red, and NIR wavelengths, respectively, have been prepared by dendrimer encapsulation.[34] Unlike semiconductor nanocrystals the absorption wavelength of these QDs is also size-dependent rendering their behavior more akin to conventional organic dye molecules, allowing for the possibility of fluorescence resonance energy transfer between particles without simultaneous excitation.

1.3. *Excitation schemes*

A number of optical schemes have been developed for the study of single molecules in low temperature matrices, on surfaces, in thin films, solutions, capillaries, and in micro-droplets.[35] The most common geometries for single molecule studies are illustrated in Fig. 2, where a high numerical aperture microscope objective lens is employed to collect as large a solid angle of fluorescence emission as possible. Under conditions for which these lenses are not designed to operate, such as those at cryogenic temperatures, alternative configurations using parabolic mirrors for fluorescence collection have been used.[36] Here we focus on room temperature single molecule methodologies, applications and results, but many of the characteristics and observables are common to most forms of single molecule

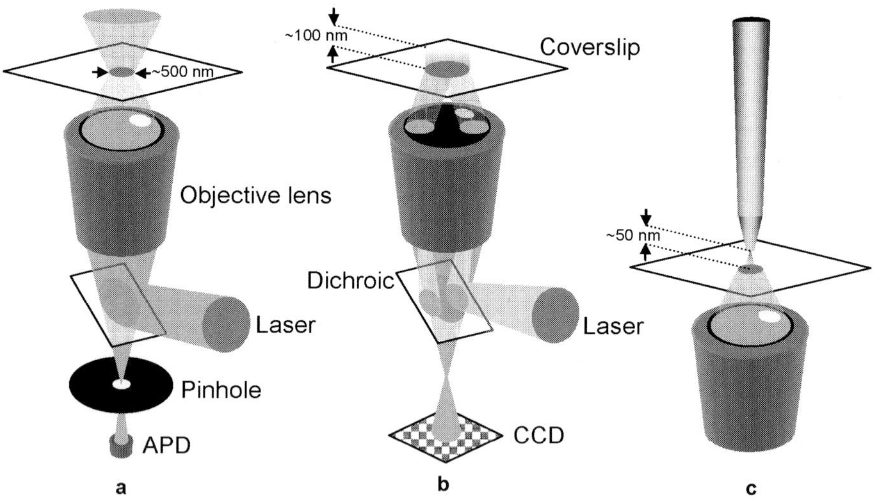

Fig. 2. Principle excitation schemes for single molecule detection: (a) Confocal fluorescence microscopy: diffraction-limited laser focus with pinhole rejection of out-of-focus fluorescence. (b) TIRFM: evanescent-field excitation by objective-type total internal reflection. (c) SNOM: sub-diffraction-limited excitation using a near-field optical fiber probe.

detection. The basic principle relies on: delivering a light source to a sample in a way that spatially constrains the excitation zone; minimizing the amount of out-of-focus fluorescence reaching the detector; maximizing the photon collection efficiency.

Experimental arrangements are broadly categorized as either confocal fluorescence microscopy, total internal reflection fluorescence microscopy (TIRFM), or scanning near-field optical microscopy (SNOM). In confocal microscopy, the objective lens is used to focus a laser beam to a diffraction-limited radius of $\sim$250 nm and a depth limited by spherical aberration to $\sim 1\,\mu$m. At a nanomolar concentration of fluorescent analyte there is then, on average, less than one molecule in the resulting $\sim$0.2-femtoliter excitation volume. A pinhole is placed in the image plane to reject any subsequent fluorescence emanating from above or below the focal plane. TIRFM uses the evanescent field produced at the interface of a dielectrically mismatched glass coverslip and sample to restrict excitation to within only 100 nm of the substrate surface. Finally, SNOM uses tapered fibers $<$100 nm or so in diameter to deliver the near-field component of a propagating beam to the sample at distances of only 50 nm or less from the sample surface.

Variations on these schemes include the use of two-photon excitation which can eliminate the need for a confocal pinhole, since out of plane fluorescence falls away more rapidly (quadratic in I_L) from point of focus than with single-photon excitation. The lateral confinement of the focus however is somewhat reduced since two photons of longer wavelengths λ_1 and λ_2 (generally derived from the same pulsed laser source) are used to match the one-photon absorption transition $\lambda = (1/\lambda_1 + 1/\lambda_2)^{-1}$. Retaining a confocal configuration with a pinhole can improve lateral resolution by up to 40%.[37] Epi-fluorescence uses optics to produce wide-field Koehler-like illumination, where the laser beam is collimated at the object plane in a similar manner to that used in objective-type TIRFM.[17] In the former case, the excitation laser is focused at the back focal plane of the objective lens along its optical axis, while in the latter the beam is focused off axis to achieve supercritical angles at the glass/sample interface. TIRFM itself can be performed in the alternative prism-type configuration in which a laser beam is totally internally reflected off the base of a prism that is positioned above the objective and within the working distance of the lens. The sample is then coupled to the underside of the prism. Recently hybrid schemes have emerged, for example confocal microscopy has been combined with TIRFM, whereby attoliter detection volumes have been produced by restricting excitation both laterally and along the optical axis, respectively.[38]

1.4. *Collection optics*

The most important element in optimizing the fluorescence collection efficiency is the objective lens. Classically the lens collects a $2\pi(1-\cos\alpha)$ solid-angle fraction of the 4π fluorescence of an isotropically emitting source, where the aperture angle α is related to the numerical aperture (NA) of the lens by $\text{NA} = n\sin\alpha$. Here n is the refractive index (RI) of the medium between lens and sample, which for $\text{NA} > 1$ is usually an immersion oil index matched to that of the glass lens, but water-immersion objectives are used increasingly in 4 Pi super-resolution and cellular imaging.[39] Recently the maximum NA has been pushed as high as 1.65 with the use of monobromonaphthalene immersion medium, but even 1.45 has now been achieved with conventional immersion oils $n = 1.52$, providing aperture angles in excess of 72° and collection efficiencies over 30%. Such high NAs allow the transmission of light at angles well above the critical angle of a glass/cell interface ($\theta_c = \arcsin(1.38/1.52) \sim 65°$) for *in vivo* TIRFM imaging, which has strongly motivated the design of these lenses.[40] High NA objectives have also proved useful in the recently developed techniques of single-molecule high-resolution colocalization (SHREC)[41] and fluorescence imaging with one nanometer accuracy (FIONA).[5] The use of a solid-immersion lens made from glasses with RIs as high as 1.85 in conjunction with a 0.55 NA objective has been shown to enhance fluorescence collection efficiencies by over four-fold that of the objective alone.[42] The principle has recently been exploited to obtain high light-field confinement and fluorescence collection in a single molecule fluorescence correlation spectroscopy (FCS) set-up modified with a solid-immersion lens.[43]

Light collected by the objective is generally filtered using first a 45° dichroic beamsplitter that, as well as directing the excitation beam through the lens in the case of confocal microscopy and TIRFM, also deflects any scattered laser light collected by the objective. Secondly, high transmission bandpass interference filters are typically used for additional filtering, particularly in multi-color imaging systems. Filter technology now allow >90% transmission in the fluorescence window with an optical density of 5–6 in the blocking regions. Furthermore multi-band filter sets now allow up to four-color excitation and detection simultaneously. Wavelength-dependent pixel-shift can result from nonparallel surfaces of the filter, a situation sometimes encountered with interference filters based on multi-layered structures with soft-coatings. Recently, however, ion-beam sputtering of hard oxide coatings on glass have allowed single-layer filters to be produced with a negligible wedge angle and hence zero-pixel shift.[44] Clearly, multi-color colocalization techniques are likely to benefit from these advances in filter technology.

1.5. *Detectors*

In detecting the small number of fluorescent photons emitted by a single molecule, the quantum efficiency (QE) and dark count or readout noise of the detector ultimately determine the signal to noise ratio. In this case, the QE defines the number of photoelectrons produced per quanta of photons incident on the detector. In point-detection schemes such as confocal microscopy and SNOM, the single-photon avalanche diode (SPAD) often provides the best detection solution.[45] These devices use thermoelectric cooling to reduce the dark count associated with thermally induced photoelectrons or charge carriers. Active quenching circuits are applied to reduce the dead-time between detected events whilst minimizing the effects of afterpulsing, a phenomenon by which erroneous signals are triggered by the release of charge carries trapped in the diode some time after the photon signal. A SPAD module with a QE $>40\%$ between 500 and 600 nm and dark count of $5\,s^{-1}$ at $-15°C$ has recently been developed with a superior response time of <60 ps.[46] While photomultipliers can provide comparable response times the efficiency of the photocathodes tend not to reach beyond 10%.

For single molecule imaging applications back-illuminated charged-coupled device (CCD) arrays offer the highest quantum efficiencies, in some cases $>90\%$ compared to $<50\%$ for front-illumination. Dark-current is reduced to <1 photoelectron/pixel/s by cooling below $-40°C$ such that readout noise, generated by the on-chip charge to voltage converter, then becomes the factor limiting the signal to noise ratio. For high spatial resolutions, slow-scan cameras that offer noise levels <10 counts at readout rates of 1 MHz are generally favored. In this case full-frame (512×512 pixels) acquisition rates are limited to 1–2 frames per second (fps) but can be improved upon by binning pixels or reading only a fraction of the array.[47] For high temporal resolutions intensified CCD cameras (ICCD) are most often employed. The intensifier combines an MCP and phosphor screen to produce substantial optical gain $>10\,000$. Signal amplification allows acquisition at video-rate (30 fps) and faster via frame-transfer or interline charge-transfer. Wide-field fluorescence lifetime imaging (FLIM) can also be achieved by gating the intensifier for periods <100 ps at different lag-times following pulsed excitation.[48] The quantum efficiency of the ICCD is inherently reduced to $\sim 50\%$ due to the combined effect of the photocathode and CCD conversion efficiencies. The spatial resolution that is generally sacrificed for high time-resolution with an ICCD is avoided to some extent in electron bombardment charged coupled (EBCCD) devices. These use the direct impact of photoelectrons on a back-thinned CCD to produce multiple charges or gain in the CCD pixels themselves. Gains, however, must be restricted to < 1000 to avoid impact damage to the CCD and the dynamic range is limited by the rate at which full charge capacity (per pixel) is reached.

Recent advances have led to the development of the electron multiplying CCD (EMCCD), which provides the necessary gain for low-light-level imaging during readout rather than during acquisition.[49] In this manner the high quantum efficiency of the CCD is fully exploited and not compromised by coupling to a photocathode. The gain is produced in the serial readout register which is extended to contain an electron multiplying region. Since dark-current is multiplied along with the signal, enhanced thermoelectric cooling to $-75°$C or more is crucial to achieve single-photon detection limits. The advantage of EMCCD cameras is that electron multiplication is independent of the readout rate allowing high frame rate acquisition with low noise. Very recently, the possibility of using an inexpensive complimentary metal-oxide-semiconductor (CMOS) process to fabricate a two-dimensional array of SPADs has been realized.[50] The CMOS process allows integration of both the photodiode and readout electronics into each active pixel sensor along with analog, digital, and timing control circuitry on a single chip. CMOS devices generally suffer from low QEs and spatial resolution by virtue of the fact that only about 30% of the sensor is photoactive. However, single molecule detection sensitivity has recently been demonstrated using a 2×2 array CMOS-SPAD device in multifocal detection and parallel FCS experiments.[51] These advances in CMOS technologies may yet herald a new era of fast, efficient, and above all low-cost arrays for single molecule detection.

2. Detection Modalities

2.1. *Single molecule signatures*

As single molecule techniques advance, the list of criteria used to test for "singleness" continues to expand. Firstly, the fluorescence intensity should be quantized at a level that does not exceed that expected for a single emitter. The intensity trajectory (I_F versus time) should exhibit photobleaching to the background count rate in a single-step process. The number of emitters observed should scale with concentration. The dispersed fluorescence spectra of individual molecules may vary over a population depending on their nanoenvironments. Moreover, spectral diffusion may be observed, where the wavelength of the absorption or emission peak changes with time. The fluorescence lifetimes of individual molecules may also vary, being subject to similar environmental factors. For a fixed or highly immobile single molecule with a well-defined absorption and emission dipole the fluorescence should be strongly polarized and display anisotropy in the emission. Correlation of the photon arrival times from a single emitting fluorophore should show evidence of antibunching, that is no two photons are emitted simultaneously. Measuring this catalog of photophysical properties requires a number of different modes of detection, acquisition and/or signal processing. Historically, separate experiments have been designed to measure individual fluorescence characteristics. However with the advent of high transmission optics and fast, high-QE detectors, single molecule multiparameter fluorescence detection (smMFD) now allows many photophysical parameters to be measured in parallel.[4] A schematic of an idealized setup incorporating the principal detection modes is presented in Fig. 3(a). In this case, a pulsed laser with a width much shorter than the lifetime of the excited state provides the trigger for time-correlated-single-photon counting (TCSPC) and, with a repetition rate much higher than the frame acquisition rate, will also act as a quasi-continuous source for wide-field imaging. These two modes represent extremes at opposite ends of a timeline spanning nine orders of magnitude, as illustrated in Fig. 3(b). Graphical representations of the single-molecule fingerprints expected over different timescales are highlighted in Figs. 3(c)–3(k).

2.2. *Photon antibunching*

On a single detector TCSPC measures the arrival time of an emitted photon following each excitation pulse. The fluorescence lifetime is then determined from the histogram of arrival times accumulated over many pulses. By splitting the emission to detect successive photons on two detectors, interphoton times can be measured. For a single molecule the distribution of these times peaks at some multiple of the period between laser pulses (i.e., 100 ns at a 10 MHz repetition rate) since only one photon can be emitted per pulse, but not every pulse will induce emission. The probability of detecting two photons quasi-simultaneously in the same pulse scales, according to Poisson

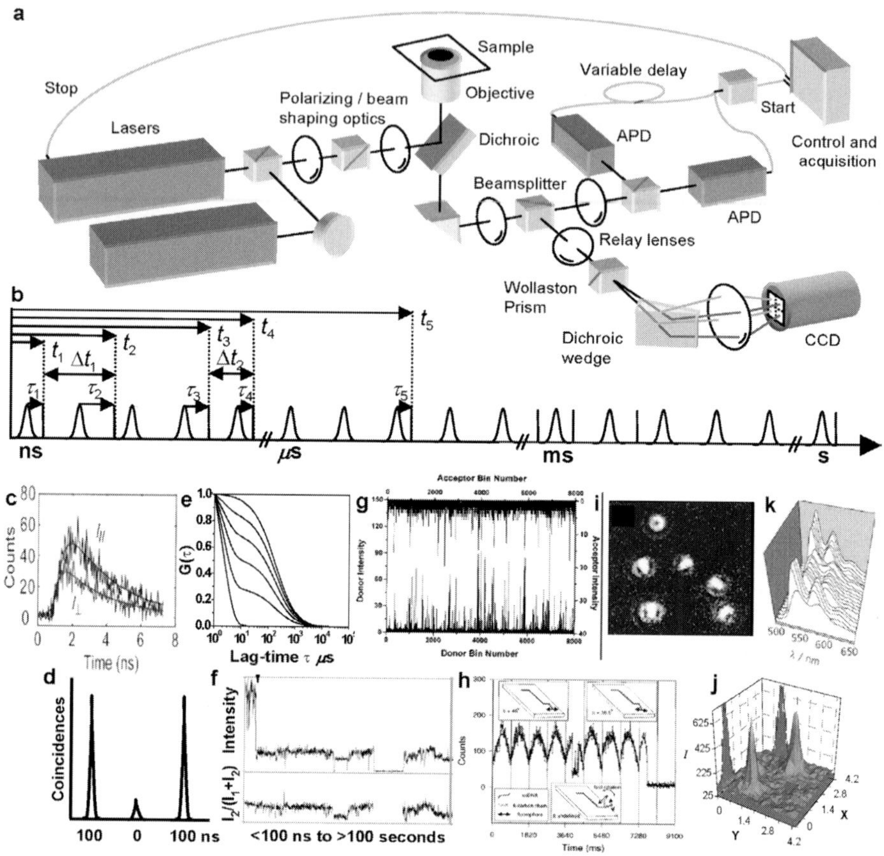

Fig. 3. (a) Experimental setup for single molecule multiparameter fluorescence detection. (b) Timeline of photon detection events showing the chronological time of arrival t, macroscopic inter-photon time Δt and microscopic photon arrival time τ following pulsed excitation. ((c)–(k)) Representations of single molecule signatures observed from different detection modalities over nine orders of magnitude in time from nanoseconds to seconds. (c) Lifetime and anisotropy. Reused with permission from Ref. 52. Copyright 2004, American Institute of Physics. (d) Photon antibunching. (e) Fluorescence correlation spectroscopy. (f) Intensity and fractional intensity trajectories of an immobilized molecule. Reprinted in part with permission from Ref. 53. Copyright 2002, American Chemical Society. (g) Fluorescence resonance energy transfer with two-color detection. Reprinted in part with permission from Ref. 54. Copyright 2000, American Chemical Society. (h) Polarization modulation spectroscopy. Reprinted with permission from Ref. 55. Copyright 1996, American Physical Society. (i) Dipole orientational imaging. Reprinted in part with permission from Ref. 56. Copyright 1999, American Chemical Society. (j) Single molecule localization. (k) Spectral diffusion. Reprinted from Ref. 57. Copyright 2004, with permission from Elsevier.

statistics, as $1 - 1/N$ for N emitters, such that in the classical limit of a molecular ensemble the interphoton-time histogram will peak at zero. In order to detect these coincident photons a suitable delay, greater than the dead-time of the acquisition electronics is introduced in one detector which is then accounted for in the histogram. As well as verifying, unambiguously, the presence of only a single molecule,[58] where more than one fluorophore is present the lack of photon-pairs can indicate the existence of nonradiative pathways by which all bar one emitting states are annihilated. For example antibunching has recently been observed in synthetic multi-chromophoric systems[59] as well as DsRed,[60] single polymeric nanostructures,[61] and quantum dots.[62] The ability to enhance photon absorption across multiple chromophores and channel excitation energy into a single emitter, via resonance transfer and annihilation processes, has been recognized as a possible route to engineering efficient, single-photon sources for quantum computing and encryption.[63]

2.3. *Fluorescence lifetimes*

The intrinsic fluorescence lifetime of a single molecule is unique to the fluorophore-type but sensitive to the fluorophore's environment, since interactions of the fluorophore with its host macromolecule or matrix can provide nonradiative routes for excited-state relaxation. Thus information offered by fluorescence lifetime measurements is essentially two-fold. For a mixture of fluorescent molecules in a reasonably homogenous environment, lifetimes can be used to uniquely identify individual species even when there is significant spectral overlap in their emissions. Conversely, for a population of identical fluorophores the distribution of individual lifetimes will reflect the static heterogeneity of the host environment. Moreover, the lifetime trajectory (τ_F versus time) of a single fluorophore should correlate, to some extent, with fluctuations in the nanoenvironment from which structural dynamics may be inferred. Since the earliest demonstrations of time-resolved single molecule detection[64,65] the technique has found applications in molecular identification in cells;[66] the measurement of heterogeneities on surfaces[67] and in the segmental dynamics of polymers;[68] conformational dynamics of DNA[69] and proteins;[70] and the photophysics of autofluorescent proteins.[71]

Conventionally, TCSPC requires a minimum of $\sim$100 photons to construct a fluorescence decay curve, limiting the temporal resolution to $\sim$10 ms per lifetime measurement at moderate detection rates of $\sim$10 kHz.

Dynamics occurring on timescales shorter than a millisecond, such as structural transformations in DNA and proteins, manifest as nonexponential kinetics in the fluorescence decay.[72] Methods of analyzing photon arrival by time-stamping each photon with a "macroscopic time" between detected photons as well as the "microscopic time" of arrival following each excitation pulse have improved time resolution. In burst integrated fluorescence lifetime (BIFL) the arrival times of photons are binned into histograms that evolve by stepping through consecutive photons in a sliding scale analysis or batches of photons in a jump-directed analysis.[69] The time-resolution is then limited by the lag-time between photons to $\sim 250\,\mu$s, depending on the rep-rate of the laser. A more recent photon-by-photon approach partitions the binning such that there is *at most* one photon per bin stamped with its microscopic time of arrival.[73] The time resolution is then essentially limited by the dead-time of the detection system, typically on the order of nanoseconds. Fluctuations in the trajectory of arrival times as a function of the chronological or macroscopic time can then be analyzed statistically by autocorrelation to obtain fluorescence lifetimes. Photon-by-photon analysis has recently been used to study photo-induced electron transfer in single flavin reductase,[73] and synthetic donor–acceptor dendritic systems.[74]

2.4. *Polarization spectroscopy*

Coupling time-resolved techniques with polarization spectroscopy allows fluorescence anisotropy to be measured, a property from which the rotational dynamics of a single molecule can be determined. For a fluorophore to be excited efficiently by a linearly polarized laser pulse its absorption dipole must be closely aligned with the polarization axis. The emission then remains highly polarized unless the molecule undergoes rotation during the lifetime of the excited state. The extent to which the emission is depolarized by rotation is determined by the difference in fluorescence intensities measured parallel $I_{||}(t)$, and perpendicular $I_{\perp}(t)$, to the excitation polarization. This difference is a measure of the anisotropy, $R(t)$, in the polarized emission and is given formally by normalizing to the total fluorescence:

$$R(t) = \frac{I_{||}(t) - I_{\perp}(t)}{I_{||}(t) + 2I_{\perp}(t)}.$$

The rotational diffusion time and zero-time anisotropy amplitude are determined from exponentials fits to the parallel and perpendicular fluorescence decay components. Here the anisotropy amplitude represents the extent to which the molecule is rotationally mobile. Fluctuations in the anisotropy amplitude and rotational time constants have been measured for single fluorescently labelled proteins T4 lysome[75] and calmodulin[52] with time resolutions as high as 10 ms. In a different application burst integrated fluorescence anisotropy has been used successfully to identify single rhodamine molecules and fluorescent proteins within a mixture.[76]

Rotational dynamics occurring on timescales longer than the excited state lifetime can be interrogated by monitoring the steady-state fluorescence as a function of the polarization angle of the excitation source or an analyzer placed in the emission path. Early measurements of the orientation of fixed dye molecules on surfaces[55,77] or slowly moving probes in host polymer films[78] were measured by simple single channel polarization modulation. In a recent modification the excitation polarization $\mathbf{E}$ is modulated, while emission is measured along two orthogonal polarization axes $\mathbf{s}$ and $\mathbf{p}$ defined by a polarizing beamsplitter. For a fixed dipole the measured intensities I_s and I_p have the same phase and full modulation depth as the field vector $\mathbf{E}$ rotates in and out of alignment with the absorption dipole $\boldsymbol{\mu}_\mathbf{a}$. However, for a rotating fluorophore, I_s and I_p will be out of phase by virtue of the fact that emission is strongly polarized along $\mathbf{E}$ for $\boldsymbol{\mu}_\mathbf{a} || \boldsymbol{\mu}_\mathbf{e}$ (emission dipole). As the excitation polarization is rotated, $I_s > I_p$ for $\mathbf{E} \parallel \mathbf{s}$ and $I_p > I_s$ for $\mathbf{E} \parallel \mathbf{p}$. Hindered rotations are then identified with phase shifts $<180°$ and $>0°$ between I_s and I_p and modulation depths $<100\%$.[79] The technique was suitably demonstrated by resolving differences in the rotational characteristics of TMR labeled DNA and the enzyme Staphylococcal nuclease (SNase) adsorbed on glass: the former exhibiting discrete reorientations; the latter more continuous but hindered rotations.[79] Steady-state polarization has since been used extensively to measure the relative orientation and mobility of the donor/acceptor molecules in fluorescence resonance energy transfer assays[80] and to track rotations in biomolecular motors such as myosin.[81]

2.5. *Wide-field orientation imaging*

Applied to wide-field imaging, polarization can offer the simultaneous mapping of the static orientations in a population of single molecules as well as the dynamic reorientations of individual probes on the millisecond to second timescale. To measure anisotropy on an imaging system, a Wollaston prism is used to split the fluorescence into orthogonal polarizations with a narrow angle of separation such that the two components are projected onto two halves of the CCD array.[82] The technique was used to directly visualize the rotational motion of singularly labeled actin filaments during translocation by myosin.[83] A further development employs excitation polarized along four axes in the sample plane to resolve ambiguities associated with the combined effects of orientation and mobility.[84] In the extreme case of a freely rotating fluorophore the polarization components along only two orthogonal axes will be equal and essentially indistinguishable from those of a molecule with its dipole fixed at 45° between the same axes. The refined technique has been used to determine various domain orientations of the motor-protein, kinesin, during binding and processive movement along microtubules.[85] Dual-polarization has been combined with dual-color imaging, using a wedged dichroic beamsplitter, in a quad-view system that has been used to track rotational motions as well as fluorescence resonance energy transfer in single TMR–biotin/Cy5–streptavidin complexes.[86]

A recent advance in visualizing single molecule orientations is dipole imaging. For a dipole emitting in free-space the intensity of photons propagating in the direction $\boldsymbol{k}$, at an angle χ from the axis of the emission dipole $\boldsymbol{\mu}_\mathbf{e}$ is proportional to $[1 - (\boldsymbol{\mu}_\mathbf{e} \cdot \boldsymbol{k})^2] = \sin^2\chi$. In dipole imaging the 3D anisotropic $\sin^2\chi$ distribution is directly resolved by introducing an aberration in the detection optics, either by imaging through a layer of liquid[87] or by defocusing the microscope objective lens by a few hundred nanometers.[88] The effect is to project the fluorescence collected at high angles by the lens along with light directed along the optical axis onto the same extended image plane. The result is a 2D spatially distributed emission pattern characteristic of the dipole orientation, defined by the polar angle θ with respect to the optical axis and in-plane azimuthal angle ϕ. For a dipole normal to the sample plane ($\theta = 0°$) a ring-doughnut structure is observed, while a dipole in the plane ($\theta = 90°$) produces a lobe-like pattern. The orientation of a molecule with arbitrary θ and ϕ can be determined directly from simulations of the dipole pattern with accuracies of 2° or more.[56] Dipole imaging has been used to follow the heterogeneous rotational dynamics of $\mathrm{DiIC_{18}}$ probes during polymer relaxation[89] and more recently to measure the orientation of the single molecule emitter in a multichromophoric dendritic system.[90] In a recent application, the dynamics of a ligand–protein complex consisting of TMR-biocytin bound to streptavidin have been interrogated by dipole imaging at video rates.[91] In this case the

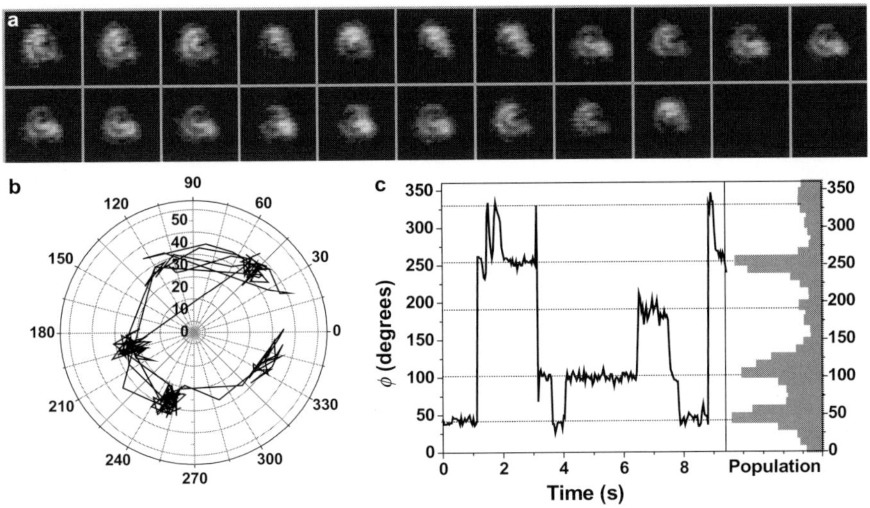

Fig. 4. (a) Evolution of the dipole emission pattern of a single tetramethylrhodamine (TMR) dye probe with time at 30 fps. The TMR is tethered to a streptavidin complex via a biotinylated linker. (b) The angular trajectory (θ, ϕ) of the dipole shows precession of the TMR in a narrow range of polar angles $\theta = 30-40^{\circ}$ about the out-of-plane axis and reorientates in the azimuthal plane at a rate of about 0.5–1 Hz. (c) The trajectory of the azimuthal angle showing the dipole adopting a few favored orientations, the angular separation of which appears conserved from the accompanying histogram of the entire population of ligand–protein complexes analyzed.

TMR dipole trajectories displayed a high degree of dynamic homogeneity as well as static order across the population of complexes analyzed (Fig. 4). The emergence of a set of favored orientations of the TMR molecule was attributed to specific interactions between the TMR probe and the protein surface that appear highly conserved throughout the population of streptavidin nanoenvironments.

2.6. *Fluorescence correlation spectroscopy*

The diffusion of single mobile fluorescent molecules through a tightly focused laser gives rise to photon bursts as individual molecules transit the excitation volume. A number of statistical methods have been developed to analyze the frequency, width and amplitudes of these bursts. The most ubiquitous of these, fluorescence correlation spectroscopy (FCS), uses autocorrelation to measure the similarity between the signal $I(t)$ and the signal $I(t+\tau)$ delayed by a lag time τ[92]:

$$G(\tau) = \frac{\langle \delta I(t)\delta I(t+\tau)\rangle}{\langle I\rangle^2} = \frac{1}{N}\left(1+\frac{\tau}{\tau_D}\right)^{-1}\left(1+\frac{\tau\omega^2}{\tau_D}\right)^{-1/2}\left(1-\beta+\beta\exp\left(-\frac{\tau}{\tau_T}\right)\right).$$

Here $\delta I(t) = I(t) - \langle I\rangle$ is the fluctuation in intensity about its mean $\langle I\rangle$. Within a single burst of fluorescence, consecutive signals will have similar intensities and their product will contribute to large correlation amplitudes for lag times shorter than the period of the burst. For longer lag times the intensities correspond to uncorrelated fluctuations in the background and the product falls away rapidly to zero. Thus the decay time of the autocorrelation curve on long timescales (100 μs to ms) corresponds to the characteristic duration of the burst or transit time τ_D of the molecule through the excitation volume. The expression given above represents the autocorrelation function for the diffusion of N molecules through a Gaussian laser focus with aspect ratio (diameter/depth) of ω. Exponential components are added to account for molecular processes that introduce rapid fluctuations during transit. For example, in the above equation β might represent the fraction of time the fluorescence is quenched during a burst due to intersystem crossing into the triplet state with a lifetime τ_T. Equilibrium and time constants for isomerization in Cy5,[93] folding kinetics in DNA hairpins[94] and quadraplexes,[95] proton transfer in GFP[96] have all been measured using kinetic models of this kind. Beyond elucidating photophysical processes, FCS has been used to identify species in multi-component systems,[97] assay PCR products,[98] and measure ligand–protein interactions.[99] Increasingly, FCS is being used *in vivo* to quantify expression levels, investigate ligand–receptor interactions and study protein trafficking.[100]

The ability of FCS to separate two fluorescent species by autocorrelation requires their diffusion coefficients to differ by at least two-fold, corresponding to a factor of about eight in their molecular weight. To address this shortcoming statistical methods of analyzing the fluctuations in fluorescence amplitude, that are more sensitive to the fluorescing species, have emerged in the form of the photon counting histogram (PCH)[101] and fluorescence intensity distribution analysis (FIDA).[102] Fluorescence-intensity multiple distributions analysis (FIMDA) is a recent adaptation that allows both diffusion time and molecular brightness to be measured concurrently by mapping the evolution of the PCH with the signal integration time.[103] The sensitivity of these techniques has been tested by successfully resolving components in mixtures of singularly and doubly Alexa488 labeled IgG antibodies[104] and separating SH2-phosphopeptide complexes from unbound peptide.[103] More recently the photon-arrival time distribution (PAID) has been developed to exploit photon-rich time intervals when a molecule is present in the detection volume.[105] This is achieved by analyzing the fluctuations in the time interval between photons rather than counting photons in evenly spaced time intervals. The success of statistical methods in fluorescence fluctuation analysis is evidenced by their increasing application in high throughput screening and drug discovery.[106]

2.7. *Spectral diffusion*

Intensity fluctuations in the fluorescence from a single molecule can be attributed to a number of processes including intersystem crossing to the nonfluorescent triplet state; conformational changes within the chromophore (isomerization) or host matrix; chemical transformation; and excited-state energy transfer. These processes affect the fluorescence cycle by altering the excitation rate due to changes in the orientation of the transition dipole; the excited state lifetime and radiative yield; the rate constants of nonradiative relaxation pathways or the excitation rate due to spectral shift in the absorption frequency. The latter, known as spectral diffusion, has been the subject of a number of room-temperature investigations of single molecules on surfaces and in thin polymer films. Single molecule emission spectra are generally monitored by projecting the dispersed fluorescence from a spectrometer onto a photodiode array or more commonly a CCD. With integration times between 170 ms and 1 min diffusion dynamics from hundreds of milliseconds to seconds have been measured. To observe higher frequency fluctuations the fluorescence can be separated by a dichroic beamsplitter to measure the intensity of the long and short wavelength components on two SPADs. The intensity at any one detector as a fraction of the total intensity $I_{\text{short}}/(I_{\text{long}} + I_{\text{short}})$ is then a measure of the wavelength shift from the emission maximum.[107,108]

Early experiments on sulphorhodamine 101 molecules spin coated on glass under a polymethylmethacrylate film revealed spectral diffusion dynamics on two distinct timescales, hundreds of milliseconds and tens of seconds. The fast component was attributed to spontaneous, thermally activated changes in the nuclear coordinates of the fluorophore, while the slow component was found to have a laser power dependence indicative of a photoinduced structural rearrangement.[109] In separate studies of single DiIC_{12} molecules absorbed on bare glass, emission spectra were found to exhibit a variety of forms, from the narrow vibronic band structure associated with low temperatures, to broad featureless spectra that highlighted the heterogeneity of the surface environment.[110] Recently the correlation between intensity fluctuations and spectral diffusion has been explicitly mapped for perlyene derivatives in a polystyrene host.[57] Molecules were observed to undergo discrete, sporadic, jumps between a set of states with markedly different spectra, which was attributed to the extrinsic conformational switching of the matrix in the proximity of the chromophore.

2.8. *Fluorescence resonance energy transfer*

The nonradiative process of fluorescence resonance energy transfer (FRET) that occurs between an excited donor fluorophore and acceptor molecule has been exploited extensively to study the structure and dynamics of biomolecules. For donor/acceptor molecules in close proximity the excitation of the donor induces the resonant oscillation of the acceptor transition dipole, which subsequently undergoes emission while the fluorescence of the donor is quenched. The strength of this dipole–dipole resonance has a $1/R^6$ dependence on the fluorophore separation R, such that the FRET efficiency E drops rapidly between about 10 and 100 Å according to the equation:

$$E = \frac{R_0^6}{R_0^6 + R^6} = 1 - \frac{\tau_{\mathrm{D(A)}}}{\tau_{\mathrm{D(0)}}} = \left[1 + \gamma \frac{I_{\mathrm{D(A)}}}{I_{\mathrm{A(D)}}}\right]^{-1},$$

where R_0 is the donor/acceptor separation at 50% transfer and has a value characteristic of the FRET-pair involved. The transfer efficiency can be measured from the ratio of the donor lifetimes in the presence $\tau_{\mathrm{D(A)}}$ and absence $\tau_{\mathrm{D(0)}}$ of the acceptor, using TCSPC techniques for example. Alternatively the intensities of the donor $I_{\mathrm{D(A)}}$ in the presence of acceptor and sensitised acceptor emission $I_{\mathrm{A(D)}}$ can be monitored on two channels. In this case the donor to acceptor intensity ratio must be corrected by a factor $\gamma = \eta_{\mathrm{A}}\phi_{\mathrm{A}}/\eta_{\mathrm{D}}\phi_{\mathrm{D}}$, where ϕ_{D} and ϕ_{A} are the quantum yields for the individual FRET-pair under interrogation and η_{A} and η_{D} are the detection efficiencies of each channel. In practice, the correction factor can be shown to equal $\Delta I_{\mathrm{A}}/\Delta I_{\mathrm{D}}$, the ratio of acceptor to donor intensity changes following acceptor photobleaching. Since fluctuations in both donor and acceptor emissions can arise from other processes, namely depolarization and spectral diffusion, it is important to account for these when determining donor/acceptor separations and inferring dynamics. The first direct measurement of single molecule FRET was between TMR and Texas Red labels of two complimentary, hybridized 20 mers of DNA immobilized on glass using SNOM.[111] A proliferation of single molecule FRET studies of proteins and RNA folding followed, many of which have been reviewed extensively elsewhere.[112,113]

Advances in single molecule FRET continue apace. For example, the folding of kinase and cold shock proteins has been studied by encapsulation in tethered vesicles to extend the interrogation time without the problems associated with immobilizing the molecule to a surface.[112,114] The power of multiparameter detection has been combined with FRET to monitor the sub-millisecond equilibrium fluctuations of HIV-1 reverse transcriptase.[115] A single molecule optical switch based on the reversible isomerization of Cy5, mediated by the controlled excitation of Cy3, has been engineered using a DNA scaffold.[116] Three-color single molecule fluorescence energy transfer (3CsmFRET) has been demonstrated on DNA constructs labeled with Alexa 488, TMR, and Cy5 from which correlated multiple conformational changes have been inferred.[117,118] Finally methodologies have been developed to make quantitative FRET measurements via two-color alternating laser excitation (ALEX)[119] and pulsed interleaved excitation (PIE)[120] without the need to photobleach the acceptor.

2.9. *Single molecule localization*

The resolution of the optical microscope in the far-field is limited by the diffraction of light at a wavelength λ to about $\lambda/2\text{NA} \sim 250\,\text{nm}$, according to the Rayleigh criterion. At the opposite end of the scale single molecule FRET measures intermolecular distances $<10\,\text{nm}$. Recent developments in single molecule image analysis have attempted to bridge the gap by localizing molecular centers with sub-diffraction limit accuracy. The spatial intensity distribution of a single molecule point-source is defined by the narrow point-spread function (PSF) of a well-focused microscope. This is found to be best-modeled by a 2D-Gaussian. The position of the Gaussian peak in x or y can then be determined with a standard error σ in its mean that is dependent on the signal N; the standard deviation s of the Gaussian ($\sim$PSF half-width); the background noise b and image pixel size a according to[121]:

$$\sigma = \sqrt{\frac{s^2}{N} + \frac{a^2/12}{N} + \frac{8\pi s^4 b^2}{a^2 N^2}}\,.$$

The error represents the sum contribution from shot noise, "pixilation noise" and background counts. In principle, the peak position can be located with an arbitrarily high precision by increasing signal to noise. For example, for a single molecule image of 10^4 photons within a PSF of half-width ~ 125 nm that is resolved on a 9 μm per pixel CCD at $100\times$ magnification, the standard error in locating the molecular center is only $\sigma \sim 1.5$ nm in any one direction for a background of 10 counts. The principle of locating molecular centers has been applied to single particle tracking and high-resolution colocalization. The first demonstration used SNOM to localize carbocyanine molecules in a PMMA matrix with an accuracy of $\sim\lambda/50$.[77] Far-field epi-fluorescence experiments have tracked TMR labeled phospholipids in supported bilayer membranes with a precision of 30 nm.[47] The positional accuracy allowed ligand–receptor binding to be measured by two-color colocalization with only a 40-nm uncertainty in their separation.[122] In a recent hybridization experiment, a TMR labeled 20 mer was immobilized directly on a streptavidin-modified glass surface and colocalized, with a Cy5 complimentary target, again with $\sim$40 nm accuracy, using the intensity weighted pixel coordinates of the molecular centers as shown in Fig. 5.[123] The tracking of GFP in polyacrylamide gels,[124] glycine receptors on the neuronal membrane of a living cell[125] and the high-resolution multi-color colocalization of single nanocrystals[126] have been demonstrated with positional uncertainties as low as 6 nm (Fig. 5).

Single molecule localization has been refined most recently in the form of FIONA, fluorescence imaging with 1 nm accuracy.[5] The technique was used to track the hand-over-hand walking of myosin V along immobilized actin filaments with a positional accuracy of between 1.5 and 3 nm. In another incarnation, single molecule high-resolution imaging with photobleaching (SHRImP) was used to measure the separation between two Cy3 dye labels on hybridized, complementary DNA strands of differing lengths.[127] Here the positions of two overlapping PSFs were resolved following the photobleaching of one dye to reveal the PSF of the second, which allowed inter-dye separations to be determined with accuracies of 1–3 nm. Finally, single molecule high-resolution colocalization (SHREC) has advanced the measurement of distances between dyes emitting at different wavelengths by achieving image registration on a two-channel, dual-view imaging system to within 3.3 nm.[41] The length of a 30 base-pair sequence of dsDNA labeled at opposite ends with Cy3 and Cy5 was measured repeatedly without photobleaching. Furthermore, myosin V labeled with Cy3 and Cy5 on its two heads was shown to walk hand-over-hand, with each head stepping 72 nm

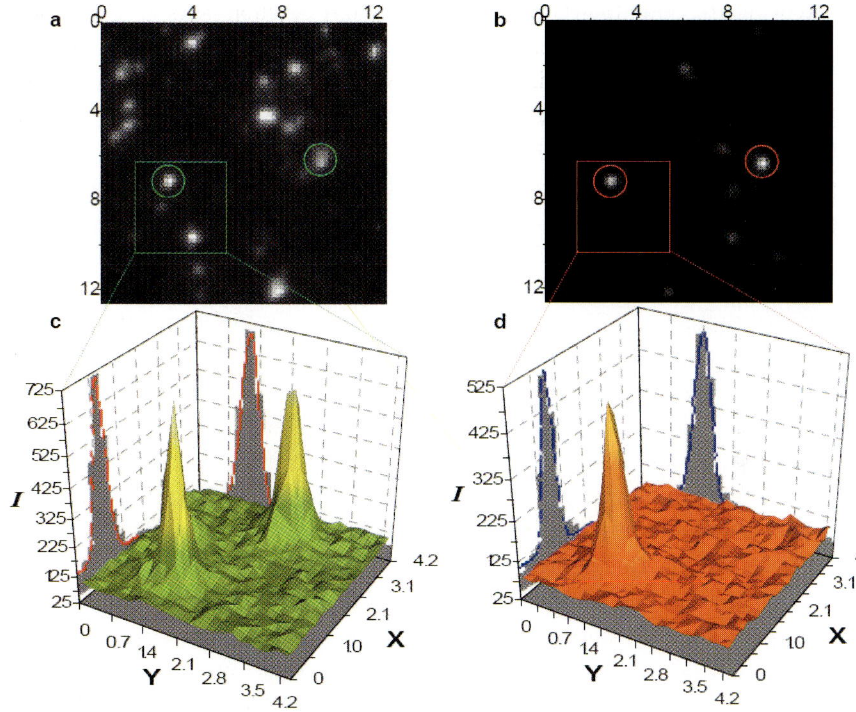

Fig. 5. Single molecule localization. Two-color excitation and detection of (a) TMR and (b) Cy5 labeled complimentary 20 mers of DNA. Colocalized molecules are circled on the TMR (green) and Cy5 (red) channels, respectively. The position of each molecule highlighted in the region of interest was measured to an accuracy of (c) $\sigma_{\mathrm{TMR}} = 25\,\mathrm{nm}$ and (d) $\sigma_{\mathrm{Cy5}} = 30\,\mathrm{nm}$ with a minimal detectable separation of $d = ({\sigma_{\mathrm{TMR}}}^2 + {\sigma_{\mathrm{Cy5}}}^2)^{1/2} = 39\,\mathrm{nm}$. The measured peak-to-peak distance of 11 nm, in this case, represents a hybridization event with a false positive of <1 in 2000 at the experimental coverage of ~ 0.1 molecules/μm^2.

along the actin filament and with the correct 36 nm separation maintained between the two heads.

3. Outlook

A renaissance in optical microscopy has been fueled by advances in photon detection technologies, compact, low-cost laser manufacture, and fluorescent probe engineering. With the sensitivity to detect and track individual molecules, fluorescence microscopy is providing new and exciting insights into the physical, chemical, and biological behavior of matter at the molecular level. Developments in excitation schemes, detection modalities and the statistical analysis of the information rich photons from individual molecules will continue to advance single molecule fluorescence technique. For example molecularly weak interactions such as protein–protein, ligand–receptor, and enzyme reactions with binding constants in the micromolar range have been studied by reducing the excitation volume to the zeptoliter scale ($>10^{-19}$ liter).[128] Such small volumes have been achieved using evanescent field excitation constrained in waveguides etched in aluminium-coated glass substrates. Engineering the point spread function using stimulated emission depletion (STED) has recently been used to further constrain the region from which emission occurs. Here a dump laser is transformed with a central node that is colocalized with the diffraction-limited focus of the pump laser, such that fluorescence at the rim of the focus is quenched by stimulated emission. A far-field resolution as high as 16 nm has recently been demonstrated using STED and single molecule probes.[129] The possibility of controlling the fluorescence cycle using STED directly on a single molecule, to reduce the probability of shelving excitation in the triplet state and hence minimize intermittent dark states, has also been proposed.[130] Most recently the excited state dynamics of a single molecule have been studied using pump–probe laser techniques, pushing the temporal resolution into the femto- to picosecond regime.[131] Continuous improvement in far-field resolution, both in time and space will, undoubtedly, open up many more avenues of investigation and fields of application for single molecule detection.

Acknowledgment

It is with immense gratitude that I acknowledge the support of the Royal Society through a University Research Fellowship.

References

1. C. Bustamante, Y. R. Chemla, N. R. Forde and D. Izhaky, *Annu. Rev. Biochem.* **73**, 705–748 (2004).
2. M. De Serio, R. Zenobi and V. Deckert, *Trend. Anal. Chem.* **22**, 70–77 (2003).

3. K. Kneipp, H. Kneipp, I. Itzkan, R. R. Dasari and M. S. Feld, *J. Phys. Condes. Matter* **14**, R597–R624 (2002).
4. S. Felekyan, R. Kuhnemuth, V. Kudryavtsev, C. Sandhagen, W. Becker and C. A. M. Seidel, *Rev. Sci. Instrum.* **76**, 083104 (2005).
5. A. Yildiz, J. N. Forkey, S. A. McKinney, T. Ha, Y. E. Goldman and P. R. Selvin, *Science* **300**, 2061–2065 (2003).
6. T. Hirschfeld, *Appl. Opt.* **15**, 2965–2966 (1976).
7. E. B. Shera, N. K. Seitzinger, L. M. Davis, R. A. Keller and S. A. Soper, *Chem. Phys. Lett.* **174**, 553–557 (1990).
8. X. S. Xie and J. K. Trautman, *Annu. Rev. Phys. Chem.* **49**, 441–480 (1998).
9. W. E. Moerner and D. P. Fromm, *Rev. Sci. Instrum.* **74**, 3597–3619 (2003).
10. P. Tinnefeld and M. Sauer, *Angew. Chem. Int. Ed.* **44**, 2642–2671 (2005).
11. W. Demtroder, *Laser Spectroscopy: Basic Concepts and Instrumentation* (Springer, Berlin, 1996).
12. D. Eisenberg and W. Kauzmann, *The Structure and Properties of Water* (Oxford University Press, New York, 1969).
13. M. Dahan, T. Laurence, F. Pinaud, D. S. Chemla, A. P. Alivisatos, M. Sauer and S. Weiss, *Opt. Lett.* **26**, 825–827 (2001).
14. Y. H. Lin, R. Weissleder and C. H. Tung, *Bioconjug. Chem.* **13**, 605–610 (2002).
15. X. Michalet, F. F. Pinaud, L. A. Bentolila, J. M. Tsay, S. Doose, J. J. Li, G. Sundaresan, A. M. Wu, S. S. Gambhir and S. Weiss, *Science* **307**, 538–544 (2005).
16. P. S. Dittrich and P. Schwille, *Appl. Phys. B — Laser Opt.* **73**, 829–837 (2001).
17. M. Tokunaga, K. Kitamura, K. Saito, A. H. Iwane and T. Yanagida, *Biochem. Biophys. Res. Commun.* **235**, 47–53 (1997).
18. J. R. Grunwell, J. L. Glass, T. D. Lacoste, A. A. Deniz, D. S. Chemla and P. G. Schultz, *J. Am. Chem. Soc.* **123**, 4295–4303 (2001).
19. N. Panchuk-Voloshina, R. P. Haugland, J. Bishop-Stewart, M. K. Bhalgat, P. J. Millard, F. Mao and W. Y. Leung, *J. Histochem. Cytochem.* **47**, 1179–1188 (1999).
20. J. Arden-Jacob, J. Frantzeskos, N. U. Kemnitzer, A. Zilles and K. H. Drexhage, *Spectrochim. Acta A Mol. Biomol. Spectrosc.* **57**, 2271–2283 (2001).
21. J. E. T. Corrie, J. S. Craik and V. R. N. Munasinghe, *Bioconjug. Chem.* **9**, 160–167 (1998).
22. K. A. Willets, O. Ostroverkhova, M. He, R. J. Twieg and W. E. Moerner, *J. Am. Chem. Soc.* **125**, 1174–1175 (2003).
23. E. Arunkumar, C. C. Forbes and B. D. Smith, *Eur. J. Org. Chem.* 4051–4059 (2005).
24. W. E. Moerner, *J. Chem. Phys.* **117**, 10925–10937 (2002).
25. R. Y. Tsien, *Annu. Rev. Biochem.* **67**, 509–544 (1998).
26. T. B. McAnaney, W. Zeng, C. F. E. Doe, N. Bhanji, S. Wakelin, D. S. Pearson, P. Abbyad, X. H. Shi, S. G. Boxer and C. R. Bagshaw, *Biochemistry* **44**, 5510–5524 (2005).

27. M. F. Garcia-Parajo, M. Koopman, E. van Dijk, V. Subramaniam and N. F. van Hulst, *Biophys. J.* **82**, 48A (2002).
28. R. M. Dickson, A. B. Cubitt, R. Y. Tsien and W. E. Moerner, *Nature* **388**, 355–358 (1997).
29. S. Habuchi, R. Ando, P. Dedecker, W. Verheijen, H. Mizuno, A. Miyawaki and J. Hofkens, *Proc. Natl. Acad. Sci. USA* **102**, 9511–9516 (2005).
30. M. Nirmal, B. O. Dabbousi, M. G. Bawendi, J. J. Macklin, J. K. Trautman, T. D. Harris and L. E. Brus, *Nature* **383**, 802–804 (1996).
31. M. Bruchez, M. Moronne, P. Gin, S. Weiss and A. P. Alivisatos, *Science* **281**, 2013–2016 (1998).
32. J. S. Steckel, J. P. Zimmer, S. Coe-Sullivan, N. E. Stott, V. Bulovic and M. G. Bawendi, *Angew. Chem. Int. Ed.* **43**, 2154–2158 (2004).
33. S. W. Kim, J. P. Zimmer, S. Ohnishi, J. B. Tracy, J. V. Frangioni and M. G. Bawendi, *J. Am. Chem. Soc.* **127**, 10526–10532 (2005).
34. J. Zheng, C. W. Zhang and R. M. Dickson, *Phys. Rev. Lett.* **93**, 077402 (2004).
35. S. M. Nie and R. N. Zare, *Annu. Rev. Biophys. Biomol. Struct.* **26**, 567–596 (1997).
36. P. Tamarat, A. Maali, B. Lounis and M. Orrit, *J. Phys. Chem. A* **104**, 1–16 (2000).
37. M. Gu and C. J. R. Sheppard, *J. Mod. Opt.* **40**, 2009–2024 (1993).
38. T. Ruckstuhl and S. Seeger, *Opt. Lett.* **29**, 569–571 (2004).
39. K. Bahlmann, S. Jakobs and S. W. Hell, *Ultramicroscopy* **87**, 155–164 (2001).
40. Y. Sako and T. Yanagida, *Nat. Cell Biol.* **5**, SS1–SS5 (2003).
41. L. S. Churchman, Z. Okten, R. S. Rock, J. F. Dawson and J. A. Spudich, *Proc. Natl. Acad. Sci. USA* **102**, 1419–1423 (2005).
42. K. Koyama, M. Yoshita, M. Baba, T. Suemoto and H. Akiyama, *Appl. Phys. Lett.* **75**, 1667–1669 (1999).
43. A. Serov, R. Rao, M. Gosch, T. Anhut, D. Martin, R. Brunner, R. Rigler and I. Lasser, *Biosens. Bioelectron.* **20**, 431–435 (2004).
44. T. Erdagon and A. Pradhan, *Biophoton. Int.* **12**, 40–44 (2005).
45. L. Q. Li and L. M. Davis, *Rev. Sci. Instrum.* **64**, 1524–1529 (1993).
46. I. Rech, G. B. Luo, M. Ghioni, H. Yang, X. L. S. Xie and S. Cova, *IEEE J. Select. Topic. Quant. Electron.* **10**, 788–795 (2004).
47. T. Schmidt, G. J. Schutz, W. Baumgartner, H. J. Gruber and H. Schindler, *J. Phys. Chem.* **99**, 17662–17668 (1995).
48. K. Dowling, M. J. Dayel, S. C. W. Hyde, P. M. W. French, M. J. Lever, J. D. Hares and A. K. L. Dymoke-Bradshaw, *J. Mod. Opt.* **46**, 199–209 (1999).
49. C. G. Coates, D. J. Denvir, N. G. McHale, K. D. Thornbury and M. A. Hollywood, *J. Biomed. Opt.* **9**, 1244–1252 (2004).
50. A. Rochas, M. Gosch, A. Serov, P. A. Besse, R. S. Popovic, T. Lasser and R. Rigler, *IEEE Photon. Technol. Lett.* **15**, 963–965 (2003).
51. M. Gosch, A. Serov, T. Anhut, T. Lasser, A. Rochas, P. A. Besse, R. S. Popovic, H. Blom and R. Rigler, *J. Biomed. Opt.* **9**, 913–921 (2004).
52. X. Tan, D. H. Hu, T. C. Squire and P. Lu, *Biophys. J.* **86**, 475A (2004).

53. K. D. Weston, M. Dyck, P. Tinnefeld, C. Muller, D. P. Herten and M. Sauer, *Anal. Chem.* **74**, 5342–5349 (2002).
54. L. Ying, M. I. Wallace, S. Balasubramanian and D. Klenerman, *J. Phys. Chem. B* **104**, 5171–5178 (2000).
55. T. Ha, T. Enderle, D. S. Chemla, P. R. Selvin and S. Weiss, *Phys. Rev. Lett.* **77**, 3979–3982 (1996).
56. A. P. Bartko and R. M. Dickson, *J. Phys. Chem. B* **103**, 11237–11241 (1999).
57. F. Stracke, C. Blum, S. Becker, K. Mullen and A. J. Meixner, *Chem. Phys.* **300**, 153–164 (2004).
58. W. P. Ambrose, P. M. Goodwin, J. Enderlein, D. J. Semin, J. C. Martin and R. A. Keller, *Chem. Phys. Lett.* **269**, 365–370 (1997).
59. S. Masuo, T. Vosch, M. Cotlet, P. Tinnefeld, S. Habuchi, T. D. M. Bell, I. Oesterling, D. Beljonne, B. Champagne, K. Mullen, M. Sauer, J. Hofkens and F. C. De Schryver, *J. Phys. Chem. B* **108**, 16686–16696 (2004).
60. G. Sanchez-Mosteiro, M. Koopman, E. van Dijk, J. Hernando, N. F. van Hulst and M. F. Garcia-Parajo, *Chem. Phys. Chem.* **5**, 1782–1785 (2004).
61. P. Kumar, T. H. Lee, A. Mehta, B. G. Sumpter, R. M. Dickson and M. D. Barnes, *J. Am. Chem. Soc.* **126**, 3376–3377 (2004).
62. B. Lounis, H. A. Bechtel, D. Gerion, P. Alivisatos and W. E. Moerner, *Chem. Phys. Lett.* **329**, 399–404 (2000).
63. B. Lounis and M. Orrit, *Rep. Prog. Phys.* **68**, 1129–1179 (2005).
64. X. S. Xie and R. C. Dunn, *Science* **265**, 361–364 (1994).
65. W. P. Ambrose, P. M. Goodwin, J. C. Martin and R. A. Keller, *Science* **265**, 364–367 (1994).
66. J. P. Knemeyer, D. P. Herten and M. Sauer, *Anal. Chem.* **75**, 2147–2153 (2003).
67. M. Lee, J. Kim, J. Tang and R. M. Hochstrasser, *Chem. Phys. Lett.* **359**, 412–419 (2002).
68. R. A. L. Vallée, N. Tomczak, L. Kuipers, G. J. Vancso and N. F. van Hulst, *Phys. Rev. Lett.* **91**, 038301 (2003).
69. C. Eggeling, J. R. Fries, L. Brand, R. Gunther and C. A. M. Seidel, *Proc. Natl. Acad. Sci. USA* **95**, 1556–1561 (1998).
70. J. Y. Tang, E. Mei, C. Green, J. Kaplan, W. F. DeGrado, A. B. Smith and R. M. Hochstrasser, *J. Phys. Chem. B* **108**, 15910–15918 (2004).
71. B. Bowen and N. Woodbury, *Photochem. Photobiol.* **77**, 362–369 (2003).
72. Y. W. Jia, A. Sytnik, L. Q. Li, S. Vladimirov, B. S. Cooperman and R. M. Hochstrasser, *Proc. Natl. Acad. Sci. USA* **94**, 7932–7936 (1997).
73. H. Yang, G. B. Luo, P. Karnchanaphanurach, T. M. Louie, I. Rech, S. Cova, L. Y. Xun and X. S. Xie, *Science* **302**, 262–266 (2003).
74. M. Cotlet, S. Masuo, G. B. Luo, J. Hofkens, M. Van der Auweraer, J. Verhoeven, K. Mullen, X. L. S. Xie and F. De Schryver, *Proc. Natl. Acad. Sci. USA* **101**, 14343–14348 (2004).
75. D. H. Hu and H. P. Lu, *Biophys. J.* **84**, 474A (2003).
76. J. Schaffer, A. Volkmer, C. Eggeling, V. Subramaniam, G. Striker and C. A. M. Seidel, *J. Phys. Chem. A* **103**, 331–336 (1999).
77. E. Betzig and R. J. Chichester, *Science* **262**, 1422–1425 (1993).

78. K. D. Weston and L. S. Goldner, *J. Phys. Chem. B* **105**, 3453–3462 (2001).
79. T. Ha, T. A. Laurence, D. S. Chemla and S. Weiss, *J. Phys. Chem. B* **103**, 6839–6850 (1999).
80. T. Ha, *Methods* **25**, 78–86 (2001).
81. S. A. Rosenberg, M. E. Quinlan, J. N. Forkey and Y. E. Goldman, *Acc. Chem. Res.* **38**, 583–593 (2005).
82. K. Kinosita, H. Itoh, S. Ishiwata, K. Hirano, T. Nishizaka and T. Hayakawa, *J. Cell Biol.* **115**, 67–73 (1991).
83. I. Sase, H. Miyata, S. Ishiwata and K. Kinosita, *Proc. Natl. Acad. Sci. USA* **94**, 5646–5650 (1997).
84. H. E. Sosa, J. G. Peterman, W. E. Moerner and L. S. Goldstein, *Nat. Struct. Biol.* **8**, 540–544 (2001).
85. E. J. G. Peterman, H. Sosa, L. S. B. Goldstein and W. E. Moerner, *Biophys. J.* **81**, 2851–2863 (2001).
86. L. Cognet, G. S. Harms, G. A. Blab, P. H. M. Lommerse and T. Schmidt, *Appl. Phys. Lett.* **77**, 4052–4054 (2000).
87. R. M. Dickson, D. J. Norris and W. E. Moerner, *Phys. Rev. Lett.* **81**, 5322–5325 (1998).
88. M. Bohmer and J. Enderlein, *J. Opt. Soc. Am. B: Opt. Phys.* **20**, 554–559 (2003).
89. A. P. Bartko, K. W. Xu and R. M. Dickson, *Phys. Rev. Lett.* **89**, 026101 (2002).
90. W. Schroeyers, R. Vallee, D. Patra, J. Hofkens, S. Habuchi, T. Vosch, M. Cotlet, K. Mullen, J. Enderlein and F. C. De Schryver, *J. Am. Chem. Soc.* **126**, 14310–14311 (2004).
91. M. A. Osborne, *J. Phys. Chem. B* **109**, 18153–18161 (2005).
92. M. Eigen and R. Rigler, *Proc. Natl. Acad. Sci. USA* **91**, 5740–5747 (1994).
93. J. Widengren and P. Schwille, *J. Phys. Chem. A* **104**, 6416–6428 (2000).
94. G. Bonnet, O. Krichevsky and A. Libchaber, *Proc. Natl. Acad. Sci. USA* **95**, 8602–8606 (1998).
95. L. M. Ying, J. J. Green, H. T. Li, D. Klenerman and S. Balasubramanian, *Proc. Natl. Acad. Sci. USA* **100**, 14629–14634 (2003).
96. U. Haupts, S. Maiti, P. Schwille and W. W. Webb, *Proc. Natl. Acad. Sci. USA* **95**, 13573–13578 (1998).
97. Z. Foldes-Papp, U. Demel and G. P. Tilz, *Proc. Natl. Acad. Sci. USA* **98**, 11509–11514 (2001).
98. M. Kinjo, *BioTechniques* **25**, 706–715 (1998).
99. L. M. Davis, P. E. Williams, H. M. Cain, D. A. Ball, C. G. Parigger, E. D. Matayoshi and K. M. Swift, *Biophys. J.* **82**, 43A (2002).
100. V. Vukojevic, A. Pramanik, T. Yakovleva, R. Rigler, L. Terenius and G. Bakalkin, *Cell. Mol. Life Sci.* **62**, 535–550 (2005).
101. Y. Chen, J. D. Muller, P. T. C. So and E. Gratton, *Biophys. J.* **77**, 553–567 (1999).
102. P. Kask, K. Palo, N. Fay, L. Brand, U. Mets, D. Ullmann, J. Jungmann, J. Pschorr and K. Gall, *Biophys. J.* **78**, 1703–1713 (2000).

103. K. Palo, U. Metz, S. Jager, P. Kask and K. Gall, *Biophys. J.* **79**, 2858–2866 (2000).
104. J. D. Muller, Y. Chen and E. Gratton, *Biophys. J.* **78**, 474–486 (2000).
105. T. A. Laurence, A. N. Kapanidis, X. X. Kong, D. S. Chemla and S. Weiss, *J. Phys. Chem. B* **108**, 3051–3067 (2004).
106. L. Zemanova, A. Schenk, M. J. Valler, G. U. Nienhaus and R. Heilker, *Drug Discovery Today* **8**, 1085–1093 (2003).
107. T. Ha, J. Glass, T. Enderle, D. S. Chemla and S. Weiss, *Phys. Rev. Lett.* **80**, 2093–2096 (1998).
108. P. Tinnefeld, D. P. Herten and M. Sauer, *J. Phys. Chem. A* **105**, 7989–8003 (2001).
109. H. P. Lu and X. S. Xie, *Nature* **385**, 143–146 (1997).
110. K. D. Weston, P. J. Carson, H. Metiu and S. K. Buratto, *J. Chem. Phys.* **109**, 7474–7485 (1998).
111. T. Ha, T. Enderle, D. F. Ogletree, D. S. Chemla, P. R. Selvin and S. Weiss, *Proc. Natl. Acad. Sci. USA* **93**, 6264–6268 (1996).
112. B. Schuler, *ChemPhysChem.* **6**, 1206–1220 (2005).
113. X. W. Zhuang, *Annu. Rev. Biophys. Biomol. Struct.* **34**, 399–414 (2005).
114. E. Boukobza, A. Sonnenfeld and G. Haran, *J. Phys. Chem. B* **105**, 12165–12170 (2001).
115. P. J. Rothwell, S. Berger, O. Kensch, S. Felekyan, M. Antonik, B. M. Wohrl, T. Restle, R. S. Goody and C. A. M. Seidel, *Proc. Natl. Acad. Sci. USA* **100**, 1655–1660 (2003).
116. M. Bates, T. R. Blosser and X. W. Zhuang, *Phys. Rev. Lett.* **94**, 108101 (2005).
117. S. Hohng, C. Joo and T. Ha, *Biophys. J.* **87**, 1328–1337 (2004).
118. J. P. Clamme and A. A. Deniz, *ChemPhysChem.* **6**, 74–77 (2005).
119. N. K. Lee, A. N. Kapanidis, Y. Wang, X. Michalet, J. Mukhopadhyay, R. H. Ebright and S. Weiss, *Biophys. J.* **88**, 2939–2953 (2005).
120. B. K. Muller, E. Zaychikov, C. Brauchle and D. C. Lamb, *Biophys. J.* **89**, 3508–3522 (2005).
121. R. E. Thompson, D. R. Larson and W. W. Webb, *Biophys. J.* **82**, 2775–2783 (2002).
122. G. J. Schutz, W. Trabesinger and T. Schmidt, *Biophys. J.* **74**, 2223–2226 (1998).
123. M. A. Osborne, C. L. Barnes, S. Balasubramanian and D. Klenerman, *J. Phys. Chem. B* **105**, 3120–3126 (2001).
124. U. Kubitscheck, O. Kuckmann, T. Kues and R. Peters, *Biophys. J.* **78**, 2170–2179 (2000).
125. M. Dahan, S. Levi, C. Luccardini, P. Rostaing, B. Riveau and A. Triller, *Science* **302**, 442–445 (2003).
126. T. D. Lacoste, X. Michalet, F. Pinaud, D. S. Chemla, A. P. Alivisatos and S. Weiss, *Proc. Natl. Acad. Sci. USA* **97**, 9461–9466 (2000).
127. M. P. Gordon, T. Ha and P. R. Selvin, *Proc. Natl. Acad. Sci. USA* **101**, 6462–6465 (2004).

128. M. J. Levene, J. Korlach, S. W. Turner, M. Foquet, H. G. Craighead and W. W. Webb, *Science* **299**, 682–686 (2003).
129. V. Westphal and S. W. Hell, *Phys. Rev. Lett.* **94**, 143903 (2005).
130. R. J. Marsh, M. A. Osborne and A. J. Bain, *Proc. Int. Soc. Opt. Eng.* **5222**, 78–86 (2003).
131. E. M. H. P. van Dijk, J. Hernando, J.-J. Garcia-López, M. Crego-Calama, D. N. Reinhoudt, L. Kuipers, M. F. García-Parajó and N. F. van Hulst, *Phys. Rev. Lett.* **94**, 078302 (2005).

Mark Osborne graduated from the University of Sussex in 1991 with a first class honors degree in Chemical Physics. He continued research there in high-resolution laser spectroscopy of molecular collisions and obtained a DPhil in 1994. Mark spent a year at the Commissariat l'Energie Atomique, Paris, France on an EU training and mobility for researchers award where he worked on the photodynamics of atomic and molecular clusters. He returned to the UK in 1996 to a postdoctoral position at Cambridge. His research in single molecule detection led to IP in a single molecule DNA sequencing technology and ultimately the formation of a spin-out, Solexa Ltd., in 2000. Mark worked as a senior scientist for Solexa until 2001 when he took up a Royal Society Research Fellowship and returned to the University of Sussex. Mark continues research in the detection and manipulation of single molecules with a view to understanding molecular interactions and transformations at the ultimate level of sensitivity. Mark and his partner Karen have twin daughters, Bronnie and Flo, aged two, and enjoy music, film and the cultural playground of Brighton.

INDEX